CAMBRIDGE LIBRARY COLLECTION

Books of enduring scholarly value

Botany and Horticulture

Until the nineteenth century, the investigation of natural phenomena, plants and animals was considered either the preserve of elite scholars or a pastime for the leisured upper classes. As increasing academic rigour and systematisation was brought to the study of 'natural history', its sub-disciplines were adopted into university curricula, and learned societies (such as the Royal Horticultural Society, founded in 1804) were established to support research in these areas. A related development was strong enthusiasm for exotic garden plants, which resulted in plant collecting expeditions to every corner of the globe, sometimes with tragic consequences. This series includes accounts of some of those expeditions, detailed reference works on the flora of different regions, and practical advice for amateur and professional gardeners.

Wood

An eminent botanist and natural historian, George Simonds Boulger (1853–1922) wrote a number of books on plant life in the British Isles. First published in 1902, this manual explores the characteristics and uses of one of the most abundant and versatile natural materials. In the first part, Boulger outlines the general biological function and uses of wood. He also describes the classification of wood, and the durability of different timbers. The second part catalogues the types of wood that are used commercially. Boulger explains the distinguishing characteristics and uses of hundreds of different kinds of timber, which are listed alphabetically. Featuring 82 illustrations, the book also includes appendices explaining some of the terminology and science of wood, and a select bibliography. Boulger's work on economic botany, *The Uses of Plants* (1889), is also reissued in the Cambridge Library Collection.

Cambridge University Press has long been a pioneer in the reissuing of out-of-print titles from its own backlist, producing digital reprints of books that are still sought after by scholars and students but could not be reprinted economically using traditional technology. The Cambridge Library Collection extends this activity to a wider range of books which are still of importance to researchers and professionals, either for the source material they contain, or as landmarks in the history of their academic discipline.

Drawing from the world-renowned collections in the Cambridge University Library and other partner libraries, and guided by the advice of experts in each subject area, Cambridge University Press is using state-of-the-art scanning machines in its own Printing House to capture the content of each book selected for inclusion. The files are processed to give a consistently clear, crisp image, and the books finished to the high quality standard for which the Press is recognised around the world. The latest print-on-demand technology ensures that the books will remain available indefinitely, and that orders for single or multiple copies can quickly be supplied.

The Cambridge Library Collection brings back to life books of enduring scholarly value (including out-of-copyright works originally issued by other publishers) across a wide range of disciplines in the humanities and social sciences and in science and technology.

PREFACE.

In an attempt, such as this, to cover a wide ground within a book of small compass perfect accuracy cannot be hoped for, completeness is impossible and originality is neither expected nor desirable. Rather, however, than burden the body of the book with constant acknowledgments of indebtedness, I have thought it better to add a bibliographical appendix, indicating those works from which I have borrowed most freely. For Figs. 1, 7, 16, 17 and 29 I am indebted to the courtesy of Mr. Francis Darwin and the Syndicate of the Cambridge University Press; for Figs. 10, 18, 21—23 and 27 to that of Professor Marshall Ward and Messrs. Kegan Paul, Trench, Trübner & Co.; for Figs. 12, 13, 15, 26 and 30 to that of Professor Somerville and Mr. David Douglas; and for Figs. 4, 28, 32, 37 and 45—66, which are photographed from nature, to Mr. D. F. Mackenzie of Morton Hall, Midlothian; whilst Figs. 2, 3, 6, 8, 9, 11, 14, 24, 25 and 38—44 have been drawn for me by Miss Emily Carter.

The plates at the end of the book are photographs from sections of common British-grown timbers magnified 3½ times. They were prepared by Arthur Deane, Esq., Assistant Curator of the Warrington Museum, and are used by his kind permission and that of the English Arboricultural Society.

I have thought it well to indicate the pronunciation of the Latin names by putting an accent over the syllables on which the stress falls; and it may be desirable to point out here that the chief symbols employed in Part II. are explained on pp. 139-140.

How incomplete my work is may be gauged by the statement that, while there are undoubtedly several thousand woods used in various parts of the world, only about 750 are here enumerated; but these include most of those which are practically known in general commerce, and to have dealt with more would have necessitated a volume fully twice as large.

G. S. B.

CONTENTS.

PART I.—OF WOOD IN GENERAL.

CHAPTER I.

CHAPTER II.

CHAPTER III.

CHAPTER IV.

CHAPTER V.

CHAPTER VI.

CHAPTER VII.

PART II.—WOODS OF COMMERCE.

APPENDICES.

PART I.—OF WOOD IN GENERAL.

CHAPTER I.

THE ORIGIN, STRUCTURE, AND DEVELOPMENT OF WOOD AND ITS USE TO THE TREE.

FEW, if any, of the products of nature are of such manifold utility as wood. Though coal has in many lands largely replaced it as fuel, and as a source of tar, though stone, brick, and iron or steel have often been substituted for it as house-building materials, and the metals last mentioned for the construction of ships, new uses are constantly arising for it, such as railway sleepers, pavements, and paper-making, so as to more than make up for the saving effected by these substitutes. In England and the United States, for example, the consumption of wood per head of the population, during the last half century, has more than doubled.

Most people are aware that for these manifold uses a great number of different woods are employed in the various countries of the world—woods that differ in colour, grain, hardness, weight, flexibility, and other properties almost as widely as the trees by which they are produced vary in foliage, flower, or fruit. It is, however, not so generally recognized that the suitability of wood of any kind for some particular purpose depends mainly upon its internal structure. This structure is determined not by man's employment of the material, but by the vital requirements of the tree when growing.

Our present concern is with wood as a material in the arts, and not with any merely botanical interest it may have, or with its cultivation as a crop by the forester. In dealing with the means of recognizing different kinds of wood we shall, therefore, not depend in any way upon characters derived from bark, leaves, flowers, or fruit—the characters, that is, of standing, or of unconverted timber; but only on those of the wood itself as it appears in the timber market. At the same time, if we are to be able to identify woods and determine their suitability for various economic applications, it is absolutely essential that we should know something of their origin, structure, development, and use to the plants that produced them.

Wood does not occur in any plants of a lower grade than ferns; and in the higher plants in which it does occur it is chiefly, but not exclusively in the stem. In the shell of the cocoa-nut or the stone of a peach it probably serves the purpose of checking premature germination of the enclosed seed by excluding damp. In stems, however, the main physiological function of wood is the mechanical one of giving strength to resist the increasing weight of the structure as it grows erect and branches. Submerged aquatic plants, buoyed up, as they are, by the water, do not form wood in their stems, nor, as a rule, do annuals, nor, at first, the succulent, flexible shoots of longer-lived plants. In ferns, even when growing into lofty trees, and in allied plants, the wood, though dense, consists largely of scattered longitudinal strands and often of cells of no great vertical length. Though there are also generally woody layers just below the surface of the stem, giving it considerable strength as a whole, this structure renders tree-ferns useless as timber.

For all practical purposes, therefore, wood is produced only by the highest sub-kingdom of the plant world, the seed-bearing or flowering plants, the *Spermatophýta* or *Phanerogámia* of botanists. This great group of plants is sub-divided, mainly by characters derived from parts other than their stems, into two divisions, the

Gymnospérmæ, or plants the seeds of which are naked, *i.e.* not enclosed in a fruit, and the *Angiospérmæ*, or fruit-bearing plants. The Gymnosperms are all perennial trees and shrubs; but of three "Natural Orders" into which they are divided, two, the *Cycadáceæ* and *Gnetáceæ*, belong almost exclusively to the Southern Hemisphere and are valueless as timber. The third Natural Order is the *Coníferæ*, so named from the general arrangement of its seeds on a series of overlapping scales arranged in a cone, but having also other general characters, one of the most conspicuous of which is the production of numerous, narrow, rigid, undivided leaves, whence they get the familiar name of *needle-leaved trees*. The members of this Order, which includes the Pines, Firs, Larches, Cedars, etc., have much-branched stems, and wood, which, though in many points, such as its arrangement in annual rings of growth, it resembles that of some other, more highly-organized plants, has, as we shall see, many peculiarities. It is, in general, of rapid growth, soft and of even texture, and very commonly abounds in resinous substances. They are, therefore, often spoken of as "*soft woods*" or as "*resinous woods*," and being, from these characteristics, both easily worked and of considerable durability, are more extensively used than any other class of woods. The Maidenhair-tree of China and Japan (*Gínkgo bilóba*) is exceptional among conifers in having broad leaves: neither this tree nor the Yew can be said to bear cones, though their seeds are naked: the Yew is destitute of resin; and the epithet "soft-wooded" applies to Willow, Poplar, Horse-chestnut, etc., as truly as to conifers.

The second and higher division of seed-bearing plants, the Angiospermæ, is divided into two Classes, which, whilst agreeing in having their seeds enclosed in fruits, differ in many characters, and in none more than in the structure of their stems. They are known botanically, from the number of seed-leaves or *cotyledons* of their embryos, as *Monocotylédons* and *Dicotylédons*. The Monocotyledons, with one such seed-leaf, comprise lilies, orchids, bananas, palms, sedges, grasses, etc. Few of these, such as Palms and Bamboos, reach the dimensions of trees, and those which do so have generally unbranched stems which do not as a rule increase in diameter after the very earliest stages of their growth, the wood in them being confined to isolated strands crowded together towards their outer surfaces. Though such stems may occasionally, like those of tree-ferns, be utilized "in the round," and veneers, cut from the outer part of the stem of the Cocoa-nut Palm (*Cócos nucífera*), and known, from the appearance of the dark-coloured woody strands in the lighter ground-tissue, as "Porcupine-wood," are used for inlaying, Monocotyledons may well be ignored as economic sources of wood.

Dicotyledons, so named from having two seed-leaves to the embryo, comprise an immense and varied assemblage of plants,

FIG. 1.—Transverse section of an Oak, 25 years old. After Le Maout and Decaisne, from *The Elements of Botany*, by permission of Mr. Francis Darwin and the Syndicate of the Cambridge University Press.

a very large proportion of which are merely herbaceous, never forming wood. In those perennial members of the Class, however, which acquire the dimensions of trees or shrubs, the stem generally branches freely, has a separable "bark," and increases in girth with age; the wood, though, as we shall see, it differs in several important but not very obvious characters, agreeing with that of conifers in being arranged in rings produced in successive seasons (Fig. 1). These rings, as they appear in a cross-section of a tree, or conically tapering sheaths surrounding the tree, as they in fact are, form on the outside of the wood of previous seasons and beneath the bark; and this type of stem, characteristic of gymnosperms and dicotyledons, is in consequence correctly termed *exógenous*, from the Greek *ex*, outside of, and *gennao*, to produce. The term *endógenous*, still sometimes applied to the structure of the stem of monocotyledons, is less accurate. Dicotyledons are commonly slower of growth than conifers, and their wood, especially that near the centre of the stem, is often much harder. They bear as a rule also broad, net-veined leaves; and are known familiarly, therefore, as "*hardwoods*," or as "*broad-leaved trees*." Such are the Oak, Beech, Ash, Elm, Teak, Willow, Alder, etc.

It is then only with the two classes of exogenous stems, those of gymnosperms or needle-leaved trees, and those of dicotyledons or broad-leaved trees, that we are concerned.

Though, as we have already said, conifers and broad-leaved trees present important differences in the structure and consequent character of their wood, their manner of growth is so nearly identical in its initial stages and broad outlines that we may well treat them at first collectively. It is, perhaps, the many branches and the numerous small leaves exposed by means of those branches to a maximum of air and light in these two groups of plants (as contrasted with the general absence of branching, and the small number and large size of the leaves in ferns and palms) that has determined the production of the progressively-enlarging, solid stem that characterizes them. It must be remembered, however, that the stem of a tree fulfils several very distinct physiological purposes. Besides bearing up the weight of leaves and flowers so as best to obtain the air and light they require, it is the means of communication between the root and the leaves. Through it the water and its dissolved gases and saline substances, taken in by the root from the soil, are conveyed to the leaves, which have been termed the "laboratory of the plant," to be built up in them, with the carbonaceous food-material taken in from the atmosphere, into those complex "*organic*" compounds of which the whole structure of the plant is composed. Furthermore, the stem serves as a reservoir in which some of these organic compounds, the "plastic material" of the plant, are stored up for use in future growth.

Every stem and every branch—and a branch is but a secondary stem, differing only in position—as long as it remains capable of elongation, is terminated, in the groups of trees with which we are concerned, by a bud. A *bud* is a growing-point protected by overlapping rudimentary leaves.

In the immediate neighbourhood of this growing-point the stem in this its initial stage is entirely made up of structures which almost completely resemble one another. Whether we cut such a growing-point across or lengthwise it presents under the microscope the appearance of a delicate mesh-work of thin membrane filled in with a viscid semi-fluid substance. These meshes, from their resemblance to honeycomb, were in 1667 named *cells* by Robert Hooke. The delicate membranes which form them, the *cell-walls* as they are termed, are composed of a definite chemical compound known as *cellulose.* It contains the three elements, carbon, hydrogen, and oxygen, in definite proportions, which the chemist represents as $C_6H_{10}O_5$, that is, in a hundred parts by weight 44 are carbon, 6 are hydrogen, and 50 are oxygen. Cellulose, like starch and sugar, belongs to a group of compounds of carbon with hydrogen and oxygen in the proportions in which those two elements occur in water, which are known as *carbo-hydrates.* It has, in fact, the same percentage composition as starch, though differing from it in many properties. It is insoluble in water, flexible, slightly elastic, permeable, but only slightly absorbent, and does not readily undergo fermentation. When treated with acid it passes into the condition of starch, as is evidenced by its then turning blue with iodine, and under certain conditions in the living plant it would seem capable of being formed from, or of passing into, sugar. Cotton-wool consists almost entirely of pure, unaltered cellulose. The viscid, semi-fluid substance contained in the cells is of far more complex chemical composition. It contains not only carbon, hydrogen, and oxygen, but also, though in far smaller proportion, nitrogen, with traces of sulphur, and, perhaps always also, phosphorus and other elements. It is probably a mixture in varying proportions of some of those substances which, from their resemblance to albumen or white of egg, are known as *albuminoid*, and, from the readiness with which they undergo chemical change or decomposition, as *proteids*. Being the substance out of which all plant-structures originate, the sole constituent of the first germs of all living beings, it is known as *protoplasm*, from the Greek *prōtos*, first, *plasma*, formed matter.

Any collection of similar cells or modifications of cells having a common origin and obeying a common law of growth is known as a *tissue*. These young cells at the apex of a stem, of nearly uniform size, and that extremely minute, with their delicate, as yet unaltered, cell-walls filled with protoplasm, form an *embryonic tissue*, one, that is, which will undergo change. Its uniform character causes it to be termed *undifferentiated*, while the various kinds of tissue to which by different changes it gives rise are known in contradistinction as *permanent tissues*. One change to which any cell is liable so long as it contains protoplasm is division into two, a partition wall of cellulose forming across it. The formation of this solid wall from material in solution in the protoplasm, and a correlative power, which, as we shall see, the living plant possesses, of dissolving a cell-wall, illustrate that interchangeability of sugar and cellulose of which we have spoken. A tissue the cells of which undergo division is termed *merismatic* or *meristem*, from the Greek *merisma*, division; so that the embryonic tissue at the apex of a stem is known as *apical meristem*.

Although its cells are all embryonic, they nevertheless at a very early stage commonly present such a degree of differentiation as to make it possible to distinguish three well-defined rudimentary tissue-systems (Fig. 2). First, there is a single layer of cells on the outside of the growing-point, with thickened outer walls and undergoing division only in planes perpendicular to the surface. If we trace this layer backwards down the surface of the shoot below its apex we shall find it continuous with similar cells which have lost their protoplasm and have even thicker outer walls. As this outer layer of permanent tissue is called the *epidérmis*, from the Greek *epi*, upon, *derma*, skin, the embryonic layer in which it originates is termed the *dermátogen* (*derma*, skin, and *gennáo*, to produce). In the middle of the growing-point is a solid column-like mass of cells which are all

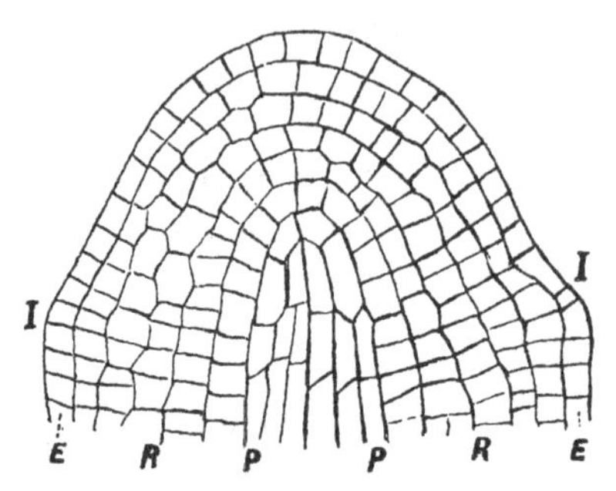

FIG. 2.—Growing-point of stem, showing apical meristem. *P*, plerome; *R*, periblem; *E*, dermatogen; *I*, rudiment of leaf. (After Leunis and Frank.)

somewhat elongated in the direction of the elongation of the stem. This is known as the *plerome* and the central axis of tissues to which it gives rise as the *stele* (Greek for a column)

FIG. 3.—Terminal bud, showing growing-point of stem, overlapped by rudimentary leaves with buds in their axils, the whole covered by dermatogen. In the centre is the stele to which descend the midribs of the leaves. (After Prantl.)

(Fig. 3). Between the outer dermatogen and the inner plerome there is a layer, or a series of layers, of cells which undergo division both in planes perpendicular to and in planes parallel to the surface of the stem. These are known as the *periblem*. On tracing them backwards down the shoot we find them continuous with tissues which immediately beneath the epidermis are commonly green, and which often have their cells much thickened in the corners in herbaceous plants or shoots, whilst still further back, on older parts of woody shoots, the green layer is often buried under one or more layers of brown cork. These tissues which thus originate in the periblem are known collectively as the *cortex*.

It is with tissues originating from the central plerome or stele that we are mainly concerned. If we cut a young shoot across, a little below its entirely embryonic apex, we shall see that, whilst there is a central whitish mass, which on being magnified exhibits a comparatively wide-meshed structure, there are round this a ring of patches of a greyer, closer tissue. These grey patches may be observed to be roundish or slightly wedge-shaped in outline, their longer diameter lying in one of the radii of the stem, and they are wider across their outer parts. They appear grey on account of the smaller diameter of their cells. Longitudinal sections show these patches to be cross-sections of long strands or bundles of cells, narrower and more elongated than those around them. The central mass of tissue is the *pith* or *medulla*, and these strands are known as *procambium* or *desmogen*.

The pith is relatively large in the stems of herbaceous plants or in young shoots (Fig. 4), but does not increase in bulk as the tree grows older. Its cells are at first full of fluid, and their walls often remain thin. Those of its outer portion, near the procambium strands, are smaller, and all its cells are often two or three times as long in the direction of the elongation of the stem as they are broad. Thus in shape they are short, polygonal, closely-packed prisms. In many cases, as in the Elder, the cells of the pith die, losing their fluid contents, shrivelling, and so completing disorganizing the entire tissue that the stem becomes hollow, or a mere line of dry powder in the centre of the innermost ring of wood marks this structural centre of the stem. In other cases, as in the Oak, the cells of the pith have their walls thickened, and turn from white to brown; but even then its relatively minute width makes it difficult to detect in a stem several years of age.

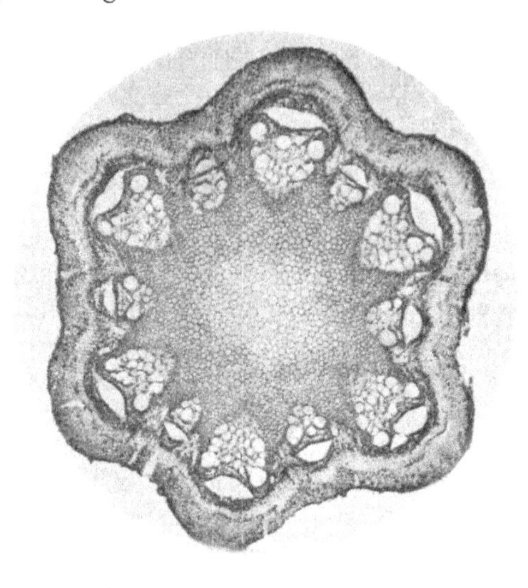

FIG. 4.—Transverse section of the stem of Traveller's Joy (*Clématis Vitálba*), showing relatively large central pith and large vessels.

The *procambium strands* extend from the rudiments of leaves near its apex right through the stem into the root. They get their name from a Latin word, *cambio*, to grow, being in a merely transitory or embryonic condition. In Monocotyledons the whole of their tissue passes into the condition of wood and bast; so that the *bundle*, as the strand in its permanent form is termed, being incapable of any further growth in diameter, is said to be *closed.* It is because it gives rise to a bundle (Greek, *desmos*, a bond) that the procambium is termed *désmogen.* In those trees, however, with which we are concerned, viz. Gymnosperms and Dicotyledons, whilst the inner portion of each strand becomes *wood* or *xylem* (Greek, *xylon*, wood) and the outer part *bast* or *phloem* (Greek, *phloios*, bark), a band between these two parts remains embryonic. This layer is called the *cambium*, or more precisely, for a reason we shall see presently, the *fascicular cambium*, the cambium, that is, within the bundle. Such a bundle, possessing such a cambium-layer, is termed an *open* one.

Between the bundles, connecting the pith in the centre with the cortex on the outside of the ring of bundles, are parts of the original or *ground-tissue* of the stem, which are known as *primary medullary rays* or *pith-rays* (Fig. 5). In Dicotyledons they are often broad and conspicuous; but in Gymnosperms they are so

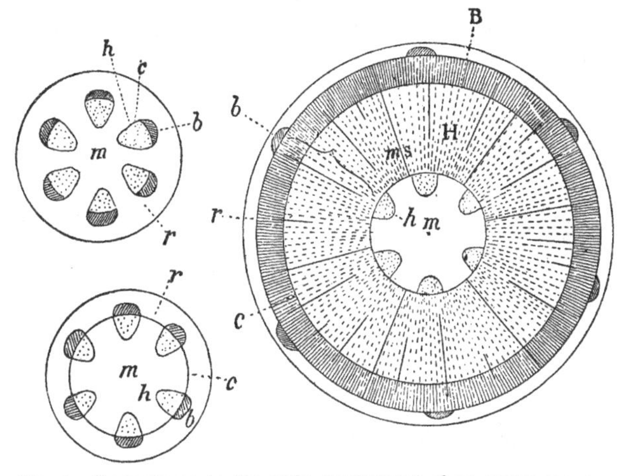

FIG. 5.—Three stages in the early development of an exogenous stem. *m*, pith; *r*, cortex; *h*, primary xylem; *H*, secondary xylem; *b*, primary phloem; *B*, secondary phloem; *c*, cambium; *ms*, pith-ray.

narrow as not to be visible to the naked eye. From the cambium-layer in one bundle to those in the bundles on either side of it the formation of cambium extends, across the primary pith-rays, so that instead of mere strips of cambium running longitudinally down the stem between the xylem and phloem of each bundle, there is now a cylindrical sheath of cambium extending from the embryonic tissue of its terminal bud downwards over the whole stem. In transverse section this sheath appears as a ring, and is accordingly sometimes called the cambium-ring. Those parts of it that extend between the bundles are termed *inter-fascicular cambium*, in contradistinction to the precisely similar tissue within the bundles. This cambium-sheath is familiar to us all as the layer of delicate thin-walled cells, full of sticky protoplasm, through which we easily tear when we peel a stick. Having what has been termed the quality of perpetual youth, it remains recognizable in a stem many years of age, and with the pith furnishes us with a convenient rough classification of all the structures of such a stem.

As we have seen, the pith, not having grown since its earliest condition, remains as a mere central line in such a stem. From this pith to the cambium-sheath is *wood* or *xylem*: outside the cambium is the *rind*, or, as it is commonly but somewhat misleadingly termed, *bark*, made up of the outer and often corky *cortex* and the inner, largely fibrous, *phloem* or *bast*.

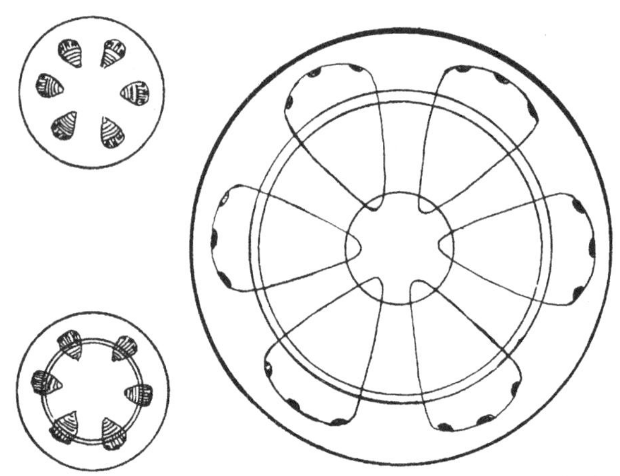

FIG. 6.—Diagrams of exogenous stem with six bundles, during the first year, at the beginning and at the close of the second year's growth, the last showing the wedge-shaped masses of primary xylem projecting into the central pith, and the formation of the first ring of secondary wood during the second year by the activity of the cambium ring.

In the first year the xylem and phloem are formed directly by the modification of the inner and outer cells respectively of the procambium-strand; but subsequently all wood, bast, and pith-rays originate in the cambium. Accordingly the xylem and phloem of the first year are termed *primary*, and that formed from the cambium *secondary* (Fig. 6).

The pith of trees seems mainly a structure of temporary utility to the plant, and the function of the cortex is chiefly protective; but as the main function of the stem is to convey liquid nourishment from the root to the leaves, and to carry back, also in a diffusible form, the material elaborated in the leaves to growing parts, it is one of the most noticeable characters of the bundles

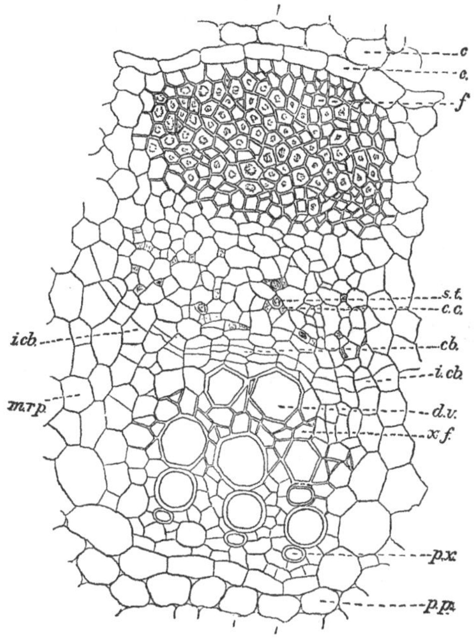

FIG. 7.—Transverse section of the stem of the Jerusalem Artichoke (*Heliánthus tuberósus*). From *The Elements of Botany*, by Mr. Francis Darwin, by his permission and that of the Syndicate of the Cambridge University Press. *c*, cortex; *f*, bast fibres; *c.c*, companion-cells; *i.cb*, interfascicular cambium; *d.v*, pitted vessel; *p.x*, spiral vessel of protoxylem; *e*, endodermis; *s.t*, sieve-tube; *cb*, cambium; *m.r.p*, pith-ray; *x.f*, wood fibre; *p.p*, pith.

that they are largely composed of *vessels*, elongated tube-like structures formed by the absorption of the transverse, or top and bottom, walls of rows of long cells placed end to end. For this reason they are often spoken of as *vascular bundles*. They also contain, however, cells which have not been thus fused into vessels, such cellular tissue, when its constituent cells are not more than three or four times as long as they are broad, being technically known as *parenchýma.*

As we have already seen, in addition to its function of conducting liquids, which necessitates these vessels or other *conducting tissue*, as it is termed physiologically, the stem has to perform the mechanical function of bearing up a considerable weight—itself, its branches, leaves, etc. To enable it to do this, both xylem and phloem are commonly accompanied by elongated elements, of which the chief characteristic is that their walls are much

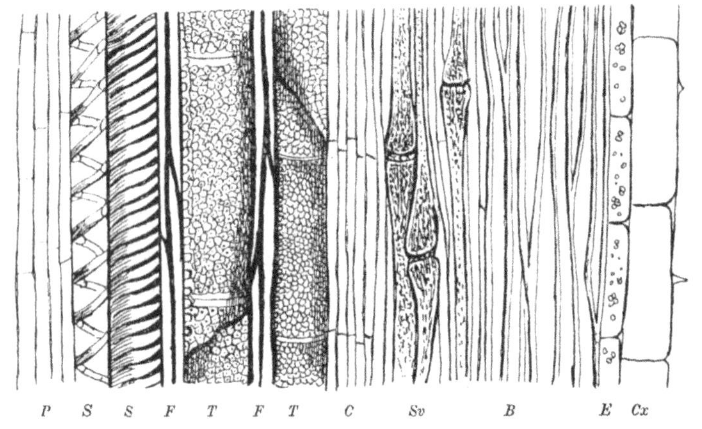

FIG. 8.—Longitudinal section of a fibro-vascular bundle in Heliánthus. (After Sachs.)
P, pith; *S*, spiral vessels of protoxylem; *F*, wood-fibres; *T*, tracheæ; *C*, cambium; *Sv*, sieve-tubes; *B*, bast-fibres; *E*, endodermis; *Cx*, cortex.

thickened and hard. The elements of this *mechanical tissue* are known as *fibres*, and from containing them the bundles are often termed *fibro-vascular bundles* (Figs. 7 and 8).

The walls of cells, fibres, and vessels in the xylem acquire mechanical strength or resistance by undergoing a change known as *lignification.* This consists in their impregnation with a substance known as *lignin.* Lignin consists of the same three elements as cellulose, viz. carbon, hydrogen, and oxygen, but in different proportions, its percentage composition being 49 per cent. of carbon, 6 of hydrogen, and 44 of oxygen. Its chemical constitution is, however, as yet unknown. It is harder and more elastic than cellulose, readily permeable by water, but not absorbent, not, that is, retaining the water. It is more soluble in acids, such as chromic acid, than is cellulose, and is recognized by turning brown when treated with Schulze's solution, a mixture of zinc-chloride, potassium-iodide, and iodine which turns unaltered cellulose blue.

The elements of the phloem, with which we are less concerned than we are with the xylem, though often variously thickened, are not lignified. They consist of *bast-parenchýma*, *sieve-tubes*, *companion-cells*, and *bast-fibres*, besides the medullary rays which traverse xylem and phloem alike. *Bast-parenchýma* consists of slightly elongated cells in vertical rows of four or six, of which the terminal cells taper. This arises from each row having been formed by several transverse divisions of a single procambium or cambium cell. They generally contain protoplasm and sometimes grains of starch or crystals. *Sieve-tubes* are the vessels of the bast, long tubes with transverse partition-walls and retaining their protoplasm but communicating through these transverse walls by the *sieve-plates* from which they take their name. The sieve-plate is a thin portion of the wall perforated by numerous pits close together. The sieve-tubes are the chief channel by which protoplasmic matter manufactured in the leaves is conveyed through the stem. *Companion-cells* occur only in angiosperms. In longitudinal section they appear as narrower cells alongside the sieve-tubes filled with granular protoplasm and with unperforated transverse walls adjoining those of the sieve-tubes. In a transverse section they appear like small corners cut off the larger sieve-tubes, and they have their name from the fact that each of them originates in this way, a longitudinal wall dividing the original cell into two unequal parts, of which the larger contributes to a sieve-tube, the smaller remains a cell. Bast-parenchyma, sieve-tubes, and companion-cells are known collectively as *soft bast* in contradistinction to bast-fibres or *hard bast*. *Bast-fibres* are extremely elongated structures, tapering at each end, containing only water or air, and with their walls so thickened as sometimes to almost obliterate the cavity or *lumen*, as it is termed. Their walls are generally at least partially lignified and give a reddish colour with Schulze's solution, and the thickening is absent from some spots on their walls. These unthickened spots are known as *pits*. Pits, which are important as occurring also on some of the elements that make up wood, are of two main classes, *simple* and *bordered*. A *simple pit* is a spot at which a cell-wall is left unthickened, generally on both sides, each successive thickening-layer leaving the same space uncovered. It appears accordingly as a bright spot on the wall; or, if in section, as a canal, the length of which depends upon the thickness of the wall. A

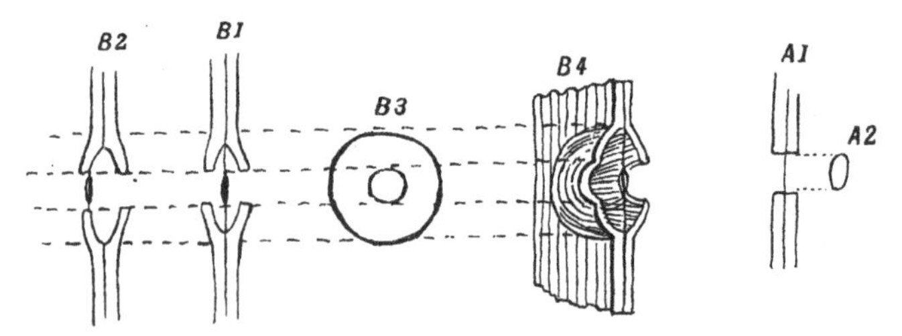

FIG. 9.—Pits. *A*, Simple pit; *A1*, in tangential longitudinal section; *A2*, in surface view. *B*, Bordered pit; *B1*, in tangential longitudinal section; *B2*, the same, with the middle lamella thrust to one side; *B3*, in surface view; *B4*, in semi-profile.

bordered pit is so-called because the bright spot appears surrounded by, or crossed by, a second circle or ellipse. The structure will be best understood from the diagrams (Fig. 9). In the thickening of the cell-wall the area of the outer circle is at first unthickened, but successive layers of thickening overlap this unthickened area more and more so as to make a short canal broad at the end near the original cell-wall and narrow at the end towards the centre of the cell. Subsequently a slight thickening termed the *torus* forms in the centre of the unthickened area. Pressure of liquid on one side of the pit-membrane often forces it against the "border," in which case the torus does not completely occupy the opening in the border or inner circle. The whole mechanism has been compared to a laboratory filter, the border being the funnel that acts as a support, the unthickened membrane, which is permeable, corresponding to a filter-paper and the torus to the small platinum cone sometimes placed in the middle of the filter to protect it from direct pressure of liquid. The bordered pits on xylem vessels in Oak have been compared to screw-heads, discs traversed by an elongated mark like the groove for a screw-driver, and the structure has been explained by the following imaginary model.[1] "Imagine a pair of watch-glasses each pierced by a narrow slit, and imagine them united face to face with a delicate circular piece of paper between them, and then fixed into a hole cut in a thick piece of card. The outline of the screw-head is the outline of the united watch-glasses where they are let into the card: the groove in the screw-head is the oblique cleft which leads into the space between the glasses." In some cases, under pressure from the cell-contents on the other side of it, the unthickened membrane in a pit bulges into the cavity of the adjoining vessel. Such projections, which are known as *tyloses*, may undergo cell-division and may even form a mass of tissue blocking up the entire lumen of the vessel. This is the case in some of the vessels of Oak and still more strikingly in the Locust or Acacia (*Robínia Pseudacácia*), in which the wood consequently appears non-porous, but, their cell-walls being thin, the tyloses appear in transverse section as light yellow spots on the dark heartwood. In Letterwood (*Brósimum Aublétii*), on the other hand, the tracheæ are filled up with tyloses, the cells of which have their walls very much thickened so that they appear dark.

We come next to the tissues which are of the greatest importance in our present study—those of the *xylem* or wood, developed on the inner side of the procambium strand and subsequently on the inner side of the cambium sheath. The development of xylem in a procambium strand begins with the conversion of one or a few cells, or vertical rows of cells, of the inner part of the strand into spirally, or occasionally annularly, thickened *tracheids* or *tracheæ*, known as the *protoxylem* or first-formed wood. This conversion consists in the loss of their protoplasmic contents, the lignification of their walls, the

[1] Francis Darwin, *Elements of Botany*, pp. 77-8.

deposit of a spiral thickening band internally, or of a series of rings, and, in the case of tracheæ, the absorption of the trans-

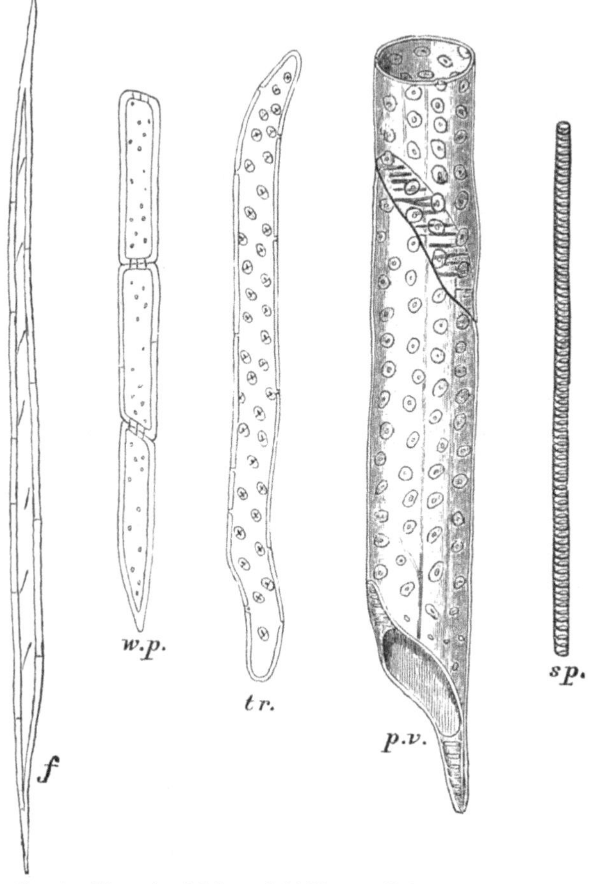

FIG. 10.—Elements of Oak wood, highly magnified. *f*, fibre; *w.p*, part of row of wood-parenchyma cells; *tr*, tracheid; *p.v*, trachea (part of); *sp*, part of a spiral vessel. From *The Oak*, by permission of Prof. Marshall Ward and Messrs. Kegan Paul, Trench, Trübner & Co.

verse walls of the vertical rows of cells. Whilst *tracheids* are elongated cells, losing their contents, generally becoming lignified and having thickened walls, so as to be adapted for the conveyance of air or water, *tracheæ* or true *vessels* differ from them only in being formed by the fusion of vertical rows of cells. In a transverse section the protoxylem is recognizable by the relatively small diameter of its tracheæ or tracheids; and, where there is a distinct pith, they may be seen projecting into the outer part of the pith in a discontinuous ring known as the *medullary sheath.* In longitudinal section the loose rings or spirals of their thickening are usually conspicuous, since, being the first vascular elements to form, they are considerably stretched by the growth in length of the adjoining fundamental tissue. The spiral or annular thickening permits, by an uncoiling in the former or a separation of the rings in the latter, a considerable amount of such stretching (Fig. 10).

The differences between the wood of coniferous trees and that of broad-leaved trees show themselves in the protoxylem and the rest of the primary wood, though they are even more important in the secondary xylem, *i.e.* that formed after the cambium-ring is complete. We will, therefore, now deal with them separately, taking the simpler type, that of the conifers, first (Fig. 11).

The xylem of conifers, both primary and secondary, consists mainly of tracheids. It contains, that is, no true vessels or tracheæ. In addition to the protoxylem the primary wood, *i.e.* that which is formed direct from the inner cells of the procambium strand, contains other wider tracheids with bordered pits between the turns of their spiral thickening.

A cross section of a Pine or Spruce shows distinct annual rings each made up of an inner, softer, light-coloured portion, the *spring wood,* and an outer, firmer, darker-coloured portion, the *summer wood.* The outer zone of the wood, that next to the bark, comprising from 30 to 50 of the most recently formed of these annual rings and from one to three or more inches across, is of lighter colours and is known as the *sap-wood.* Many of its cells are still in a sufficiently active state of vitality to store up starch, at least in winter, though growth is confined to the outermost layer of all, the cambium. The inner rings are darker and constitute what is known as the *heart-wood*, the cells of which are physiologically dead and serve only a mechanical function, that of supporting the weight of the tree and resisting the lateral strain of the wind. The darker colour of this heart-wood is due to infiltration of chemical substances into the cell walls, but not, in pine, as is sometimes supposed, to any greater thickening, lignification, or filling up of the cells than there is in the sap-

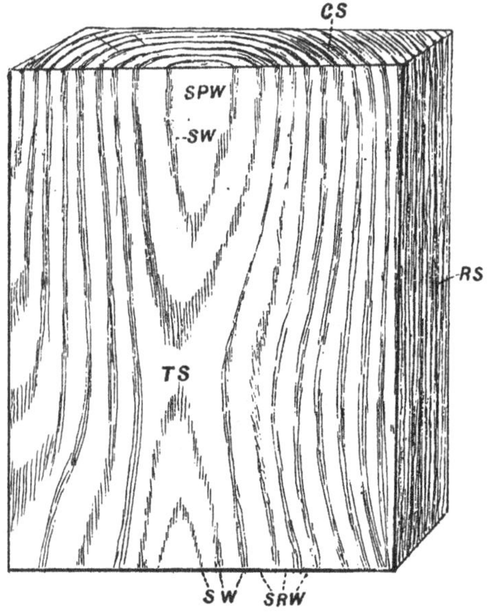

FIG. 11.—Coniferous wood, about natural size. *TS*, tangential section; *RS*, radial section; *CS*, cross section; *SPW*, spring wood; *SW*, summer wood. (After Roth.)

wood. The proportion of sap-wood to heart-wood is always considerable, but it varies in width even in different parts of the same tree, the same year's growth being sometimes sap-wood in one part and heart-wood in another. The width of the annual rings varies from half-an-inch or more near the centre of very quick-grown trees to one-eighth or one-sixth of an inch (3-4 mm.), common widths for the twenty innermost rings in deal, one-twelfth of an inch, a general average width, one-thirtieth

(0·7 mm.), an average for the twenty outermost rings, and even a minimum of one two-hundredth of an inch (0·2 mm.).[1] Many local causes, especially exposure to wind, produce excentricity of growth, few trees presenting a truly circular cross-section or a truly central pith, though this is more common among pines than among other trees. Branches almost always present an excentrically oval section, the pith nearer to the upper surface. The summer-wood in each ring being darker, heavier, and denser, its relative proportion to the spring-wood largely determines the weight and strength of the wood, so that colour becomes a valuable aid in distinguishing heavy, strong pine wood from that which is light and soft. Whilst on a cross-cut or transverse section the annual growths appear as rings, on a longitudinal radial section they are represented by narrow parallel stripes alternately light and dark, and on a longitudinal but tangential section by much broader alternating and less parallel stripes with some V-shaped lines (Fig. 12).

Under the microscope a transverse section of coniferous secondary wood presents regular straight radial rows of apparently four-sided meshes or openings, the transverse sections of tracheids. These are as broad in a radial as in a tangential direction in the spring wood, but much narrower radially in the summer wood of each ring. The cell-walls also are thicker in the summer wood. The radial walls have bordered pits, and in some cases such pits also occur on the tangential walls. Scattered through the summer wood are numerous irregular grayish dots, which on being magnified are seen to be the cross sections of

[1] Poplars grown in moist ground may reach a diameter of 14 inches in 8 years. Laslett records (*Timber and Timber-trees*, ed. 2, pp. 44-5) exceptionally fine English Oak and Elm, and an average drawn from several specimens of Canadian Oak and Elm which gave the following number of rings at 6, 12, 18, and 24 inches diameter :

	6 in.	12 in.	18 in.	24 in.
English Oak, - - -	13	19	24	30
Canadian Oak, -	49	105	160	216
English Elm, - -	10	16	25	36
Canadian Elm, - - -	80	156	252	—

relatively large spaces, the *resin-passages*, each surrounded by a layer of thin-walled cells, the *resin-epithelium* (Fig. 13). These

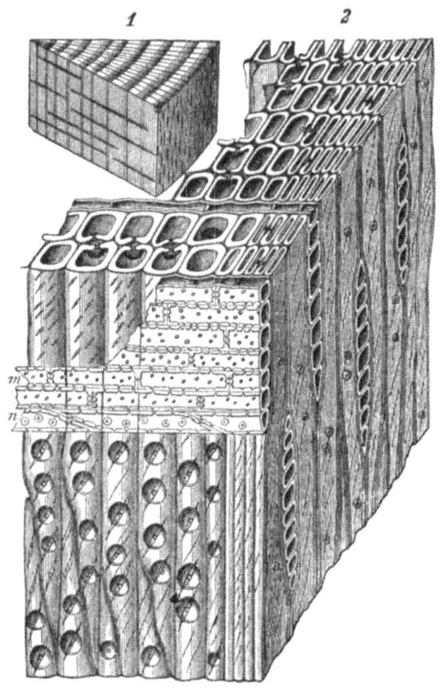

FIG. 12.—1. Piece of wood of Spruce (*Picea excélsa*) with the bark removed, natural size. 2. A portion from the nearest upper outside angle of 1, showing wood near the outside of an annual ring, magnified 100 times. From Hartig's *Timbers and how to know them*, by permission of Dr. Somerville and Mr. David Douglas.

resin-passages are not cells or vessels, but intercellular spaces, into which the resin oozes from the surrounding epithelium (Fig. 14). They generally occur singly, though sometimes in groups, and are most readily detected on a very smooth surface, or are often more easily seen on radial or tangential sections. On these they appear as fine lines or scratches running longitudinally. The whole mass of xylem is traversed radially by *pith-rays*, most of which appear in the transverse section of the stem as only one cell in width and made up of cells elongated radially. In a longitudinal and radial section (Fig. 15) it appears that the tracheids are from $\frac{1}{20}$ to $\frac{1}{5}$ inch long, 50-100 times as long, that

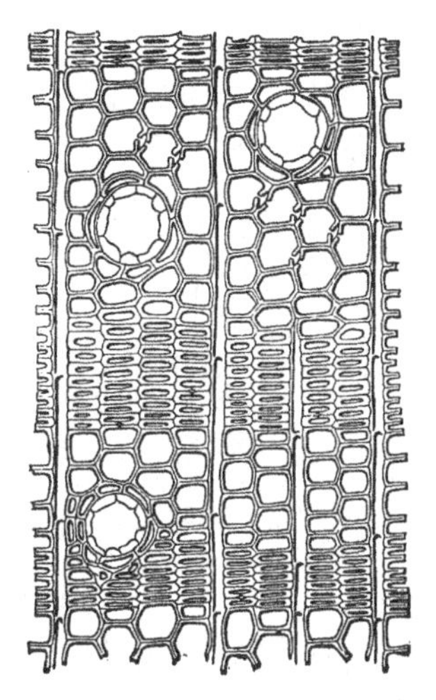

FIG. 13.—Transverse section of Spruce (*Pícea excélsa*), magnified 100 times, showing narrow rings, thin walls and three resin-ducts. From Hartig's *Timbers and how to know them*, by permission of Dr. Somerville and Mr. David Douglas.

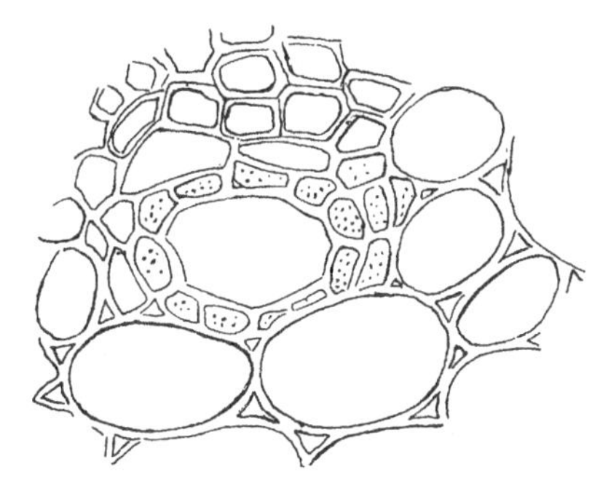

FIG. 14.—Resin-duct in coniferous wood, in transverse section, highly magnified, showing the epithelial cells surrounding the duct.

is, as they are wide; that they have their bordered pits in a single row down their radial walls; and that they are closed at their ends by a tapering to one side like the cutting edge of a carpenter's chisel. The pith-rays in longitudinal sections are seen to extend only a short way longitudinally, each appearing on radial sections as a band of 8 to 10 rows of cells elongated at right angles to the elongation of the tracheids like bricks in a wall 8-10 bricks high, with bordered pits on the cells of the

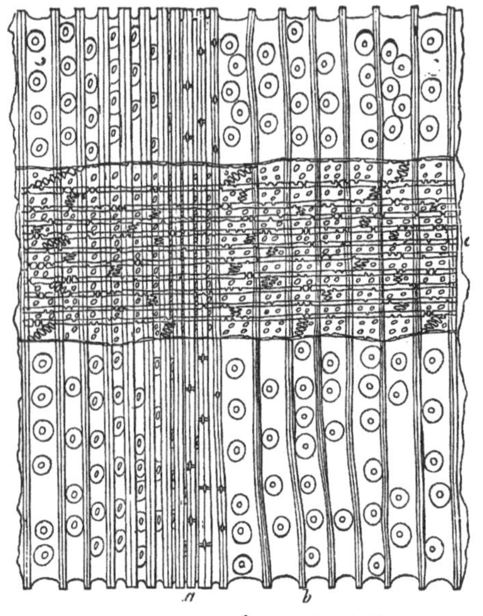

FIG. 15.—Radial section of Silver Fir (*Ábies pectináta*), showing a medullary ray, with simply pitted, parenchymatous cells, crossing wide tracheids of spring wood, and narrower ones of autumn wood, with bordered pits. Magnified 100 times. From Hartig's *Timbers and how to know them*, by permission of Dr. Somerville and Mr. David Douglas.

upper and lower rows, in Pines and Spruces, and simple pits on the others. On tangential sections the rays appear as vertical series of 8-10 pores tapering above and below. In Pines there are some larger pith-rays containing horizontal resin-passages.

The development of this comparatively simple type of wood from the cambium can be readily traced. The cambium is a cylindrical sheet of very thin-walled cells, each of which is rectangularly prismatic, broader in a tangential direction and tapering above and below to a radially-directed chisel-edge.

These cells contain protoplasm. After they have grown somewhat in a radial direction, partition walls form across them in the longitudinal tangential direction, so that each cell gives rise to two radially placed towards one another, and, this process being then repeated in one or both of the resultant cells, a radial

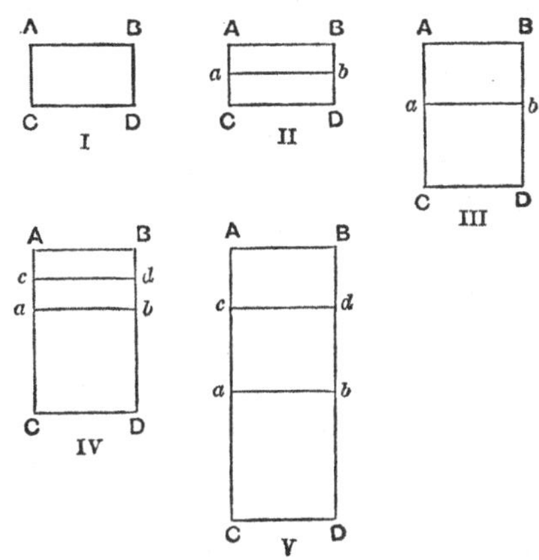

FIG. 16.—Diagram illustrating merismatic tissue. I, a merismatic cell ABCD; II, a cross-wall *ab* has appeared; III, A*ab*B has grown and again equals ABCD in size, whilst *a*CD*b* has also grown; IV, A*ab*B has been divided by a cross-wall *cd*; V, A*cd*B has again grown: it equals ABCD in size and is ready again to divide. Meanwhile *cabd* and *a*CD*b* have increased in size considerably. From *The Elements of Botany*, by Mr. Francis Darwin, by his permission and that of the Syndicate of the Cambridge University Press.

row is formed (Fig. 16). After several such divisions the innermost and earliest-formed of these cells ceases to divide, and uses up its protoplasmic contents in lignifying and thickening its walls, except at certain spots which become pits. It has, in fact, become a water-and-air-conducting tracheid. A cambium cell in the same radial row as a pith-ray undergoes transverse division into 8-10 superposed cells which elongate radially and retain protoplasmic contents, thus continuing the pith-ray (Fig. 17). In spring, when there is little heat, light, or activity of root and leaf to supply material, and when the bark, split by winter, may exert but little pressure, tracheids are produced with relatively thin walls and wider radial extension, constituting the spring wood; but in summer heat, light, and physiological activity, thicker walls are produced, whilst increased pressure of new bark allows less radial extension. As winter comes on, the active growth and division of the cambium cells cease, and its recommencement to form large thin walled tracheids in the following spring, after being dormant for several months, produces the sharp contrast between compressed summer tracheids and larger spring ones that marks a new annual ring.

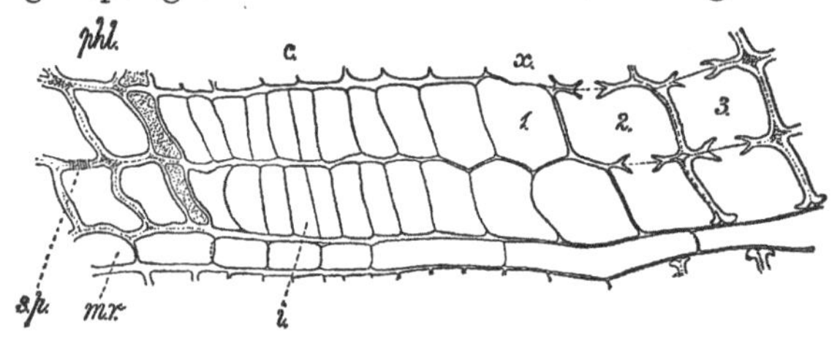

FIG. 17.—Transverse section of Scots Fir (*Pinus sylvéstris*). After Strasburger. From *The Elements of Botany*, by permission of Mr. Francis Darwin and the Syndicate of the Cambridge University Press.

phl, phloem; *s.p*, sieve-plate; *m.r*, pith-ray; *c*, cambium; *i*, initial cell of cambium; *x*, xylem; *1*, *2*, *3*, successive stages in the development of bordered pits.

The simple uniformity of structure in coniferous wood contributes largely to its great technical value.

Space does not permit any detailed discussion of the physiological uses of the different parts of such a stem as that of a conifer to the growing tree. The following recapitulation must suffice. The vitality of the pith of trees is generally confined to the very earliest stages of their existence, and the spirally-thickened elements of the protoxylem also only serve as conducting tissue when all the xylem is young. Heart-wood has ceased to have any active functions, serving merely for strength. Whilst cortical tissue serves to protect from external action, damp, etc., and to check transpiration, the sieve-tubes of the phloem appear to be the chief carriers of the food-materials elaborated by the leaves to the growing parts of the stem; and the formation of new phloem and xylem is the one function of the cambium. In the sap-wood of conifers consisting, as it does, so largely of tracheids, it is these tracheids, communicating as they do by the bordered pits on their radial walls, that convey water and air from the roots to the leaves, though they also store up starch in autumn and winter. The pith-rays being

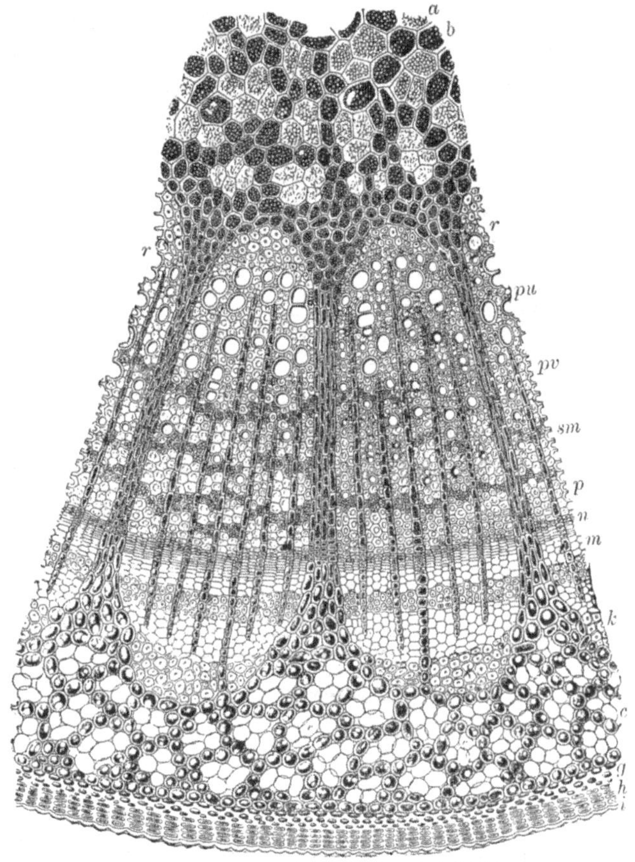

FIG. 18.—Transverse section of part of young stem of Oak, highly magnified. *ab*, pith; *c*, cortex; *i*, epidermis; *h*, periderm; *g*, collenchyma; *r*, spiral vessels forming protoxylem; *pv*, pitted vessels (tracheæ); *sm*, secondary pith-rays; *p*, wood-parenchyma; *n*, *m*, cambium; *k*, bast-fibres. After Hartig, from *The Oak*, by permission of Professor Marshall Ward, and Messrs. Kegan Paul, Trench, Trübner & Co.

elongated radially, retaining their protoplasm, forming starch, and communicating through their pitted walls with phloem and even cortex as well as xylem, undoubtedly play an important part in the transfer of formative material from one part of the stem to another.

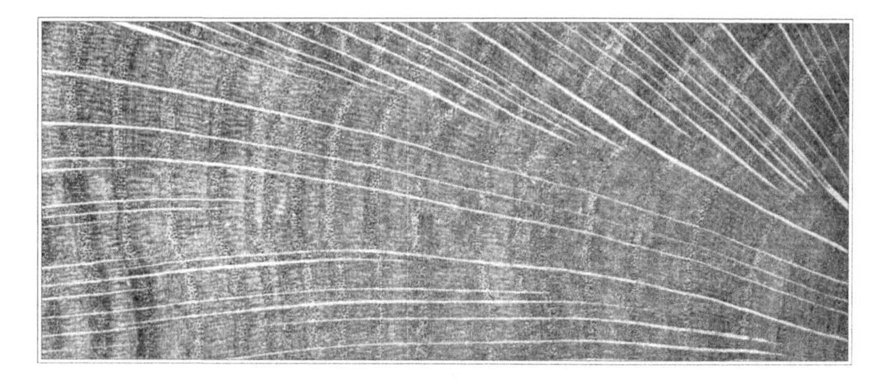

Fig. 19.—Transverse section of Oak, photographed direct from nature.

When we examine the stem of a broad-leaved tree, such as an oak, we find, with the same general exogenous arrangement of

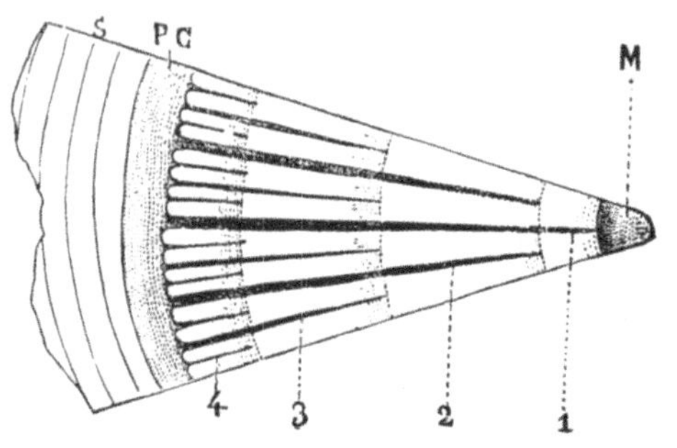

Fig. 20.—Part of transverse section through a branch of Cork Oak (*Quércus Súber*), 4 years old. After Le Maout and Decaisne, from *The Elements of Botany*, by permission of Mr. Francis Darwin and the Syndicate of the Cambridge University Press.
M, pith; PC, phloem and cortex; S, cork; 1, primary pith-ray, running from pith to cortex; 2, 3, and 4, secondary pith-rays formed in successive years.

pith, bark, heart-wood, sap-wood, and annual rings, considerably greater complexity in the variety and grouping of the elements of which the tissues are built up (Fig. 18). The pith presents considerable variety among broad-leaved trees, so as to be used to some extent in discriminating woods seen in complete cross-sections. Thus in its proportion to the area of the wood in cross-section it may vary from equality, *i.e.* being as wide as the

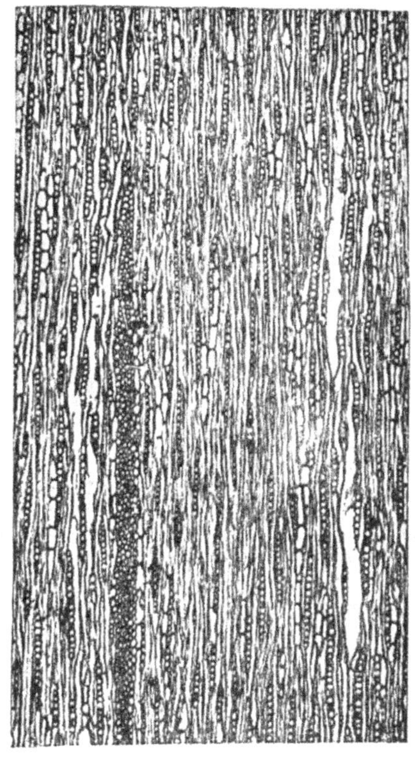

FIG. 21.—Tangential longitudinal section of Oak, magnified 50 diameters, showing transverse sections of pith-rays. After Müller, from *The Oak*, by permission of Prof. Marshall Ward, and Messrs. Kegan Paul, Trench, Trübner & Co.

xylem, as in three-year old shoots of Elder, to $\frac{1}{250}$, as in shoots of the Cork-Elm of the same age. In outline it may be pent-angular or hexagonal, as in Oak, Spanish Chestnut, Black Poplar, or White Willow; triangular, as in Birch, Beech, and conspicuously in Alder; ovoid, as Linden, Plane, Holly, Ash, and

Maples; or nearly circular. In the last-mentioned case the projections of the primary xylem into the pith may give the pith a

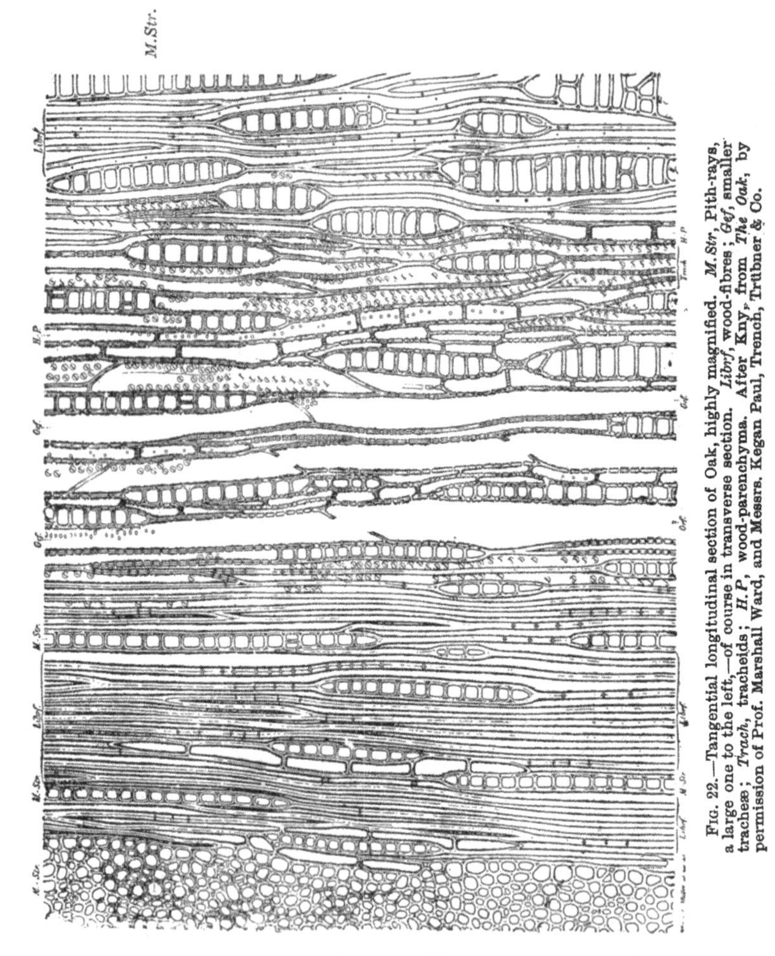

FIG. 22.—Tangential longitudinal section of Oak, highly magnified. *M.Str*, Pith-rays, a large one to the left,—of course in transverse section. *Librf*, wood-fibres; *Gef*, smaller tracheæ; *Trach*, tracheids; *H.P*, wood-parenchyma. After Kny, from *The Oak*, by permission of Prof. Marshall Ward, and Messrs. Kegan Paul, Trench, Trübner & Co.

wavy or *crenate* outer margin, as in Hawthorn, Rowan, Laburnum, Horse-Chestnut, or Elder; or this margin may appear even, as in Elm, Hazel, and Dogwood (*Cornus*). In the Walnut the pith has an interrupted or chambered structure: in the Elder it soon dies and disintegrates, leaving the stem hollow; whilst in young stems of Elm the inner portion of it has thin walls and loses

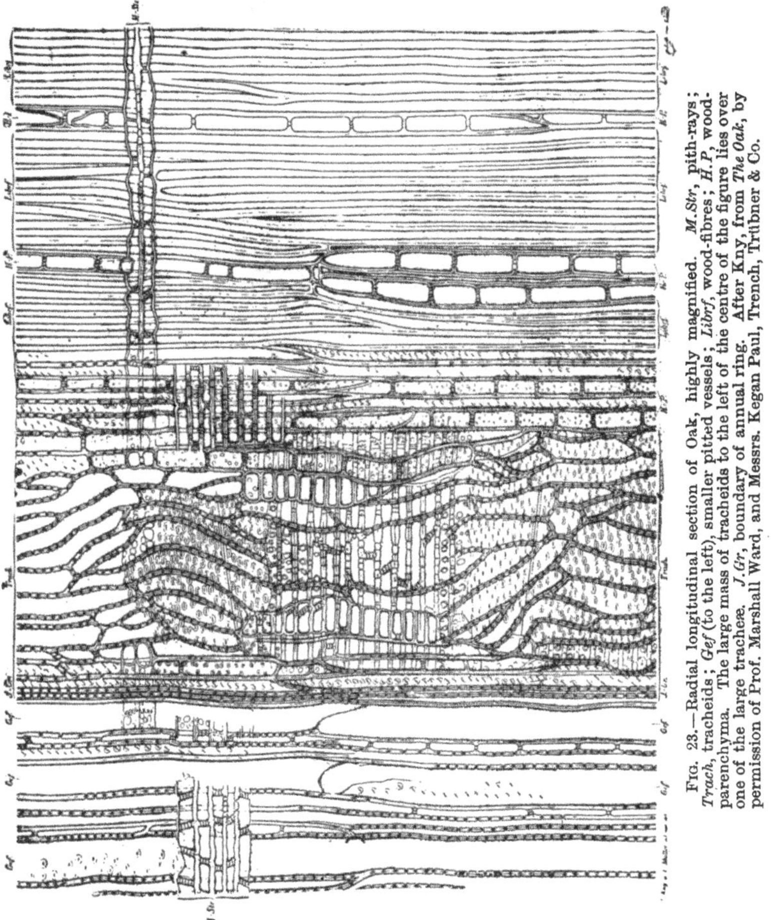

Fig. 23.—Radial longitudinal section of Oak, highly magnified. *M.Str*, pith-rays; *Trach*, tracheids; *Gef* (to the left), smaller pitted vessels; *Librf*, wood-fibres; *H.P*, wood-parenchyma. The large mass of tracheids to the left of the centre of the figure lies over one of the large tracheæ. *J.Gr*, boundary of annual ring. After Kny, from *The Oak*, by permission of Prof. Marshall Ward, and Messrs. Kegan Paul, Trench, Trübner & Co.

its protoplasm, whilst the outer part becomes thick-walled but retains its cell-contents.

The pith-rays of broad-leaved trees are in general far more

conspicuous than those of conifers. In Oak the large primary pith-rays extending from pith to cortex are often twenty or more cells in width, appearing as long, clearly defined, grayish lines in a transverse section of the stem (Fig. 19). The secondary pith-rays are much narrower as well as shorter (Fig. 20). In a tangential section (Figs. 21 and 22) the primary rays may be several hundred cell-rows, *i.e.* upwards of an inch, in height, and, however wide at the middle, taper to one cell at each end. On a radial

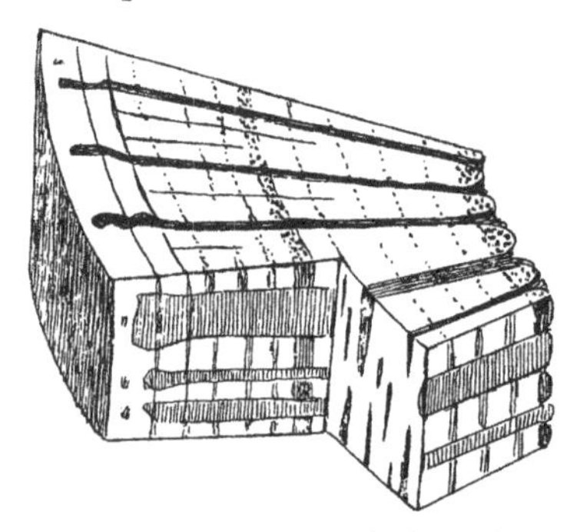

FIG. 24.—Two annual rings of wood and the bark of the Oak, the upper surface in transverse section, part of the inner ring (unshaded) in tangential, and the front view of both rings in radial section. The medullary rays are shown black in transverse, shaded in radial section. (After Hough.)

section they appear as broad, shiny bands, the "mirrors" or "silver grain," so that they are conspicuous on any section, in whatever plane it may be. In Oak they constitute 16-25 per cent. of the wood (Figs. 23 and 24).

The protoxylem of broad-leaved trees differs from that of conifers mainly in that its spirally-thickened elements are *tracheæ* or true *vessels*, owing to the absorption of the transverse walls of a vertical row of tracheids. But it is in the elements of the secondary xylem that we find the greatest complexity and variety. This may contain from three to five of the following six kinds of elements: tracheids, vessels, woody fibres, fibrous cells with thick or with thin walls, and wood-parenchyma. The *tracheæ* or true vessels vary considerably in transverse diameter, some of them being the widest pores seen in a transverse section of wood and being sometimes specially conspicuous in the spring-wood. Some of them, in young wood, have net-like thickening, but most of them have bordered pits, as have also the tracheids. The chief differences in fact between these two kinds of elements are the smaller diameter and lesser length of the tracheids. As they are each formed from a single cambium-cell, these *tracheids* have no transverse divisions; whereas in the vessels there are much-perforated or partially absorbed partitions inclined towards the pith-rays, indicating the origin of the vessels from the fusion of a chain of cells. *Woody fibres* may be as long as, or longer than, the tracheæ, and are often more pointed, but their distinctive characteristic is their much-thickened, lignified walls, marked with few simple pits, often oblique and narrow. This thickening of their walls sometimes almost obliterates the cell-cavity or lumen, and, together with their early loss of all contents but water and air, serves to indicate their main function to be that of mechanical support. *Fibrous cells* only differ from fibres in retaining their protoplasmic contents. Their walls sometimes remain thin. Both thick-walled fibrous cells and woody fibres sometimes become *chambered* by the formation of delicate transverse walls. *Wood parenchyma* consists of vertical groups of short cells, the upper and lower cell of each group tapering to a point, each group originating, in fact, from the transverse division of one cambium-cell. They retain their protoplasm and become filled with starch in autumn. Their walls are not much thickened, but are lignified and pitted, having bordered pits where in contact with tracheæ or tracheids, but simple pits elsewhere. Wood-parenchyma is commonly grouped in narrow circles round the vessels, appearing in longitudinal sections as cloudy margins to them. It may expand from such circles laterally into wings forming a spindle-shaped patch with the vessel in the centre, and these wings may widen until they meet others, so forming straggling oblique lines, long wavy streaks, or concentric circles ("*false rings*"). These transverse lines of tissue may be very narrow, as in Ebonies, or broad and conspicuous. Wood-parenchyma much resembles the pith-rays, especially in tangential longitudinal sections; but its walls are not elongated radially.

As has been said, the wood of broad-leaved trees may contain from three to five of these different elements. Vessels are always present, but in some cases tracheids are absent. The wood of Plane, Ash, and *Citrus* (Orange, Lemon, etc.), for example, consists of vessels, woody fibre, thin-walled fibrous cells and wood-

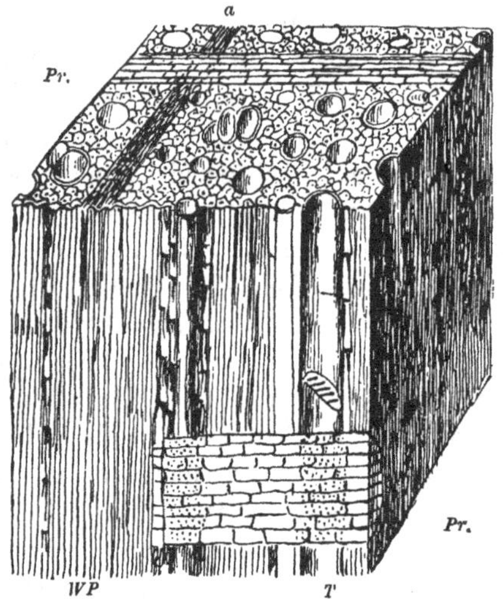

FIG. 25.—A piece of dicotyledonous wood, magnified about 100 times. A transverse section is shown above, with a pith-ray (*Pr.*) crossing the zone of autumn wood (*a*) which forms the outer boundary of an annual ring. In front is a radial longitudinal section showing wood-parenchyma (*WP*), some large tracheæ (*T*), and much wood-fibre, crossed by another pith-ray. The tangential section is in shadow.

parenchyma only. That of Holly, Hawthorn, and *Pyrus* (Apple, Pear, Rowan, etc.) is made up of vessels, tracheids, and wood-parenchyma: that of Maples, Elder, Ivy, *Euonymus*, etc., contains also thick-walled fibrous cells. The wood of *Berberis* (Barberry) consists exclusively of vessels, tracheids, and thin and thick-walled fibrous cells; and that of Oaks, Hornbeams, Plum, and Buckthorn of vessels, tracheids, woody fibre, and wood-parenchyma (Figs. 25 and 26). The most common type of structure, however, occurring in Willows, Poplars, Alder, Birch, Walnut,

Linden, *Magnólia*, *Ailanthus*, *Robínia*, etc., contains vessels, tracheids, woody fibre, thin-walled fibrous cells, and wood-parenchyma.

The distinctive features of woods, however, depend rather upon the proportions in which these elements are present and upon their arrangement than upon the absence of any of the six kinds of elements. There is, as a rule, among the woods of broad-leaved trees no such regularity of radial arrangement of

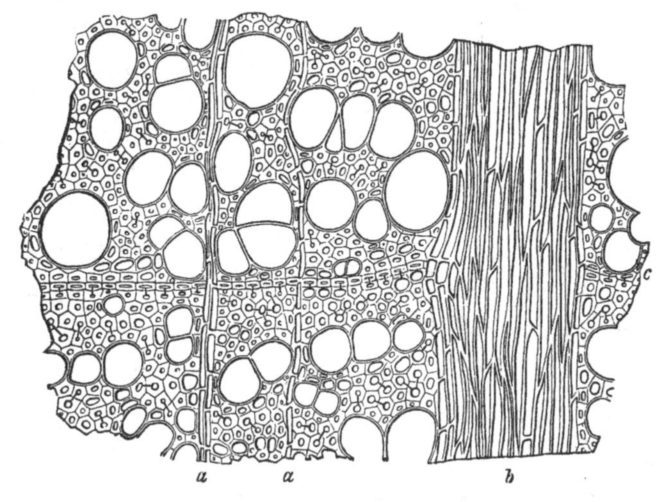

FIG. 26.—Transverse section of Beech (*Fágus sylvática*). Magnified 100 times. *a*, narrow pith-ray; *b*, broad pith-ray; *c*, boundary of an annual ring. The large pores are transverse sections of vessels (tracheæ). The thick-walled elements with narrow lumina are wood-fibres; those with thinner walls and wider lumina, wood-parenchyma or tracheids. From Hartig's *Timbers and how to know them*, by permission of Dr. Somerville and Mr. David Douglas.

elements as characterizes the simple wood of conifers. In the cambium region, it is true, owing to the repeated regular tangential divisions, the cells not only appear rectangular in a transverse section, but are also in regular radial rows; but in the xylem itself this regularity is disturbed by the different diameters attained by the various elements as they become fully formed. In Oak, for example, the annual rings are marked in a cross-section by the large and conspicuous pores, or sections of the vessels, which occupy the greater part of the spring wood of each ring (Fig. 27). On a radial section the layers appear as parallel

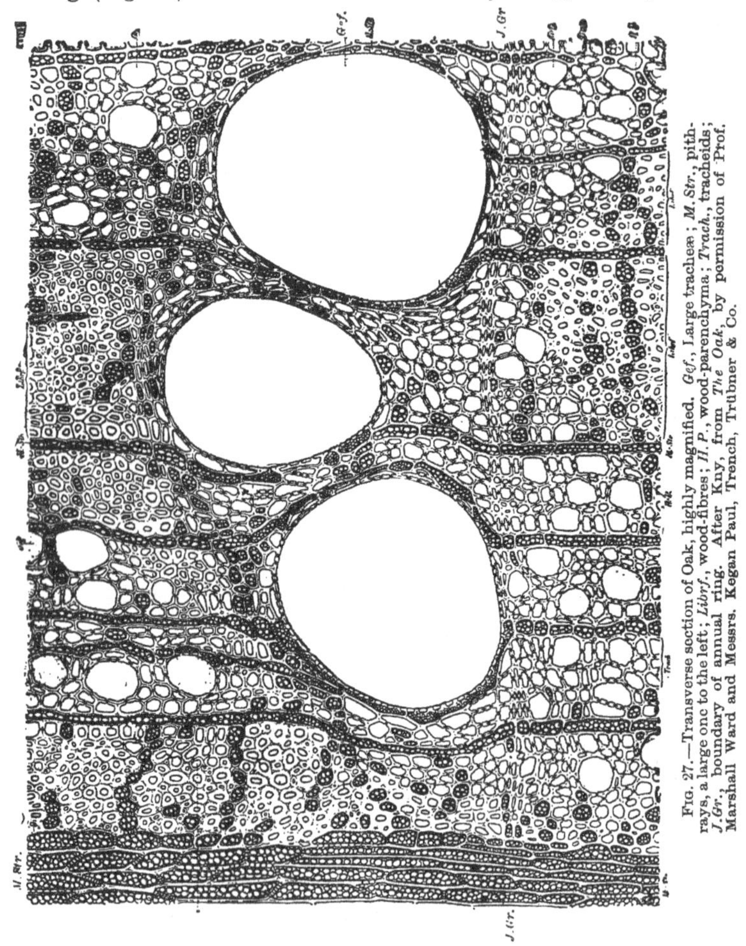

Fig. 27.—Transverse section of Oak, highly magnified. *Gef.*, Large tracheæ; *M.Str.*, pith-rays, a large one to the left; *Librf.*, wood-fibres; *H.P.*, wood-parenchyma; *Trach.*, tracheids; *J.Gr.*, boundary of annual ring. After Kny, from *The Oak*, by permission of Prof. Marshall Ward and Messrs. Kegan Paul, Trench, Trübner & Co.

stripes, and on a tangential one as broader and less parallel stripes; but, whilst in coniferous woods the dark bands were denser summer wood, in this case the darker parts are produced by the vessels in the spring wood, the more uniform fibres of the summer wood appearing lighter. Vessels, tracheids, and fibres formed in spring have larger diameters and thinner walls than those formed in autumn, which fact produces much of the distinctness of the annual rings. In timbers with well marked rings the distinctness of these rings may either be due, as in Oak, Ash, Teak, etc., to the contrast between wood with numerous large vessels and that with fewer or smaller ones; or, as in Birch, Maple, Horse-chestnut, etc., to the fibres being smaller across and thicker-walled in one part of each ring, whilst the vessels may be evenly dispersed through the whole wood. Woods differ widely as to the circularity of their rings. In not a few cases they are distinctly wavy; and, whilst in Beech and Hornbeam the crests of the waves—as seen in a cross-section—bend inwards at the primary pith-rays, in the Barberry they bend outwards. In evergreens, to which type belong the bulk of tropical broad-leaved timbers, where there is not the check to physiological activity produced by the "fall of the leaf," we do not, as a rule, find such well-marked annual rings. Sometimes, however, the annual rings are replaced by less completely concentric zones, often stretching as wavy, pale, bar-like markings from one primary pith-ray to another, and sometimes running into one another. These "false rings," as they have been termed, which are seen in the wood of Figs, She-oaks (*Casuarína*), Padouk (*Calophýllum*), etc., will be found on microscopic examination to be mainly produced by zones of wood parenchyma.

The grouping of the vessels also affords some useful distinctive characters. Thus in Box and in Quince they usually occur singly; in Hazel and Holly in groups of from 5 to 12; in Hornbeam in long sinuous radial lines between the pith-rays; in Elms in concentric bands like false rings; and in Oaks, Chestnut and Buckthorn, from 20 to 50 together, in flame-like groups (Fig. 28).

The elements of the wood are generally parallel in direction to the axis of the stem or limb in which they occur—*i.e.* the wood is *straight-grained*; but they may be spirally twisted round the stem, or oblique, in which latter case if successive layers lie in opposite directions the wood is *cross-grained.* A slightly wavy longitudinal course in the elements of the wood produces the condition known as *curly grain*, frequent in Maple; whilst slight projections or depressions repeated on the outer surface of successive annual layers produce the *bird's-eye* and *landscape* varieties in the same wood. The presence of undeveloped buds or knots, as in "burrs," produced on many trees by the attacks

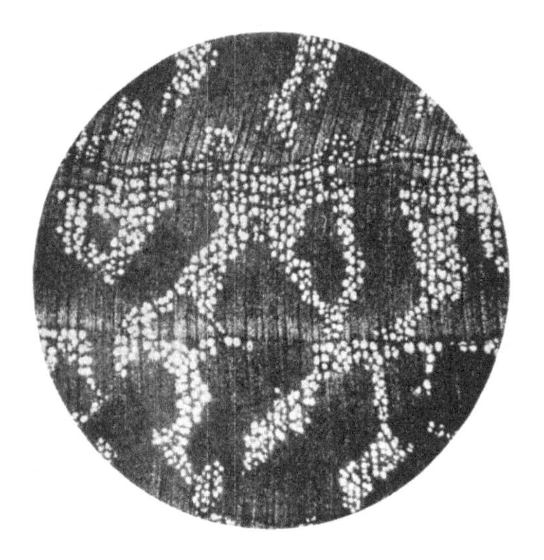

Fig. 28.—Transverse section of Buckthorn (*Rhámnus cathárticus*), showing flame-like groups of vessels.

of mites (*Phytóptus*), causes similar ornamental wavings of the grain

One main cause of the elements not being vertical is their growth in length and in diameter after leaving the cambium stage. Such growth in length causes the tips of the fibres to crowd in between those above and below, and become interlaced and oblique in direction. This adds to the toughness of the wood and makes it less easy to split, and may produce a visible twisting of stems or branches.

Up to a certain age the segments or chambers (original cells) of the vessels, the tracheids and the fibres, gradually increase each year both in length and diameter.

The pith-rays—as seen in cross-section—afford a very useful distinctive character, varying much, as they do, in number and in width. In Willow, Horse-chestnut and Ebony, as in Conifers, they are either only one cell in width, or are at least so inconspicuous as to require a lens for their observation, whilst in Oak and in the so-called She-oaks (*Casuarína*) they are conspicuous to the naked eye. They vary in width from ·005 millimetre to a millimetre; and in number from 20 or less in a breadth of 5 millimetres, as in *Labúrnum* and *Robínia,* to 64 in the same space, as in Oak, or even 140 in the case of *Rhododéndron máximum.*

Another character of some value in discrimination is the occurrence of *pith-flecks*, or *medullary spots*, dark rust-like patches, which occur in Alder, Birch, Hazel, Hawthorn and some species of Willow, Poplar and *Pýrus.* They are supposed by some authorities to originate in passages bored by the larvæ of a species of *Típula* (wire-worm) which live in the cambium, these passages becoming filled up immediately with cellular tissue; but their origin requires further investigation. We will postpone the consideration of such characters of woods as weight, hardness, colour and odour—characters that depend little, if at all, upon structure—to a subsequent chapter. It may be noted here that, while it is the lignified elements of woods, especially their tracheids and fibres, that give them their chief technological value, it is the stored up nitrogenous and other more complex, and therefore more chemically unstable, substances that are the most combustible, *i.e.* the most readily oxidized, and also the most readily decomposed by the attacks of fungi. It is these substances, therefore, that have to be eliminated, or at least taken into account, in the processes of seasoning or preserving timber, and it is their presence which renders sapwood generally less durable than the physiologically inert heartwood.

CHAPTER II.

THE RECOGNITION AND CLASSIFICATION OF WOODS.

NOT only carpenters and other workers in wood, but engineers, surveyors and timber-merchants at present recognize the timbers with which they are familiar as to kind, and even largely as to quality, by methods obviously and confessedly empirical, mere "rule of thumb." From this it results that, though woods may be accurately discriminated generically, as oak, ash, birch or pine, the species are seldom correctly distinguished, and, as a consequence, the best wood for any particular purpose is very often not obtained.

In these empirical identifications such more obvious but variable, and therefore less trustworthy, characters as weight, hardness, colour and odour are often more used than most of the structural characters described in the previous chapter. In attempting a more thorough-going discrimination we cannot ignore these more obvious characters; but it is important to recognize their variability and consequently merely secondary importance. Details as to the testing of weight and hardness will form the subject-matter of a subsequent chapter: we are here only concerned with rough approximations.

Weight of wood.—The weight of wood depends mainly upon two things, its compactness and its moisture. Compactness signifies the amount of woody or other solid matter in a given bulk, and this will generally be greater in slow-growing than in quick-growing species, greater in heartwood than in sapwood. Moisture is so far more variable in amount in the same wood, according to the extent to which it has been naturally or artificially seasoned, that no comparison of the weights of different woods can be of any value unless the samples have been kiln-dried, and even by this method it is difficult to secure a uniform elimination of moisture. Many tables have been published giving the density or specific gravity of various woods, their weight, that is, as compared to that of water, to three or even four places of decimals. A more useful form of statement, however, is perhaps the weight of a standard cube, either that of a cubic foot in pounds or that of a cubic decimetre in grams.[1] Thus, while water weighs 62·321 lbs. per cubic foot, timbers range from 13 to 85 lbs. per cubic foot.

They may be grouped from this point of view in the following six grades:

	Approximate weight of 1 cubic foot in pounds.	Specific gravity.
1. Very Light, - - - - -	not exceeding 24	·26—·4
Spruce, Willow and most Poplar.		
2. Light, - - - - - - -	24—30	·4—·5
Northern Pine, Hemlock Spruce, Linden, Chestnut.		
3. Medium, - - - - - - - - -	30—36	·5—·6
Pitch-pine, Douglas Spruce, Sycamore.		
4. Heavy, - - - - - - - - -	36—42	·6—·7
Most Birch, Beech and Walnut.		
5. Very heavy, - - - - - -	42—48	·7—·8
Hornbeam, Hickory and good Ash and Elm.		
6. Heaviest, - -	Above 48	Above ·8
Some Oak, most Teak, Mahogany, Jarrah, Mora and Greenheart.		

Whilst such kiln-dried weights as those employed here range from a specific gravity of ·26 in *Fícus aúrea*, or 13 lbs. to the cubic foot in *Erythrína suberósa*, to 1·3 in Black Iron-wood (*Condália férrea*), or 85 lbs. to the cubic foot, as in Anjan (*Hardwíckia bináta*), none of the native woods of temperate

[1] To facilitate the conversion of one measure into the other it may be noted that 1 cubic foot = nearly 28⅓ (28·315) cubic decimetres, and 1 pound avoirdupois = 453½ (453·592) grams.

latitudes are, when dry, as heavy as water.[1] Most of the woods in grade 6 of the above table grow within the tropics.

Hardness of wood.—Though in testing woods for engineering purposes various resistances, such as stiffness or elasticity and compressibility, have to be ascertained, hardness, or resistance to indentation, is often estimated roughly. It may be expressed with precision by the number of kilograms required to sink a punch one centimetre square to the depth of 1·27 millimetres ($\frac{1}{20}$ of an inch) perpendicularly to the fibres of the wood, or by the number of pounds per square inch to produce such an indentation. Here too we may, perhaps, group all woods roughly into six grades:

1. Hardest, such as the Iron-wood of India, *Mésua férrea*, which turns the edge of almost any tool, and Lignum-vitæ (*Guáiacum*), which requires 793 kilograms to produce the standard indentation.
2. Very hard, requiring more than 3200 lbs. per square inch, such as Hickory and good Oak and Elm.
3. Hard, requiring from 2400 lbs.-3200 lbs., such as Ash, Walnut, Beech, Holly and Sycamore.
4. Medium, requiring from 1600 lbs.-2400 lbs., such as Douglas Spruce.
5. Soft, requiring less than 1600 lbs., such as the majority of coniferous woods, Pine, Spruce, Cedar, Poplar, Linden and Chestnut.
6. Very soft, such as the so called Cotton-tree of India (*Bómbax malabáricum*), which is so soft that a pin can be readily driven into it with the fingers.

Hardness and density or weight to a great extent vary together. They also increase from the base of a stem up to its first branch, and decrease from that point upward.

Colour of wood.—The colour of the heartwood affords in many cases a useful aid in identification, while mere differences of tint are often indicative of quality or soundness. The black duramen of the Persimmon (*Diospýros virginiána*), of other species

[1] This is true when the contained air is not eliminated. For more precise estimates see Chapter VII.

of *Diospýros* known as Ebonies, and of Laburnum (*Cýtisus Labúrnum*), the dark brown of the Walnuts, the purplish-red of Logwood (*Hæmatóxylon campechiánum*), the lemon-yellow sapwood and bluish-red heartwood of the Barberry (*Bérberis vulgaris*), the narrow yellow alburnum and greenish duramen in Lignum vitæ (*Guáiacum officinálé*), or the mottling of dark and light browns in the Olive (*Olea europǽa*), are obvious distinctions.

The Northern Pine (*Pínus sylvéstris*) presents numerous variations in the colour of its wood, as well as in its mode of branching, dependent probably in part upon the conditions under which it is grown, and the superiority of "red deal" to the more resinous honey-yellow varieties is well-known in trade. Northern hill-grown wood is commonly redder than that of the south grown in plains, the finest being that of the Riga pines, with a close pyramid of ascending branches, including the timber from Smolensk, Vitebsk, Tchernigov, and Volhynia.

The Locust or False Acacia of the United States (*Robínia Pseudacácia*) includes at least four varieties of wood. The most durable, most beautiful, and most valuable is the red: the commonest the green, a greenish-yellow wood (apparently the only kind imported), is next in value; the black is only recorded in the Western States; and the white is the least valuable.

In West Virginia three varieties of the Tulip-tree (*Liriodéndron tulipíferá*) are distinguished as "White," "Blue," or "Yellow Poplar," of which only the last-named is commonly shipped to this country. Grown only for ornament in Europe, in America this tree is largely used for rafters, wainscots, roof-shingles, boxes, furniture, and turnery, and increasing quantities now arrive at Liverpool from New York under the names of American or Yellow Poplar, American Whitewood or Canary Whitewood. These names and that of "Tulip-wood" are nearly all objectionable, as previously applied to very different woods, or as suggesting a connection between the tree, a member of the Magnolia family, and the Poplars. The yellow variety of its wood comes from moist low-lying ground, and is valued for staining or polishing, by cabinet-makers, shop-fitters, and coach-builders.

Exposure to air or light darkens the colour of most wood, as is well seen in freshly felled, as compared to seasoned, Mahogany. Moisture carries this darkening deeper into the wood, whilst the black of Oak and the dark brown of Yew after prolonged immersion in bogs are well known. The translucency of all sound timber when in thin slices gives it a characteristic lustre, whilst incipient decay renders it dull and opaque. Any local departure from the natural colour peculiar to the species is an indication of incipient decay. The deterioration that sets in directly growing timber passes maturity generally shows itself first by a white colour at the centre of the butt end of the log. This is not a serious defect; but the yellowish-red tinge subsequently assumed indicates a loss of toughness and tenacity, and suggests that the log is not well fitted for constructive work. So too spots of discoloration scattered through a log, especially at its butt end, are liable to prove centres from which serious decay, caused or accompanied by parasitic fungal moulds, may spread. This remark does not apply, however, to the so-called *pith-flecks* or *medullary spots*, which are often numerous in woods when perfectly sound. The reddish-brown tinge known as *foxiness* is a clear sign of advanced decay, unfitting wood for any purposes requiring strength; but Oak is very often much prized by cabinet-makers when in this condition, merely on account of its colour.

Odours and resonance of woods.—The odours of woods, such as the resinous smells of Deal or Teak, the fragrance of Cedars, Toon, or Sandal-woods, the characteristic perfume of Camphor-trees and the unpleasant smells of the Stinkwood (*Ocotéa bulláta*) of South Africa and the Til (*Oreodáphné fœ́tens*) of Madeira, may sometimes be of use in discrimination, as, to an educated ear, may the notes given out by different woods when struck by a hammer. In the manufacture of musical instruments the wood must be of uniform structure, even-grained, free from knots, well seasoned, and unbent, so that each fibre may vibrate freely. The notes emitted will vary in pitch directly with the elasticity, and indirectly with the weight of the wood.

Silver Fir (*Ábies pectináta*), imported as "Swiss Pine," is employed for the sounding boards of pianos and the belly of violins, whilst Maple, a dense wood, is used for the back and ribs of the latter instrument.

Classification of woods.—Obviously these "rule of thumb" characteristics are generally made use of in practice, not separately, but together. This will also be the case in the classification which we are about to propose, which refers mainly to the appearance of transverse sections, including both heartwood and sapwood.

For ready identification and comparison of timbers, considering even the great variety that are used in the arts in various parts of the world, it is obviously necessary to have some system of classification. Botanists group trees, as they do other flowering plants, in accordance with the characters of their flowers, fruit, and leaves, a method which is undoubtedly the best for the purpose of indicating the genetic affinities of the various species. As we have seen, for instance, timber-yielding trees fall naturally into two main groups, conifers and dicotyledonous angiosperms, of which the first is generally distinguished by needle-like leaves and seeds borne exposed on the inner surfaces of scales arranged in a cone, whilst the second group has generally broad leaves and the seeds inclosed in a fruit. For the practical study of timber, however, we require a scheme of grouping based upon the wood itself; and, having often to deal with converted timber, it is well to be as independent as we can of characters derived from bark, or even from pith. Speaking of this problem in his excellent work, *Timber and some of its Diseases*, Professor H. Marshall Ward writes: "It may be doubted whether all the difficulties are likely to be surmounted. . . . In any case, while allowing that it is as yet impossible so to arrange a collection of pieces of timber, that all the kinds can be recognized at a glance, it must be admitted that the attempt to do so at least aids one in determining many kinds."

In describing the many valuable timbers of India, Mr. J. S. Gamble makes use of eight classes of characters: (i) the size of the trees; (ii) whether they are evergreen or deciduous; (iii) the bark; (iv) the wood, its colour, hardness, and grain; (v) the annual rings; (vi) the pores or vessels; (vii) the pith rays; and (viii) other miscellaneous characters, such as concentric markings or false rings. Of these, the first three are not available to the student of converted timber. The annual rings by their width indicate the rate of growth, a character of great importance as to quality, if not of great distinctive value. More than 12 rings to the inch, giving, as it does, 6 feet of girth in 134 years, may be termed *slow* growth; from 12 to 6 rings to the inch, which would mean 6 feet of girth in from 134 to 67 years, *moderate*; and less than 6 rings to the inch, or 6 feet of girth in 67 years, *fast* growth.

The absence of pores or vessels is characteristic of coniferous woods. As to their size, Mr. Gamble classifies them in 7 groups: extremely small, as in Box; very small, as in *Ácer píctum*; small, as in Haldu (*Adína cordifólia*); moderate-sized, as in Mahwa (*Bássia latifólia*); large, as in Siris (*Albízzia Lébbek*); very large, as in *Erythrína suberósa*; and extremely large, as in many climbers (Fig. 4, for instance).

So too the pith-rays, as distinctive characters, are grouped under seven types: extremely fine, as in *Euónymus lácerus*; very fine, as in Ebony (*Diospýros Melanóxylon*); fine, as in Siris (*Albízzia Lebbek*); moderately broad, as in *Dillenia pentagyna*: broad, as in Plane (*Plátanus orientális*), in which case they measure $\frac{1}{5}$ mm.; very broad, as in some Oaks, in which they reach 1 mm.; and extremely broad, as in *Sámara robústa*. The number and distance apart of the pith-rays are also characters of consequence. When further apart than twice the diameter of the pores they may be termed *distant*.

There are some of these microscopic characters that are eminently distinctive of large groups, such as the Natural Orders into which botanists group plants. The *Cupulíferæ*, for instance, that great group to which the Oaks, Beeches, Chestnuts and Hornbeams belong, have their pores in wavy radial lines or queues: in the *Ebenáceæ*, or Ebony tribe, and the *Sapotáceæ*, a closely-allied tropical Order, including the Bullet-woods (*Mímusops*), the pores are in short, wavy lines, and there are wavy false rings; but whilst the *Ebenáceæ* have white, grey or black wood, that of the *Sapotáceæ* is reddish. So too the tropical Order *Anonáceæ*, or Custard-Apple family, which includes the Lancewood of the West Indies, has regular ladder-like transverse bars on its woods that are very characteristic.

Several of the characters used in the classification of woods, such as weight per cubic foot, hardness and amount of ash left on combustion, not only vary together, but also differ according to the age of the tree and the distance of the sample from the root. Weight, for instance, increases from the butt to the lowest branch, and decreases from the latter point upward.

Among minor characters sometimes of use in discriminating woods may be mentioned the colour of a solution obtained by boiling the wood in water or in alcohol, its reaction when treated with a solution of iron sulphate or perchloride, and the colour of the ash produced in burning. Jarrah, for instance, yields a black cindery mass, whilst the only less valuable paving wood Karri gives a white ash.

Unfortunately trees of the same Order, or even of the same genus, by no means always have similar woods. Mr. Gamble, for instance, cites the important genus *Dalbérgia*, three Indian species of which—the Blackwood (*D. latifólia*), Sissoo (*D. Síssoo*), and *D. lanceolária*—have hard, dark-coloured, heavy woods; whilst other species have only white and often soft sapwood, not forming any duramen. When, however, we compare heartwoods microscopically they do as a rule resemble one another in allied species.

In many cases a knowledge of the locality from which a timber comes may aid us in identifying it. Thus, save by this means, it is apparently impossible to distinguish the woods of *Cupréssus Lawsoniána* from Oregon, *C. Thyóides* from the Eastern States, *Thúya gigántea*, the Canoe Cedar or Red Cedar of the West, and *T. occidentális*, the Arbor-vitæ of the North-east, all of them being known to American timber-merchants as White Cedar.

The following table is by no means exhaustive, few Asiatic or Australian woods being, as yet, classified in it. It has seldom been possible to carry the discrimination further than genera. Though obvious naked-eye characters have been largely employed, use is also made of those seen only in microscopic sections. For this purpose it is only necessary to take a single shaving, across the grain, with a well sharpened plane, put it at once into methyl blue or some other dye, and then mount it as an ordinary microscopic slide. The first character to be observed is the presence or absence of "pores," or the transverse sections of large tracheæ. If they are absent, which practically means that the wood is coniferous, we next look for conspicuous resin-canals, and for the

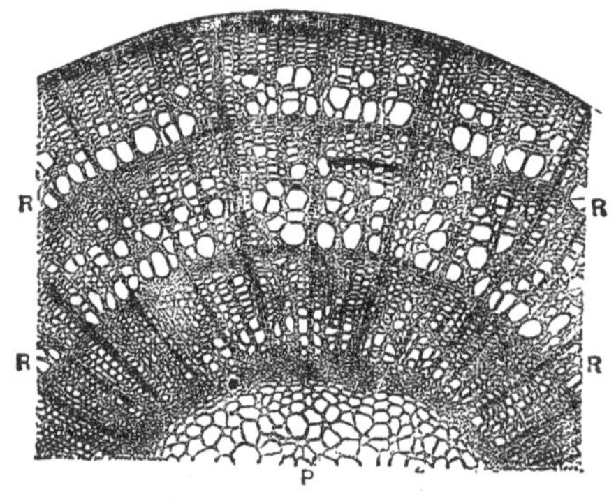

FIG. 29.—Transverse section of Linden, a ring-porous wood, showing three annual rings. After Van Tieghem, from *The Elements of Botany*, by permission of Mr. Francis Darwin and the Syndicate of the Cambridge University Press.

presence of heartwood defined by a distinct colour. The outlines of the annual rings, the hardness, colour, weight, taste and smell of the wood then afford further means of identification; whilst such microscopic characters as the presence of tracheids in the pith-rays, or of spiral thickening in the tracheids are only requisite as a last resource. Where, on the other hand, the presence of "pores" indicates that the wood is that of a broad-leaved tree, we first note whether there are, or are not, distinct annual rings, or whether "false rings" of wood-parenchyma are present; then whether the "pores" are so collected in the inner or spring portion of each ring that we should class the timber in question as "ring-porous" (Fig. 29), or whether they are so scattered that we may call it "diffuse-porous." The grouping of the pores, the prominence of the pith-rays, the weight, hardness and colour here again furnish subsidiary characters.

I. CONIFEROUS OR NON-POROUS WOODS.

No visible or conspicuous pores on a transverse section, even when magnified, the wood containing no tracheæ or true vessels, except immediately round the pith. Resin-canals often present in the autumn wood. Annual rings generally sharply marked by denser, dark-coloured autumn bands. Pith-rays very fine and numerous, invisible to the naked eye.

A. Without conspicuous resin-canals.

1. No distinct heartwood: rings well rounded.
   - *a.* Yellowish-white, soft: no tracheids in the pith-rays. *Abies.* The True or Silver Firs, *e.g. A. pectináta* of Central Europe, *A. Webbiána* of the Himalayas, *A. balsámea*, the Balsam Fir of the North-Eastern United States, and *A. grándis*, *A. cóncolor*, *A. amábilis*, *A. nóbilis*, and *A. magnífica* of the Western States.
   - *b.* Reddish, soft, brittle: pith-rays with tracheids. *Tsúga.* The Hemlock Spruces, including *T. canadénsis* of North-east, and *T. Mertensiána* of North-west America.
2. Heart-wood present and contrasting in colour.
   - *a.* Heavy, hard, non-resinous, dull. Heart-wood brownish or orange-red: sapwood lemon-colour. Rings excentric, wavy and sinuous. *Taxus.* The Yews, including *T. báccata* of Europe and Northern Asia, and *T. brevifólia* of North-west America.
   - *b.* Light, soft—medium hard, usually aromatic. Heartwood rose, yellowish or brownish red: sapwood yellowish white. Rings wavy and sinuous. Pith-rays very fine. The "Red Cedars," *Junίperus.*

Heartwood rose to brown red. *J. virginiána*.

Heartwood yellowish-brown. *J. commúnis* and *J. Oxycédrus.*

c. Very light, very soft, odourless. Heartwood light-red, turning brownish: sapwood narrow, amber-coloured. Rings regular. Pith-rays very distinct, especially on the radial section. Resin-canals in a single row, or absent. "Redwoods," *Sequoía.*

d. Medium heavy and hard, often camphor-scented. Heartwood rich brown, often mottled with darker brown or yellow: sapwood narrow, white. Rings wavy. "Cypress Pines," etc. *Cállitris.*

e. Light, moderately hard, or soft, fragrant. Heartwood yellowish or reddish-brown. Rings well rounded. Resin-ducts few and narrow. *Cédrus.*

3. Heartwood present, but differing only in shade from the sapwood, of a dull yellowish or greyish brown.

a. Odourless and tasteless. *Taxódium.*

b. Light, soft, with slight resinous odour, tasteless. Rings finely and coarsely wavy. Pith-rays very fine but distinctly coloured. "White Cedars," including *Thuya occidentális* and *T. gigántea, Cupréssus thyoídes* and *C. Lawsoniána.*

c. Light, soft, with resinous smell and peppery taste. Incense Cedar, *Libocédrus.*

Near here belong apparently the Huon Pine and allied species, *Dacrýdium*, etc.

B. Resin-canals present, at least in autumn wood.

1. Heartwood not distinctly coloured, white: resin-canals few, very narrow: rings imperfectly rounded; tracheids in pith-rays. Spruces, *Pícea.*

2. Heartwood distinct.

a. Resin-canals not numerous, nor evenly distributed.

(i) Canals solitary or here and there in pairs; tracheids without spirals. Heartwood reddish-brown, sapwood yellowish. Knots irregularly distributed. Larches or Tamarack, *Lárix.*

(ii) Canals in groups or lines of 8—30: tracheids with spirals, otherwise resembling Larch. Douglas Spruce, *Pseudotsúga.*

b. Resin-canals numerous, evenly distributed. Knots in regular whorls. *Pinus.*

(i) Wood tolerably hard and firm: transition from spring to autumn wood abrupt: resin-canals more numerous

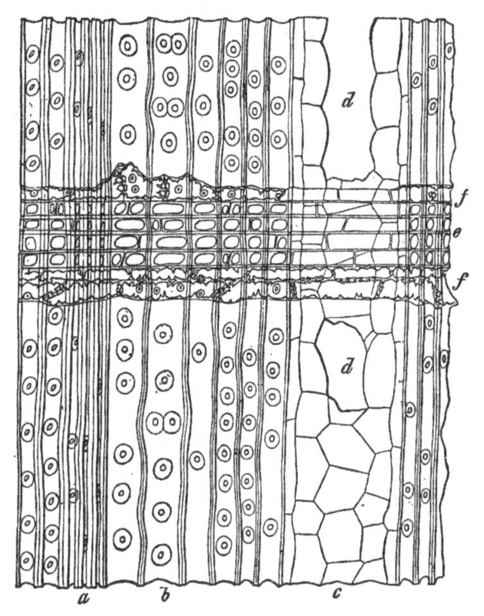

FIG 30.—Radial section of Scots Fir (*Pínus sylvéstris*). Magnified 100 times. *a*, narrow tracheids of autumn wood with small bordered pits on their radial walls; *b*, broad spring tracheids; *cd*, resin-duct lined with epithelium; *e*, parenchyma of pith-ray with large simple pits; *f*, tracheids of pith-ray with small bordered pits and dentate projections. From Hartig's *Timbers and how to know them*, by permission of Dr. Somerville and Mr. David Douglas.

in autumn wood: heartwood reddish: tracheids of pith-ray with dentate projections, when seen in radial section (Fig. 30). Hard Pines.

* 1 or 2 simple pits on radial wall of each tracheid of pith-ray. "Norway pine" of U.S.A., *Pinus resinósa.*

** 3 to 6 such pits.

† Wide rings. Loblolly and Short-leaf Pines of U.S.A., *P. tǽda* and *P. echináta*; Northern, Black Austrian, and Cluster Pines of Europe, *P. sylvéstris*, *Larício*, and *Pináster*.

†† Narrower rings. Longleaf Pine of U.S.A., *P. palústris*; Dwarf Pine of Europe, *P. montána*.

(ii) Wood soft and light: transition from spring to autumn wood gradual; autumn wood narrower and with fewer resin-canals: tracheids of pith-ray without dentate projections. Soft Pines.

* Rings rather narrow, circular: resin-ducts very large and numerous: wood yellowish. *P. Cémbra.*

** Rings broad: wood redder. Weymouth and Sugar Pines, *P. Stróbus* and *P. Lambertiána* of U.S.A.; and probably the Aleppo Pine, *P. halepénsis.*

II. LEAF-WOODS, HARD-WOODS, OR POROUS WOODS.

Pores visible on transverse section, either to the naked eye or when magnified, often characteristically grouped, especially in spring-wood. Pith-rays either all fine or some broad.

A. Without distinct annual rings, though sometimes with false-rings or partial zones of wood-parenchyma. Mostly tropical.

1. *With false rings.*

a. Some pith-rays broad. Indian Oaks, *Quércus lamellósa*, etc.

b. All pith-rays narrow.

(i) False rings very distinct.

* No distinct heartwood: wood moderately hard and dense, greyish. Banyan, *Fícus bengalénsis*, Myrobalans, *Terminália belérica*, and various Asiatic Acacias and other *Leguminósæ*.

** Dark heavy heartwood. *e.g.* the very hard, tough purplish-brown Jhand, *Prosópis spicígera.*

(ii) False rings obscure: wood dense, heavy, red, brown, purple or black. Including the chief hardest woods of India and other tropical countries, such as the Ebonies, *Diospýros*, Ironwood, *Mésua férrea*, Pynkado, *Xýlia dolabrifórmis*, Anjan, *Hardwíckia bináta*, Rosewoods, *Dalbérgia*, *Pterocárpus*, etc., Babul, *Acácia arábică* and other species, such, perhaps, as the Australian Myall, *A. homalophýlla*, Saj, *Terminália tomentósa*, Bandara, *Lagerstrœ́mia parvifólia*, Lignum vitæ, *Guáiacum*, etc. Olive (*Ólea europǽa*), a close, compact, yellow wood, characteristically mottled with brown, with uniformly scattered vessels, may, perhaps, be classed here.

2. *Without false rings.*
   a. Soft, with no distinct heart. Silk-cotton, *Bómbax*, Mango, *Mangífera*, etc.
   b. Harder, denser, usually with distinct heart. Siris, *Albízzia Lébbek*, Eng, *Dipterocárpus tuberculátus*, etc.

B. With distinct annual rings.

1. *Ring-porous*: vessels in spring wood large or numerous, those in summer wood small or few and scattered.
   a. Vessels in the spring wood larger.
   (i) Vessels in tree-like or dendritic groups, or in circles, often scattered in the inner part of the rings.
   * Slightly dendritic or concentric: pores in summer wood minute, regularly distributed, singly or in groups, or in short peripheral, but never radial lines.
   ‡ Pith-rays minute, scarcely distinct.
   § Wood heavy and hard: vessels in summer wood not in clusters, or 2-4 together.
   (a) Heartwood not yellow in radial section; continuous zone of pores in spring wood. Ash, *Fráxinus*.

Vessels in summer wood in peripheral lines. White and Green Ash, *F. americána* and *F. víridis.*

Vessels in summer wood not united in peripheral lines. English, Black and Red Ash, *F. excélsior* (Fig. 31), *F. sambucifólia*, *F. pubéscens.*

(b) Heartwood yellow, very heavy and very hard. Osage Orange, *Maclúra.*

§§ Wood light and soft: vessels in summer wood in groups of 10—30. *Catálpa.*

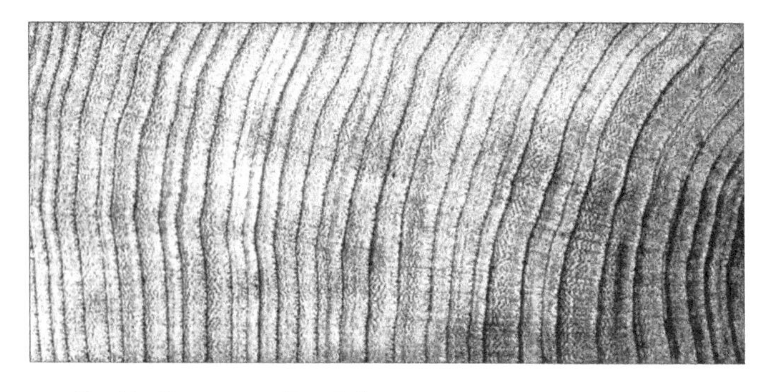

FIG. 31.—Transverse section of Common Ash (*Fráxinus excélsior*), photographed from nature.

‡‡ Pith-rays very fine, but distinct: heartwood reddish brown: sapwood yellowish white: vessels in summer wood single or in short lines: odour. *Sássafras.*

‡‡‡ Pith-rays fine, but distinct.

§ Very heavy and hard: heartwood dark yellowish brown: sapwood yellow: vessels 1—10 together, filled, so appearing as yellow dots. Black Locust, *Robínia.*

§§ Heavy: moderately hard or hard.

(a) Vessels in summer wood very minute, usually in small clusters of 1—8, open: heartwood yellow to light orange-brown, reddening on exposure to light: sapwood yellowish white: odourless. Mulberry. *Mórus.*

(b) Vessels in summer wood small or minute, usually solitary: heartwood cherry-red. Coffee Tree, *Gymnocládus.*

‡‡‡‡ Pith-rays fine, but very conspicuous to the naked eye: heartwood rose-red to brownish: sapwood pale lemon or greenish white; vessels open. Honey-Locust, *Gledítschia.*

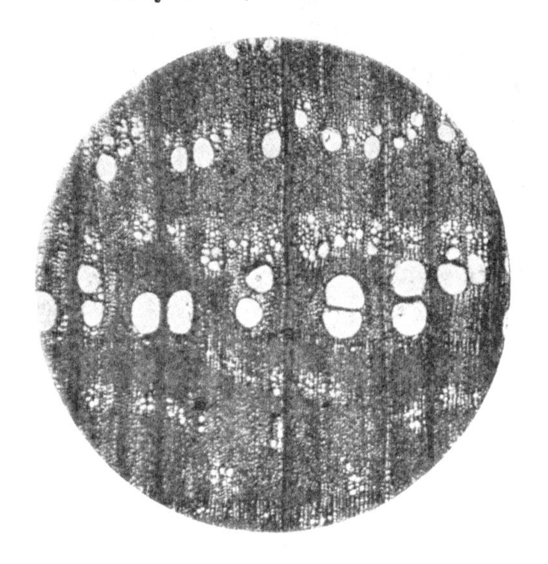

FIG. 32.—Transverse section of Nettle-tree. (*Céltis austrális*).

‡‡‡‡‡ Pith-rays rather coarse, lustrous: heartwood brownish or greyish orange: sapwood broad, yellowish: broad zone of very large open pores in spring wood: vessels in autumn wood 1—5 together in segments of circles. *Ailánthus.*

** Strikingly dendritic: pores in summer wood minute or small, appearing as finely feathered hatchings on tangential sections.

† Vessels 1-8 together: pith-rays fine, but distinct.

‡ Heartwood yellowish or greenish brown to black, hard: sapwood narrow, yellowish. *Labúrnum.*

‡‡ Heartwood greenish or yellowish white, hard, heavy: sapwood not differing. Hackberry, *Celtis* (Fig. 32).

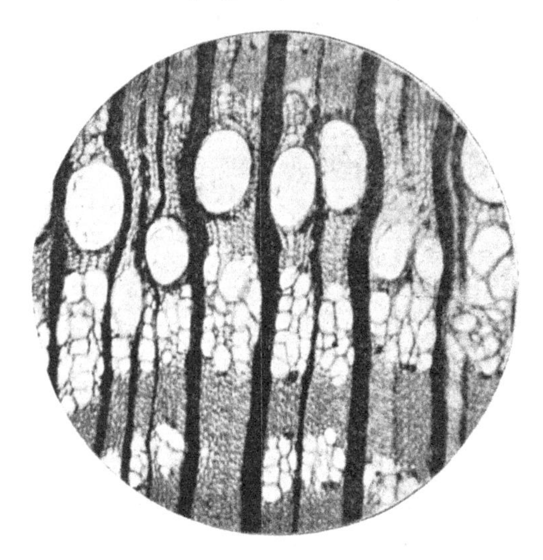

Fig. 33.—Transverse section of Common Elm (*Ulmus campéstris*).

†† Vessels 1—several dozen together, in wavy peripheral lines in autumn wood: heartwood brown, hard, heavy: sapwood yellowish-white. Elm, *Úlmus.*

‡ Pores of spring wood forming a broad band of several rows. English, Scotch and Red or Slippery Elm, *Úlmus campéstris, montána* (Figs. 33 and 34), and *fúlva.*

‡‡ Pores of spring wood in a single row, or nearly so. White, Rock, Winged and Cedar Elms of U.S.A., *Úlmus americána, racemósa, aláta* and *crassifólia.*

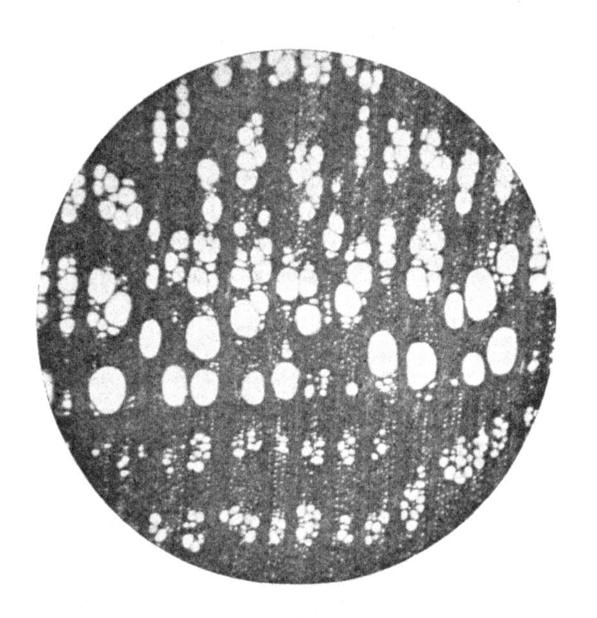

FIG. 34.—Transverse section of Wych Elm (*Úlmus montána*).

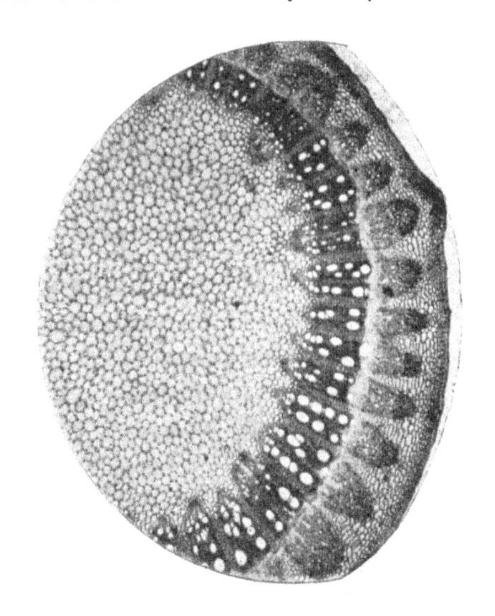

FIG. 35—Transverse section of Common Vine (*Vítis viniféra*).

*** Vessels in radial lines or queues, wavy or branched the branches often uniting.

† All the pith-rays very broad.

‡ Wood beset with large pores: heartwood reddish-brown. Vine, *Vítis* (Fig. 35).

‡‡ Wood sulphur yellow, hard: zone of vessels narrow. Barbery, *Bérberis.*

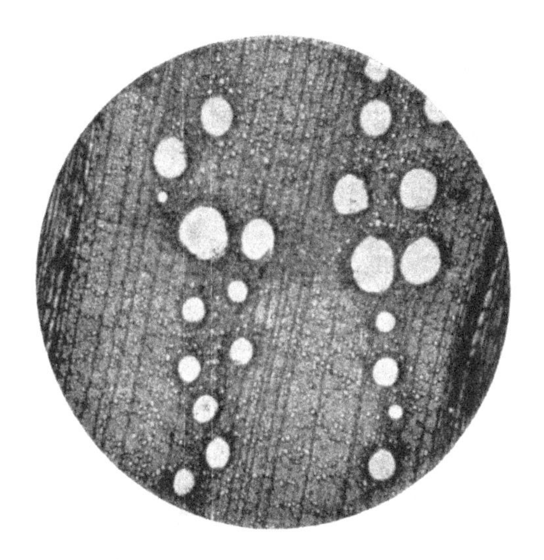

Fig. 36.—Transverse section of Cork Oak (*Quércus Súber*).

†† Pith-rays so narrow as to be hardly perceptible: heartwood oak-brown: zone of vessels very broad and vessels large, but less crowded than in Oak. Chestnut, *Castánea.*

††† Some of the pith-rays very broad and easily visible to the naked eye. Oaks, *Quércus.*

‡ Pores in summer wood very fine, numerous and crowded: heartwood light brown. White Oaks, *Q. álba, bícolor, palústris, obtusilóba,* etc., in U.S.A. and *Róbur* in Europe.

‡‡ Pores in summer wood fewer but larger: heartwood dark brown. Red and Black Oaks of U.S.A., *Q. rúbra, tinctória.*

‡‡‡ Pores few, gradually but slightly diminishing across the entire ring: wood very dense and heavy. Live Oaks, *Q. vírens* of U.S.A., *Ílex* and *Súber* (Fig. 36) of Europe.

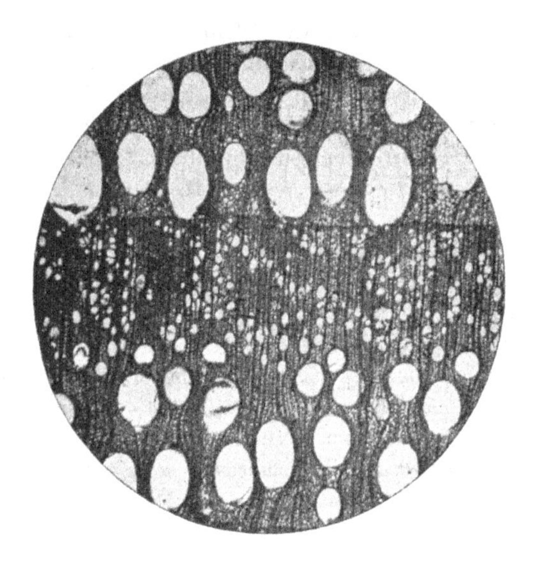

Fig. 37.—Transverse section of Judas-tree (*Cércis Siliquástrum*).

†††† Pith-rays moderately broad and distinct: vessels in spring wood very large, those in summer wood much smaller, 1—8 together: heartwood brownish yellow: sapwood white. Judas-tree, *Cércis* (Fig. 37).

**** Vessels in summer wood mostly but little smaller than those of the spring wood, scattered, solitary, or few together. Mostly hard, heavy woods.

† Fine peripheral lines of wood-parenchyma: pith-

rays fine: zone of vessels interrupted: summer wood reddish nut-brown. Hickories, *Hicória.*

†† Similar, but with blackish heartwood. Persimmon, *Diospýros virginiána.*

††† Vessels distinct and large, sometimes filled with white phosphate of lime: pith-rays fine, distinct, light-coloured: wood brownish-red. Teak, *Tectóna.*

†††† Vessels equally distributed: pith-rays fine, distinct: wood a warm red brown, often beautifully figured. Mahogany, *Swieténia Mahágoni.*

††††† Vessels very large, open or partly filled with a brown resin: pith-rays distinct: heart-wood cinnamon-brown, very soft, fragrant. Honduras Cedar, *Cedréla odoráta.*

b. Vessels in the spring wood not larger, but generally more numerous and crowded than in the autumn wood.

(i) Pith-rays distinct.

* Heartwood reddish-brown, zone of vessels in spring wood lighter coloured: vessels 1—4 together: hard, heavy. Plum, *Prúnus doméstica.*

** Heartwood yellowish-brown, with greenish streaks, fragrant: vessels 1—8 together: hard. Mahaleb Cherry, *Prúnus Máhaleb.*

*** Heartwood yellowish-brown, with an unpleasant odour at first: sapwood yellowish-white: vessels 1—8 together: moderately hard. Bird Cherry, *Prúnus Pádus.*

**** Heartwood blackish-brown, with pith-flecks: sapwood reddish: vessels 1—5 together: hard. Blackthorn, *Prúnus spinósa.*

***** Heartwood yellowish-brown: sapwood reddish-white: vessels minute, 1—13 together: hard, heavy. Cherry, *Prúnus Cérasus.*

****** Sapwood yellowish-white: heartwood slightly

browner: vessels 1—8 together: hard: pith very large. Elder, *Sambúcus.*

(ii) Pith-rays not at all, or scarcely visible.

* Heartwood orange-red: sapwood yellow: vessels about 50 together in branching flame-like groups: hard, heavy. Buckthorn, *Rhámnus cathárticus* (Fig. 28).

** Similarly coloured; but vessels 1—7 together, not in flames, but equally distributed and minute: soft. Berry-bearing Alder, *Rhámnus Frángula.*

*** Heartwood greyish green, autumn wood in darker zones: sapwood narrow, yellowish-white: soft. Stag's horn Sumach, *Rhús typhína.*

**** Heartwood greenish to golden: sapwood narrow, white: vessels 1—7 together: harder. Venetian Sumach, or Wig-tree, *Rhús Cótinus.*

***** Heartwood light brown, touched with red or violet: sapwood narrow yellowish-white: hard, heavy. Lilac, *Syrínga vulgáris.*

2. *Diffuse-porous*: vessels numerous, usually minute, but neither larger nor more numerous in the spring wood: rings sometimes rendered distinct by closer texture of the elements of the autumn wood.

a. Vessels large, open, but few.

(i) Wood soft and light: heartwood light reddish-brown. Butternut or White Walnut, *Júglans cinérea.*

(ii) Wood hard and heavy: heartwood chocolate-brown: pith-rays fine: fine peripheral lines of parenchyma vessels 1—4 together: with darker wavy zones. Common Walnut, *Júglans régia.*

(iii) Similar, but darker. Black Walnut, *Júglans nígra.* The Indian Sal (*Shórea robústa*) belongs here.

b. Vessels minute.

(i) Broad pith-rays present.

* Pith-rays numerous, mostly broad, crowded: rings bending outwards at the rays: reddish-white or

light brown: hard, moderately heavy. Plane, Buttonwood or Sycamore, *Plátanus occidentális.*

** Only some of the pith-rays broad.

‡ Broad rays numerous: rings bending inwards at the rays: reddish-white or light brown: hard. Beech, *Fágus.*

‡‡ Broad rays few, light-coloured: rings very sinuous, bending inwards at the rays: yellowish-white: hard, heavy, tough. Hornbeam, White or Blue Beech, *Carpínus.*

‡‡‡ Broad rays few: rings almost circular: reddish-white, soft. Hazel, *Córylus.*

‡‡‡‡ Broad rays few: rings bending inwards at the rays: white, becoming brownish-red, with brown pith-flecks, soft. Alder, *Alnus.*

(ii) No broad pith-rays.

* Pith-rays narrow but quite distinct to the naked eye.

† Wood hard.

§ Pith-rays with a decided satiny lustre. Maples, *Acer.*

‡ Rings perfectly circular.

¶ Wood white, hard and heavy: pith-rays straight: Sycamore or Plane. *A. Pseudoplátanus.*

¶¶ Similar; but with winding pith-rays. *A. opulifólium.*

‡‡ Rings slightly wavy.

¶ Wood reddish, very hard, sometimes with curled, bird's-eye or blister figures.

(a) Sometimes with pith-flecks. Field Maple, *A. campéstré* and Moose-wood, *A. pennsylvánicum.*

(b) Without pith-flecks. Rock or Sugar Maple, *A. barbátum.*

¶¶ Wood reddish, but lighter, hard, with very fine but conspicuous pith-rays.

(a) With distinct, dark-coloured heartwood. Red Maple, *A. rúbrum.*

(b) Without distinctly coloured heartwood. Norway or Plane Maple, *A. platanoídes.*

¶¶¶ Wood light-coloured, reddish or yellow, lighter and softer.

(a) Red-tinged, sometimes curled. Silver or Soft Maple, *A. saccharínum.*

(b) Yellowish, with very broad rings: vessels minute, numerous. Box-elder, *A. Negúndo.*

§§ Pith-rays very fine, but distinct, not markedly satiny: rings circular: wood white or greenish: vessels minute. Holly, *Ilex.*

†† Wood soft or very soft.

§ Pores crowded, occupying nearly all the space between the pith-rays.

‡ Yellowish-white, often darker or greenish in the heartwood. American White-wood, Yellow-wood or Yellow Poplar, *Liriodéndron tulipífera*, and Cucumber tree, *Magnólia acumináta* and allied species.

‡‡ Sapwood greyish-white: heartwood light to dark reddish-brown, heavy, but soft. Sweet Gum, Bilsted or Red Gum of U.S.A., *Liquidámbar styracíflua.*

§§ Pores not crowded, occupying not more than one-third of the space between the pith-rays: brownish or reddish-white to light brown; only slightly silky; pith-rays less distinct and less lustrous than in the Maples: light. Linden, Lime or Basswood, *Tilia.*

** Pith-rays not distinct to the naked eye.

† Wood hard: distribution of vessels uniform, or sometimes in wormlike lines.

§ Vessels 1—3 together.

‡ Wood flesh-coloured, with pith-flecks. Hawthorn, *Cratǽgus Oxyacántha.*

‡‡ Yellowish-white. Spindle-tree, *Euónymus europǽus.*

‡‡‡ Greenish. Bladdernut, *Staphýlea pinnáta.*

§§ Vessels 1—4 together.

‡ Without pith-flecks.

¶ Heartwood flesh-coloured. Dogwood, *Córnus sanguínea.*

¶¶ Brownish-red, no distinct heartwood. Pear, *Pýrus commúnis.*

¶¶¶ With a dark red-brown heartwood. Apple, *Pýrus Málus.*

‡‡ With pith-flecks.

¶ Sapwood reddish-white: heartwood reddish-brown: pith-flecks few, near centre. White Beam, *Pýrus Ária.*

¶¶ Brownish-yellow: pith-flecks numerous. Wild Service-tree, *Pýrus torminális.*

§§§ Vessels 1—5 together: pith-flecks numerous: reddish. Rowan or Mountain Ash, *Pýrus Aucupária.*

§§§§ Vessels 1—8 together, minute: sapwood whitish: heartwood reddish, with satiny lustre. Birch, *Bétula.*

§§§§§ Light yellow, very compact and fine-grained, almost horny: rings scarcely visible: heavy. Box, *Búxus.*

†† Wood soft.

§ Creamy white, yellowish or reddish, light: vessels 1—7 together, indistinct: rings wide. Horse-chestnut or Buckeye, *Ǽsculus.*

§§ Sapwood white or reddish: heartwood light red to dark brown, sometimes lustrous, light: rings sometimes angular: vessels in worm-like groups. Willow, *Sálix.*

§§§ Sapwood white: heartwood light brown, lustrous: rings angular: vessels 1—5 together. Black Poplars or Cottonwoods, *Pópulus nígra, monilífera, balsamífera.*

§§§§ Rings circular: vessels 1—7 together.

‡ Without pith-flecks: heartwood yellow-brown. White Poplar, *Pópulus álba.*

‡‡ With white pith-flecks: white, with no distinct heartwood. Aspen, *Pópulus trémula.*

CHAPTER III.

DEFECTS OF WOOD.

IN every stage of their growth trees are liable to mischances, from defects of soil or climate, from accident, or from the attacks of fungi, of insects, or of other animals. Some of these mischances have permanent and important effects upon their wood. Although, in healthy surroundings and in the absence of external injury, there is no very definite limit to the longevity of any species of tree, after it has reached maturity a certain deterioration generally shows itself at the centre of the trunk, which will subsequently manifest itself as decay. After felling, shrinkage in the process of drying and the attacks of species of fungi, mostly differing from those that injure growing trees, develop further defects in timber of the very gravest practical import to the consumer.

The attacks of insects or of fungi upon the leaves of trees, though they may prove fatal to seedlings, have generally in later stages of growth merely the effect of injuring the nutrition of the plant. They may thus diminish the amount of wood formed in the season, and may, therefore, be of first-rate importance to the forester or timber-grower, but do not in general concern the timber-user.

Cup-shake.—When, however, the caterpillars of some moths, such as *Tórtrix viridána*, entirely destroy the young leaves of the Oak in June, though the tree may put out new leaves in July and August, it will only do so at the expense of wood-forming reserve-materials, and there may possibly result so complete a check to the nutrition of the tree that the wood of one year may fail to cohere to that of the preceding season, a *cup-shake* or *ring-shake* being produced (Fig. 38).

Such a separation between successive annual rings—a defect seriously interfering with the conversion of timber into planks—is, however, undoubtedly produced for the most part by various other causes, and may be briefly here described. It occurs in

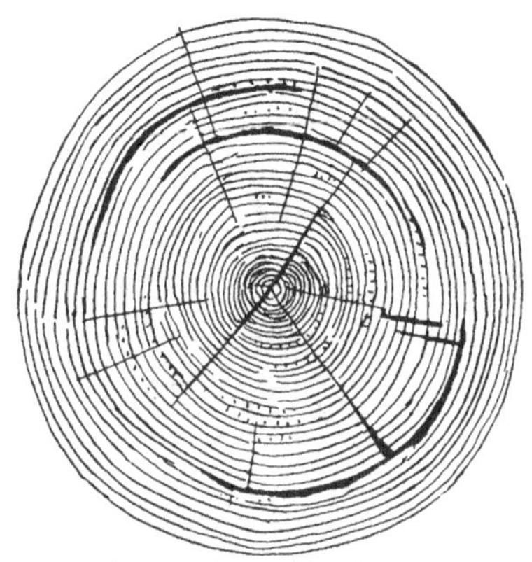

FIG. 38.—Cup and heart shake.

various species, such as Hazel, Oak, Poplar, Pitch Pine, and Lignum Vitæ, and seems to some extent local. The oaks of Sicily, for instance, a variety of our British species, *Quércus Róbur*, and those of the Forest of Dean (*Q. Róbur*, var. *sessiliflóra*) are peculiarly subject to this defect, which in the latter case has been doubtfully ascribed either to the rocky character of the soil or to the swaying to and fro of the tall trees by strong winds. This action of wind bending the rings of wood alternately in opposite directions, in a manner obviously calculated to tear them apart, may well explain the occurrence of this form of shake in Poplars. Cup-shake has also been attributed to frost, the rings of sapwood and heartwood in a living tree containing varying proportions of water and the outer layers being most likely to freeze first. The explosive rending of trees by frost, the noise of which disturbs the stillness of night in the forests of North America may in this way be sometimes concentric in its action. This may explain the prevalence of this defect in the swamp-loving Pitch Pine (*Pínus austrális*) of Virginia. Frost cannot, however, be the cause of the frequency of cup-shake in the tropical Lignum Vitæ; but in this case the sun may have produced an effect similar to that which sometimes occurs when part of the cambium ring at the base of a stem is injured by a forest fire. Lastly, in some Pines this defect

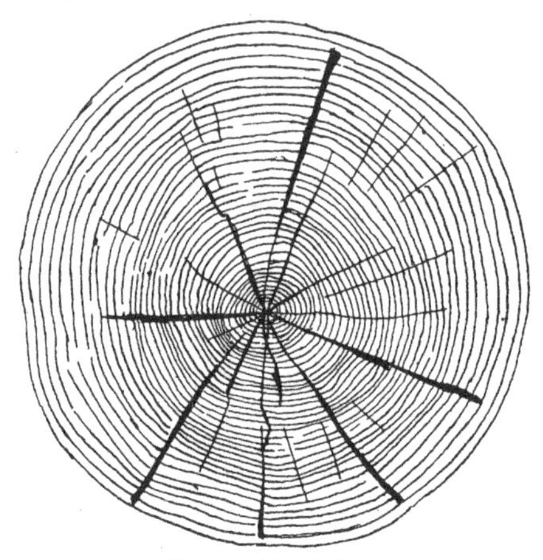

FIG. 39.—Star-shake.

is the result of the attacks of certain fungi (*Tramétes*), the "spawn" or "mycelium" of which spreads as a felted mass of colourless mould especially in the cambium. Cup-shake occurs most frequently at the base of the stem: when of long standing, it is often accompanied by traces of rot, and in many cases it is also associated with *star-shake*.

Star-shake.—*Star-shake* consists in clefts radiating from the pith along the planes of the pith-rays and widening outwards (Fig. 39).

It occurs in many species and in trees of all ages. The clefts may only extend a small distance and be so slightly open when the tree is newly felled as to be scarcely perceptible. In such a case they generally widen during seasoning, from the more rapid drying of the outer layers, their sides becoming darker in colour than the rest of the wood. In other instances the clefts may have extended to the circumference of the stem, in which case they may have been so overgrown by new wood as to form a longitudinal rib down the exterior of the bark, a sure sign of the defect to the experienced timber-surveyor. Such extreme cases at least seem to be always the result of frost or sun, the latter being specially frequent in the case of smooth thin-barked species, such as Beech and Hornbeam, in which lines of the cortex are killed by sun-burn.

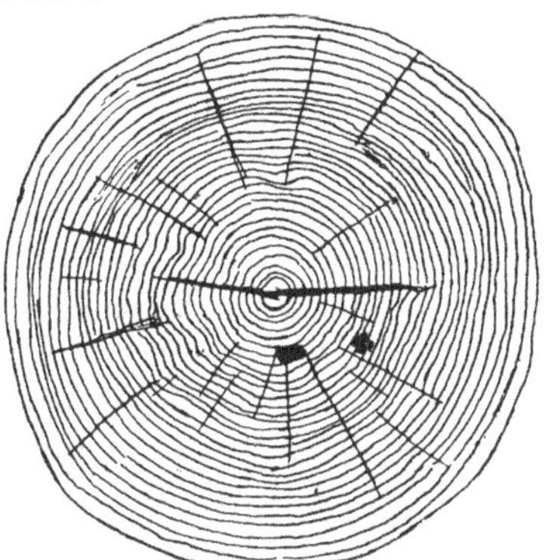

FIG. 40.—Heart-shake.

Heart-shake.—More common than either cup-shake or star-shake is *heart-shake*, one or two clefts crossing the central rings of the stem and widening towards the centre (Fig. 40).

This may occur in almost every kind of timber, whether coniferous or broad-leaved, and seems to be quite independent of soil or situation. Among species least affected by it Mr. Laslett[1] mentions the so-called African Oak or Teak (*Oldfieldia africána*), Sabicu, Spanish Mahogany, Common Elm, Dantzic Fir or Redwood (*Pínus sylvéstris*), Canadian Red Pine (*Pínus resinósa*), and, somewhat less free from it, Canadian Yellow Pine (*Pínus Stróbus*);

[1] *Timber and Timber-trees*, ed. ii. p. 54.

whilst as exceptionally liable to the defect he mentions the true Indian Teak (*Tectóna grándis*), the Australian Tewart (*Eucalýptus gomphocéphala*), the Riga and Swedish varieties of *Pínus sylvéstris*, and *P. austrális*, the Pitch Pine of the southern United States. Greenheart (*Nectándra Rodiœ́i*) commonly develops two crossing heart-shakes for two or three feet up the butt end of the log. One of the worst forms of this defect is when, owing to spiral growth, the shake shifts its direction as we trace it up the stem. It may in this way sometimes be nearly at right angles at one end of the tree to its direction at the other,—thus rendering the conversion of a log into plank well nigh impossible.

It is this hindrance to the conversion of timber into plank that constitutes the main practical importance of all forms of shake, as they do not at first involve any decay, and consequently do not much interfere with the employment of the logs in bulk. Heart-shake, however, is probably in itself an indication of that incipient decay that comes when timber has passed its maturity and the older layers shrink more than the outer.

Rind-gall.—Somewhat allied to cup-shake is the local defect known as *rind-gall*. This originates from the destruction of part of the bark of a growing tree, whether by another tree falling against it, the scorching of a forest-fire, the knawing of an animal, or even the cutting of initials by some misguided youth. If the cut has penetrated to, exposed, and destroyed the cambium, there may, in spite of the gradual overgrowth of layers of new wood from the margins of the injury, be a local want of cohesion between the exposed wood and that subsequently formed over it. This defect may entirely escape detection from the outside of an unconverted log.

Decay.—Bright-looking wood is generally of better quality than that which is dull; while any departure from the usual colour of the timber of the species is commonly, as we have already stated, an indication of at least incipient local decay. Discoloured patches, such as occur on the exterior of the butt ends of some masts of the Kauri (or Cowdie) Pine of New Zealand (*Ágathis austrális*), will generally be found to be relatively brittle. They are usually white at first and are then of small extent or consequence; but when they are yellowish-red, the mischief has gone further; and a decided red or *foxy* colour indicates a wide-spread decay so serious as to disqualify the timber for purposes of construction. Oak, however, in an advanced state of foxiness and decay is in request for cabinet-work. Such decay is a gradual oxidation, or combustion of the wood, similar to that which it undergoes when exposed to alternations of dry and damp air. In old Beeches, and other trees, it appears to begin in the pith and spread outwards, such wood being known in France as *bois rouge*; but it very frequently originates in a broken branch, a rind-gall, or a star-shake reaching the surface, so that air, damp and fungi find access to the wood of the tree. It is this decay spreading from the pith that gradually hollows out old trees; but this hollowing occurs much earlier in pollards where water and rotting leaves may accumulate in the fork of the crown, or in trees in which broken limbs or other injuries have been neglected. The breaking of a small branch may set up decay, and yet such a *druxy knot*, as it is termed, may gradually be covered up with sound wood, so that only a slight swelling may indicate the defect at the surface of the stem. Any such excrescence should be removed directly a tree is felled; as, though the healing over, by excluding further damp, may have checked the mischief, there is no telling from the outside how deep it may have extended, and such a patch of decayed wood, if left to itself, is certain on being laid bare in the process of conversion to absorb more atmospheric moisture and so enlarge itself.

Fungal attack.—All such decay is immensely hastened, and undoubtedly in many cases originated, by the action of fungi. Some of these plants confine their attacks to living trees, others to timber after it is felled; and of the first-mentioned class some are *true parasites*, attacking the roots of living and otherwise healthy trees, whilst others are *wound-parasites*, the minute spores or reproductive germs finding their way into the tree through some wound not produced by the fungus. Holes bored by insects, excoriations of the bark by animals of any kind, and branches broken by wind or badly pruned, afford wounds suitable for the attacks of these last. When the disease caused by a wound-parasite manifests itself first in the cortical and cambium tissues it is termed a *canker*. Some fungi are confined to single species of trees, others attack conifers only, others hard woods only, whilst some seem capable of attacking trees of all kinds alike. The fungi most destructive to timber belong to the more highly organized subdivisions of the class, the *Peziza*, which produces the canker in the Larch, being, for instance, one of the *Ascomycétes*, whilst many others known as "wet rot," "dry rot," etc., are members of the order *Hymenomycétes*, that to which the mushrooms belong.

One of the most generally destructive of these last is the toad-stool *Agáricus* (*Armillária*) *mélleus*, clusters of the yellow fructifications of which are often seen near the base of unhealthy Beech, Spruce, Oak, or other trees in autumn. The upper surface of its tawny cap is shaggy with hair: the gills on the under surface run down on to the stalk, round which there is a well marked torn ring; and the spores, when ripe, are white. Underground, instead of the delicate white "spawn" or mycelium, resembling cobweb, which is common among fungi, this species produces stout, purplish-black strands, which may extend, at a depth of six or eight inches below the surface, to a distance of several feet. These strands are known as *rhizomorphs*, from their root-like appearance. They have growing points capable of penetrating the cortex of living tree roots, and, when they have done so, extend into the cambium and send off branches into the pith-rays and the wood. When this parasite attacks a resinous tree, such as Spruce, a quantity of the resin flows from the pierced root, and the fungal threads travel partly along the resin-passages. In these cases the fungal threads commonly exude a fermentative secretion, by means of which they soften and dissolve the walls of cells or vessels: on penetrating cells containing protoplasm, starch, etc., they readily absorb such substances; but they also destroy cellulose and lignin itself, at first producing various discolorations of the wood, and ultimately reducing it to the condition of "touchwood" or "punk." "It will readily be understood that all these progressive changes are accompanied by a decrease in the specific gravity of the timber, for the fungus decomposes the substance much in the same way as it is decomposed by putrefaction or combustion, *i.e.* it causes the burning off of the carbon, hydrogen, and nitrogen, in the presence of oxygen, to carbon-dioxide, water, and ammonia, retaining part in its own substance for the time being, and living at its expense.[1]

Another true parasite, *Tramétes radicipérda*, only attacks conifers. Its spores, which can be readily conveyed in the fur of mice or other burrowing animals, germinate in the moisture around the roots: the fine threads of "spawn" penetrate the cortex and spread through and destroy the cambium, extending in thin, flat, fan-like, white, silky bands, and, here and there, bursting through the cortex in white oval cushions, on which the subterranean fructifications are produced. Each of these is a yellowish white felt-like mass, with its outer surface covered with crowded minute tubes or "pores" in which the spores are produced. The wood attacked by this fungus first becomes rosy or purple, then turns yellowish, and then exhibits minute black dots, which surround themselves with extending soft white patches.

The many pores in the fructification of *Tramétes* indicate its kinship with the genus *Polýporus*, many species of which are well-known as "shelf-funguses," projecting like brackets from the stems of trees, and having their pores on their under-surfaces. Most of these are wound-parasites. One of the commonest, the yellow cheese-like *Polýporus sulphúreus*, occurs on Oak, Poplar, Willow, Larch, and other standing timber, its spawn-threads spreading from any exposed portion of cambium into the pith-rays and between the annual rings, forming thick layers of yellowish-white felt, and penetrating the vessels of the wood, which thereupon becomes a deep brown colour and decays.

The ravages of such wound-parasites are often the result of

[1] *Timber and some of its Diseases*, by Prof. H. Marshall Ward, F.R.S., to which work I am particularly indebted in the present chapter.

neglect, broken branches being left untrimmed as a lodgement for the spores of the fungus. We have known an Elm-tree to be divided in this way by a broad zone of touchwood, originating from the attack of a *Polýporus* on a snag, so that, though sound timber both above and below, the tree snapt readily in half in a slight gust of wind.

Another species of *Polýporus*, *P. vaporárius*, though it acts as a wound-parasite on coniferous trees, frequently develops and does its chief mischief in stacked timber. It is then commonly confused with the true dry rot, of which we shall speak presently. Its spores (which are, as in most fungi, extremely minute and produced in myriads) fall into cracks of wood, whether the result of injuries to timber when standing, or "shakes" developed after the tree is felled and barked. As their spawn-threads develop in the timber and gradually decompose and absorb its substance, the wood shows deep red or brown streaks, warps and cracks up, and becomes thoroughly rotten, and is penetrated by thick snowy-white ribbons of the felted fungus. In stacked timber this rot frequently develops mainly in the lower, less ventilated, layers of a stack.

Some of the diseases that show themselves conspicuously in the cortex and are known as *cankers* may be set up by frost, by sun, or by insect attack; but in Oak, Beech, Maple, Hornbeam, Alder, Lime, and Larch, canker is the result of wound-parasite fungi. The spores of most, if not all, of these fungi are incapable of penetrating sound cortex; but how many are the chances that bring about small ruptures of this layer! In the case of that most destructive of cankers, the Larch disease, it has been shown that the fungus which produces it, *Pezíza Willkómmii*, is far less common and less deadly in the drier colder air of Alpine heights where the Larch is indigenous; but that late frosts attacking the more advanced and sappy trees in the moist air of the lowlands kill many a shoot and form wounds by which the spores can enter. The moister and warmer air at the same time is more favourable to the growth of the fungus. Its spawn-threads ramify in all directions through the wood, turning it brown and drying it up; while resin flows out at the wound in the bark, which enlarges yearly as the tissues surround it with successively wider-gaping lips of cork in the futile effort of the tree's vitality to heal it over. Round the margins of the wound appear the little orange cup-shaped fructifications of the *Pezíza* scattering their spores so as to infect other trees; whilst the ultimate effect is that each tree is ringed by the destruction of its cortex and then generally succumbs.

Many of the fungi which attack standing timber are so ruinous in their action that the wood of the affected trees will never reach the hands of the timber-merchant; but the wood-worker is more seriously interested in those diseases which attack converted timber. Of these the most important is "dry rot" (*Merúlius lácrymans*). The spores of this fungus germinate on damp wood, provided some alkali is present, such as the ammonia fumes in stables. Then, under the influence of warm, still air (*i.e.* the absence of ventilation) its spawn-threads spread not only in all directions through the wood, forming greyish-white cords and flat cake-like masses of felt on its surface, but even over surfaces of damp soil or brickwork, and thus to other previously uninfected timbers. Feeding upon the elements of the wood, getting its nitrogen from cells which retain their protoplasm, such as those of the pith-rays, but its carbonaceous and mineral substances from the walls of the tracheids and other fibrous elements, the fungus destroys the substance of the timber, lessening its weight, and causing it to warp and crack; until, at length, it crumbles up when dry into a fine brown powder, or, readily absorbing any moisture in its neighbourhood, becomes a soft, cheese-like mass. At an earlier stage the affected timber appears dark-coloured and dull; and, long before its total disorganization, it will have lost most of its strength. Imperfectly seasoned timber is most susceptible to dry rot: the fungus can be spread either by its spawn or by spores, and these latter can be carried even by the clothes or saws of workmen, and are, of course, only too likely to reach sound wood if diseased timber is left about near it; but on the other hand dry timber kept dry is proof against dry rot, and exposure to really dry air is fatal to the fungus. If only the ends of properly seasoned beams which are inserted in brick walls are previously creosoted, it will prove a most effective protection.

Burrs.—Another class of malformations of considerable interest to the timber-merchant are the gnarled and warty excrescences known as *burrs* or *knauers*. These are sometimes due to some mechanical injury to the cortex, at other times apparently to the sudden exposure of a previously shaded stem to the light, as by the felling of a neighbouring tree. They consist of a number of adventitious buds, capable of growing in thickness and putting on wood, but insufficiently nourished to grow in length. In course of years they may grow several feet across, their wood being very irregular, and, owing to its slowness of formation, very dense. The cross-sections of these bud-axes, as in the "bird's-eye" variety of the Hard Maple (*Ácer barbátum*), the Elm, the Yew, the Walnut, and other species, furnish beautiful veneers.

Injurious animals.—Brief mention must be made here of three classes of enemies to both living and converted timbers, viz. the ship-worms or Teredos, the termites (erroneously known as "white ants"), and various insect-larvæ known generally as "worms." *Terédo navális*, the ship-worm, and its allies, are bivalve mollusks, which bore into most kinds of timber when immersed in sea-water, some very dense species, and especially those with pungent resinous secretions, being proof against them. On the other hand, creosoting is by no means always sufficient to keep off their attacks. Shipworms occur in all seas: they generally bore with the grain, lining their burrows with a layer of calcareous matter, and carefully avoiding one another's burrows. They will sometimes completely riddle timber within four or five years. In Australia they are known as "cobra."

The termites belong to the Neuroptera, an entirely distinct Order of the insect class from that to which the true ants belong. They occur in a great variety of species throughout the Tropics, but especially in South America, living in societies of prodigious numbers, and, no doubt, fulfilling a useful function in the economy of nature, by disintegrating, removing, and destroying wood that is already decayed, just as the ship-worms rid the seas of much derelict timber. The termites will, however, attack most species of wood after conversion, sometimes eating their way upward from the foundations of a house to its rafters until all its timbers are reduced to a mere shell, or completely destroying wooden articles of furniture. The pungent resinous secretions which repel the teredo seem also generally effective as a protection against termites.

The large and voracious larvæ of some moths are most destructive to growing trees, and sometimes attack converted timbers. Very generally their eggs are laid in the bark, and the grubs generally bore downward through the sapwood. The Goat-moth (*Cóssus lignipérda*), for instance, specially attacks aged and already unsound Willows, Ash, and Elm; but will attack converted as well as living wood. The Wood-leopard (*Zeuzera æsculi*) specially attacks living fruit-trees and Horse-chestnuts, and its Australian congener, the Wattle Goat-moth (*Z. eucalypti*) frequent the various species of Acacia. Such insects are most destructive; but their large galleries are only too obvious in converting timber. Of the wood-boring beetles, on the other hand, many only attack unhealthy trees: others, such as *Scólytus destrúctor*, the Elm-bark beetle, tunnel in and under the bark, especially of fallen logs, only occasionally penetrating a small depth into the outer wood. Others are far more destructive, in many cases mainly attacking sound converted timber. The widespread Death-watch beetles, for instance, (*Anobium domesticum*, *A. tessellatum*, and allied forms,) the chief cause in England of the familiar "worm-holes" in Oak, frequently entirely destroy the timbering of roofs, and still more commonly riddle our smaller articles of furniture. In the Tropics and warmer Temperate regions their place is largely taken by the numerous family *Bostrýchidæ*, some of which attain far larger dimensions.

CHAPTER IV.

SELECTION, SEASONING, STORAGE, AND DURABILITY OF WOODS.

Selection of wood.—The wood-worker must, of course, determine first what kind of wood is best suited for his purpose, and then take steps to secure that the wood he obtains is a sufficiently good sample of its kind.

It cannot unfortunately be at all assumed that the botanical determination of the species will prove a guarantee of the quality of a timber. Experience shows that *Pínus sylvéstris* or *Quércus Róbur* from different parts of Europe, or even from different situations in one country or *Tectóna grándis* from different districts of India may be a very different thing from the same species of Pine, Oak, or Teak from elsewhere. Botanical identification, therefore, though a most important preliminary, will not obviate other tests. For many purposes, such as mere temporary hoardings, crates, packing-boxes, or the carcases of low-priced furniture, cheapness may be a consideration paramount to all others.

Speaking generally, warm countries, sunny exposures, and dry, elevated land produce heavier, harder, and stronger timber.

It is important that timber should be selected for felling when mature, when the quantity of sap-wood is small and the heart-wood nearly uniform, hard, compact, and durable. After this stage, wood may become brittle, inelastic, discoloured, and perishable, while before maturity, when the sap-wood is in excess, it will seldom be durable. Oak, for instance, for building, should not be less than 50 nor more than 200 years old, and Teak not less than 80 years of age.

Autumn or winter-felled wood, owing to the lower temperature, splits less in drying, and for this reason, and on account of the season being less favourable to fungus-growth, is more durable than that felled in the spring or summer.

Shakes, knots, especially if disposed in a ring round a stick, upsets, *i.e.* fibres crippled by compression, or cross-grain are all defects which reduce the strength of timber. Both butt and top should be close, solid, and sound, any sponginess near the pith, discoloration at the top, rind-gall, worm-holes, or splits produced in seasoning being indications of weakness. Bright-coloured and smooth-working wood is generally better than any that is dull or works with a rough surface; and heavier wood is in all respects stronger than lighter wood of the same species.

Where lightness and stiffness are desirable, coniferous wood is generally preferable; and, where a steady load has to be supported, the denser coniferous woods equal those of broad-leaved trees, which are costlier and heavier. Where, however, moving or jarring loads have to be sustained, the tougher hard woods should be used.

Conversion of timber.—Split wood is straighter in grain and more easily seasoned than sawn timber; and, when sawn, timber will prove stronger and more durable, will season better and will warp less if sawed as nearly as possible along the radii of the annual rings, or, as it is termed, "quarter" or "rift" sawed. This method is more expensive than tangent sawing; but a little consideration will show how it secures—in flooring boards, for instance—a more even exposure of the grain (*i.e.* the hard bands of summer wood) on the surface. It must be borne in mind that in a squared beam with the pith in its centre, whilst we have some complete annual cones of wood appearing as rings at the butt end and tapering to a point or to smaller rings at the top, we shall also have other imperfect cones represented by rings at the top but presenting tangent or "bastard" faces on the sides of the beam and not represented at the butt. These different "structural aggregates" differ materially in strength, the central cone with its numerous knots being the weakest part, whilst the strongest is the hollow cylinder formed of cones that occur as

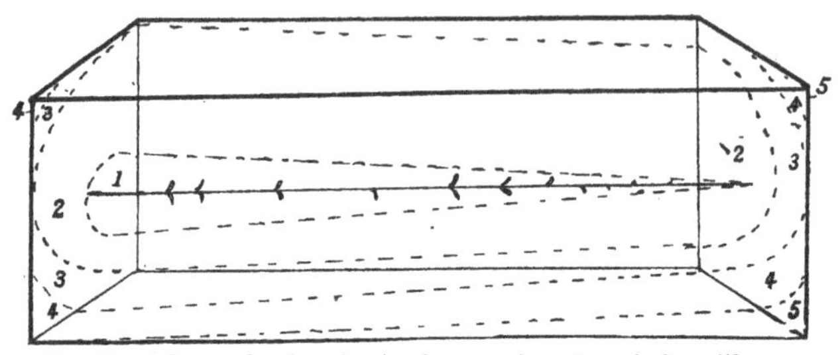

FIG. 41.—A beam, showing structural aggregates. 1, central or pith cone; 2, cylinder of rings continuous throughout; 3 and 4, partial cylinders, making "bastard faces" on the sides. (Modified from Roth.)

rings both at butt and top (Fig. 41). Quarter-sawing secures the most advantageous uniformity in the proportion of each of these aggregates in every plank.

In ordinary tangent-sawed timber it is, as pointed out by Mr. Laslett, important to notice that there is an outside and an inside to every board, and that it is desirable in construction to leave the

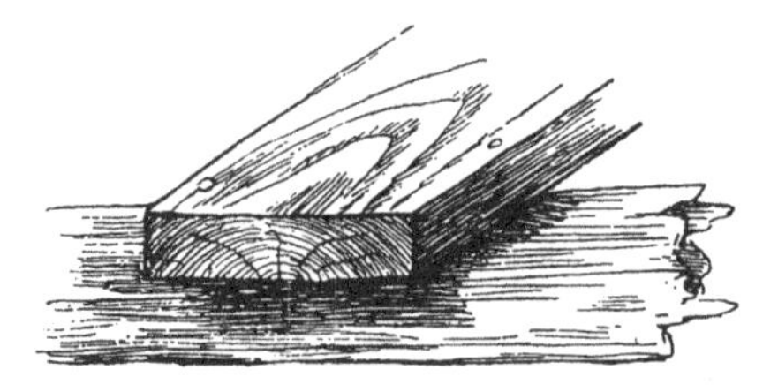

FIG. 42.—Plank well laid, with inside, or inner rings, downward. (After Laslett.)

outside exposed, as shown in Fig. 42, since otherwise (Fig. 43) the inner rings of wood soon shell out.

Seasoning.—All wood when first felled contains a large quantity of moisture, and this, together with the readily decomposable organic or protoplasmic matter also present, furnishes (especially at temperatures between 60° and 100° F.) the most favourable conditions for the growth of those fungi which are the main causes of decay. If completely submerged, or buried, or when once dried and kept so, timber may last indefinitely. The piles in the Swiss lake-dwellings must be many centuries old: wood of *Juníperus Oxycédrus* buried in the island of Madeira has remained undecayed and fragrant for four hundred years; and Spruces three to four feet in diameter have been observed in the moist forests of north-west America growing on the prostrate but still sound trunks of *Thúya gigántea*.

By girdling standing timber the process of seasoning is to a great extent anticipated. Thus, in order to float the timber,

FIG. 43.—Plank badly laid, with the inside, or inner rings, upward. (After Laslett.)

which in its green state is at least as heavy as water, it is the general practice in Burmah to cut a complete ring through the bark and sapwood of the Teak three years before it is intended to fell it. This stoppage of all ascending sap kills the tree in a few weeks: the heat of the climate helps the seasoning process; and, as usually about a year elapses between the felling of the timber and its delivery in England, it is then fit for immediate use. It has, however, been objected to this process that it causes or intensifies heart-shake, and, by drying the wood too rapidly, renders it brittle and inelastic.

Seasoning of some kind is, in all other cases, rendered imperative by the changes in volume, irregular shrinkage, or warping, that all green woods undergo under the influence of changes in atmospheric temperature and moisture, especially in their cross sections. So important is it to avoid this warping in furniture, wheelwright's work, etc., that it is a common practice to block out work roughly and let it season a little longer before finishing.

The strength of many woods is nearly doubled by seasoning, hence it is very thriftless to use it in a green state; as it is then not only weaker, but is liable to continual change of bulk and form. The longitudinal fibres of the wood being, as it were, bound together by the radiating pith-rays, as the wood shrinks it finds relief by splitting radially from the centre along the pith-rays. When a log is sawn into four quarters, by passing the saw twice through the centre at right angles, the outer annual rings shrink the most, so that the two flat surfaces of each quarter of the log cease to be strictly at right angles to one another. In tangent-sawn timber, however, the same shrinkage causes the centre plank to contract in thickness at its edge, whilst planks cut from the outside will shrink in breadth, their edges curving away from the centre of the tree.

It seems to be generally agreed that *natural* or *air seasoning* gives the best results. Firewood should be dried rapidly; but in other cases slow drying in cool air—a process difficult to effect in the tropics—is most desirable in order to reduce the amount of cracking. The timber should be squared as soon as cut, and even halved or quartered, for the rate of drying depends largely on the shape and size of the piece, an inch board drying more than four times as fast as a 4-inch plank, and more than twenty times as fast as a 10-inch timber. The wood is then piled in the seasoning yard so as to be protected as far as possible from the sun and rain, but with air circulating freely on all sides of each log. Bad ventilation is sure to cause rot; but at the same time exposure to high wind is likely to cause unequal drying, and is, therefore, to be avoided. One of the most fertile causes of decay is the leaving of logs to sink into soft ground where they are felled, often in the immediate neighbourhood of rotting stumps or dead twigs, the most fertile source of infection by fungus-spores than can be imagined. Timber should therefore be stacked, or at least skidded a foot off the ground, as soon as possible and protected by a roof. Experience is against the stacking of timber vertically or at an angle, as this only produces unequal drying; but planks may be stacked flat or on edge. Laslett gives the following table of the times required for seasoning Oak and Fir in a shed:

		Months.		Months.
Pieces 24 ins. and upwards square,	Oak about	26,	Fir,	13
Under 24 ins. to 20 ins.	,, ,, ,,	22	,,	11
,, 20 ,, 16 ,,	,, ,, ,,	18	,,	9
,, 16 ,, 12 ,,	,, ,, ,,	14	,,	7
,, 12 ,, 8 ,,	,, ,, ,,	10	,,	5
,, 8 ,, 4 ,,	,, ,, ,,	6	,,	3

For planks half or two-thirds of these times would be requisite, according to their thickness. Too prolonged seasoning will cause an undue widening and deepening of the shakes that open at the surface during drying.

Kiln-drying or hot-air seasoning is a more rapid, but more expensive process, suited only to boards or other small material. It is a common practice to first steam the timber, which reduces

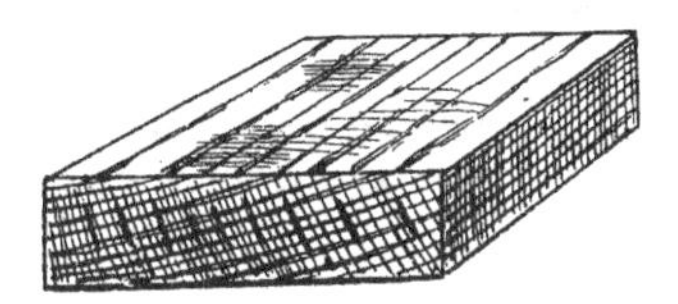

FIG. 44.—"Honeycombed" board, splitting along the pith-rays. (After Roth.)

its hygroscopicity and, therefore, its warping. This, however, is said to reduce the strength, if not also the durability of the wood. If not steamed, the ends of boards should be clamped before kiln-drying to prevent splitting and warping. Neither hygroscopicity nor shrinkage of wood can be altogether overcome by drying at temperatures below 200° F.; but as a rule only the first shrinking is likely to cause splitting, so that any timber which has had from three to six months air-drying may be safely kiln-dried. Too rapid kiln-drying, however, is apt to produce "case-hardening" in Oak and other hard woods, the drying and shrinking, that is, of an outside shell followed by "honeycombing," or splitting of the interior along the pith-rays (Fig. 44). Previous air-drying or steaming will obviate this. Various temperatures are employed in kiln-drying; but it is stated that at 100° to 120° F., Oak, Ash, and other hardwoods can be seasoned in dry kilns without any of the loss of strength often alleged to result from artificial heat. Poplar planks are dried in kilns in America at 158° F. to 180° F.; but Oak, Ash, Maple, Birch, Sycamore, etc., are first air-seasoned for three to six months, and are then exposed to these temperatures for six to ten days for 1-inch stuff. Pine, Spruce, Cypress, and Cedar of the same dimensions are dried for four days immediately after being felled and sawn up. Such temperatures are more than sufficient to kill and prevent fungus growth, and the fact that well-ventilated seasoned wood is seldom attacked shows that the amount of moisture then left in the wood is insufficient to support fungus growth. Walnut gun-stocks are desiccated in the rough by a current of air at 90° or 100° F., passing over them at such a rate as to change the whole volume of air every three minutes, and it is found possible in this way to save a year of seasoning. Temperatures of 250°-300° F. are almost certainly detrimental to the wood. Such desiccated timber must not be exposed to damp before being used or it will re-absorb moisture, and coloured woods are said to lose colour and lustre under this treatment.

Seasoning by passing the smoke-laden products of combustion from the furnace through the timber pile has been found successful, and has an important preservative effect. A modification of this, known as M'Neile's process, consists in exposing the wood to a moderate heat in a moist atmosphere charged with the products of the combustion of fuel.

Boiling timber in water has much the same effect as steaming, but is costly, and probably weakening in its effects.

Seasoning by immersion in water is a slow method that answers well for wood to be used in water or in damp situations. It reduces warping, but renders the wood brittle and less elastic. It is important that the submergence be total, as otherwise there is great danger of fungus attack along the water-line. Two or three weeks' water-seasoning is often a good preparation for air-seasoning, and it must be remembered that foreign timbers have often had some weeks or months of such treatment while being transported by water to the port of shipment. It is important that wood seasoned in this way be thoroughly dried before use, otherwise dry rot will set in. In Mauritius, Ebony, which is perfectly sound when freshly cut, is immersed immediately for 6 to 18 months, and then, on being taken out, is secured at both ends of the logs with iron rings and wedges. Soaking timber or burying it under corn were methods of seasoning practised by the ancient Romans, who also steeped wood in oil of cedar to protect it against worms.

Salt water makes wood harder, heavier, and more durable; and the rules of Lloyd's add a year to the term of classification of a ship if she is "salted" during construction, having her timbers, that is, packed with salt. Salt water cannot, however, be applied to any timber intended for use in ordinary buildings, as it gives the wood a permanent tendency to attract moisture from the air.

Boiling in oil is an effective and strengthening, but costly, method of seasoning, employed in making wooden teeth for mortice gears. The wood is roughed out in blocks little more than the size of the finished work, and the oil kept at a temperature not exceeding 250° F.

In Australia the abundance of hardwood, its great weight, and the high price of labour, has led to a general total neglect of seasoning, which has had a very deleterious effect upon the reputation of Australian timbers in the markets of the world. Though admittedly too costly for general use, the modification of the oil process adopted by Mr. J. H. Maiden, curator of the Technological Museum of New South Wales, for museum specimens of timber is interesting. The logs are stood on end and the upper end is soaked with boiled linseed oil, and a day or two later covered with a cream of white lead. Iron bands are then put round each end of the logs and hammered to their outline, the ends of the bands being turned out at right angles and bored for a screw bolt, by means of which the bands can be tightened up every few days.

Carbonizing, or charring the outer surface of wood, destroys all fungus-germs at the surface; and, charcoal resisting the solvents of fungi, this process renders the wood little liable to subsequent infection. Thus it is stated that the stakes found in the bed of the Thames, near Weybridge, and supposed to have been used to oppose the invading Romans, and the piles upon which the city of Venice was built, had alike been charred. M. de Lapparent, who introduced this process into the French dockyards forty years ago, held that the durability of carbonized timber is secured by the absence of fermentation in the juices of the interior of the wood. The results are satisfactory, but care must be taken not to cause surface splitting. M. de Lapparent's process is carried out by means of a jet of gas.

Another important series of methods of seasoning are those which may be termed *impregnation-methods*, which all depend upon the principle that the sap may be replaced by some substance that is antiseptic or poisonous to fungus-germs. The most primitive of these is merely to paint the substance, such as tar, as thickly as possible over dry wood and leave it to soak in, and this undoubtedly has a great preservative effect, even on sapwood or wood very imperfectly dried; but the chief drawback to this, and the chief difficulty in several other impregnation processes, is the very small distance that the liquid soaks, so that slight cracks expose unprotected wood to fungus attack. Whilst it is comparatively easy to inject sapwood in a longitudinal direction, it is far more difficult to inject heartwood; and it is vastly easier to force liquids through wood tangentially than radially.

An improvement on any painting process is to submerge the timber in a bath of the preservative, which may be tar, sulphate of iron or of copper, chloride of zinc or creosote; and in these processes the replacement of the air and sap in the wood by the liquid will generally be hastened by heat. Penetration is, however, slight, and long submergence renders the timber brittle.

Of the materials employed for impregnating timber, the most effective is corrosive sublimate (mercuric chloride), the use of which is known, from its inventor, Kyan, as kyanizing. It forms insoluble compounds in the wood, and is, therefore, permanent; but its costliness and dangerously poisonous character are against it. Zinc chloride, introduced by Sir William Burnett, is cheap and effective against both insects and fungi, but less so than creosote. Copper sulphate, sometimes used for sleepers in France, is cheap; but is deposited in crystals in the wood, rendering it brittle, and, owing to its solubility, is as easily washed out as it is injected. Creosote, originally suggested by Bethell, and now very largely employed in various ways, is cheap, lasting in its effects, and useful in rendering the wood damp-proof. The more expensive carbolic acid and ferric tannate have also been used.

To force the antiseptic solution into the wood, M. Boucherie, who first employed copper sulphate, proposed placing it in an elevated reservoir connected by a pipe with the lower end of a log; but this requires the log to have its bark on, and is thus a wasteful process.

A more complicated and costly, but very successful, process consists in the use of air-tight chambers, in which the converted timber is placed. The air is then partially exhausted, so as to draw out some of that in the vessels of the wood, and the antiseptic solution is then forced in by pumps, preferably with steam or heat, the whole process occupying less than an hour. About 75 lbs. of creosote, however, are required for the impregnation of an ordinary railway sleeper, and various attempts to reduce this quantity by the use of some liquid solvent have failed. Herr F. Seidenschnur proposes that the timber be first steamed under pressure, the air then exhausted by reduced pressure, and then an emulsion of 15 per cent. of creosote, in a resin soap to which water is added, forced in under a pressure of seven atmospheres. The latest process, known as the Nodon-Bretonneau method, is electrical. The timber is placed on a lead plate, connected with the positive pole of a dynamo, in a tank filled with a solution of 10 per cent. of borax, 5 per cent. of resin, and ·75 per cent. of carbonate of soda, a second lead plate, connected with the negative pole, being on top. The circuit is completed through the wood; and, within from 5 to 8 hours, the sap rises to the surface of the bath, the aseptic borax and resin solution replacing it in the pores of the wood. Artificial drying, or a fortnight's natural seasoning in summer weather, will then complete the process.

Such impregnation-methods double or treble the life of railway sleepers. On the other hand, it should be remembered that paint prevents not only the entrance of moisture, but also its exit; so that if applied to imperfectly seasoned wood it merely protects the dry rot which finds a sufficiency of moisture in the wood. Even perfectly seasoned wood, if not protected by tar or paint, requires good ventilation if it is to last. Warm, moist, stagnant air or draught, and partial contact with moist earth or water are the most unfavourable conditions for the durability of timber. A process has recently been introduced for rendering wood "non-flammable" by extracting the resin and volatile matter under hydraulic action and then injecting chemicals. This process is applicable to wood up to 6 inches in thickness and to hard woods as well as soft woods, and dries them effectually at the same time, so that there is no subsequent change of form; but it adds about 25 per cent. to the cost of soft woods and more to that of hard woods. It has been largely used in the United States navy.

Storage.—For the storage of seasoned timber much the same precautions are requisite as for that which is undergoing air-seasoning, viz. thorough ventilation, absence of contact with moist earth, and preferably some protection from rain and sun. If logs are stacked with their butt-ends outward and slightly lower than their tops, if every log or scantling be so separated by small packing billets that it can be removed without disturbing the remainder, and if each tier of timber is set back a few inches so as to obviate the use of a ladder, it will render the stock not only safe but accessible.

CHAPTER V.

THE USES OF WOODS.

So multifarious are the uses to which wood is applied that it is well nigh hopeless to attempt to classify or enumerate them. Still less is it possible here to mention all the different kinds of wood locally employed for each purpose, or to describe the methods in which they are treated. We must be content with a rough catalogue mainly confined to species widely used or known in general commerce, with occasional mention of less known kinds of timber for which we believe there may be a demand in the near future.

The term "timber," from the Old English "timbrian," to build, is strictly applicable only to felled and seasoned wood fit for building, as distinguished from "fancy" or furniture-woods, dye-woods, etc. Undressed trunks without branches are termed "round timber"; or, if of young trees, "spars"; hewn logs are called "square timber"; or when quartered, "billets"; when split, "staves" or "lathwood"; or when sawn, "deals," "battens," "planks," "boards," and "scantling."

Some very strong timbers, such as Teak, Sal, and Padouk, are specially designated as "Ordnance woods."

Shipbuilding.—There is, perhaps, no purpose for which timber has been, and requires to be, more carefully tested and selected than for shipbuilding. From this point of view we have a full account of most timbers so employed in the late Mr. Thomas

Laslett's *Timber and Timber-trees*, originally published in 1875, of which a new edition by Professor Marshall Ward appeared in 1894. The requirements of the dockyard are, however, very varied, durability being generally necessary; but great strength, even if accompanied by weight, and freedom from decay on contact with metal, being important for armoured vessels; resistance to ship-worms or termites for those not metal-sheathed; lightness for boats; freedom from splintering for planks; extreme toughness for blocks; evenness of growth and great resistance to strain for masts; flexibility for oars. For general purposes, among the heavier woods, Teak (*Tectóna grándis*) is taken as a standard, and is far more used than the Oaks, whether European or American, of former days, valuable as these are, however, especially for exposed and compass timbers. The Indian Jarul (*Lagerstrœ́mia Flos-regínæ*) and Thingan (*Hópea odoráta*), the Greenheart of Demerara (*Nectándra Rodiǽi*), the Angélique (*Dicorýnia guianénsis*), African Oak or Teak (*Oldfiéldia africána*) from West Africa, Stinkwood (*Ocotéa bulláta*) and Sneezewood (*Pteróxylon útilé*) from the south of the same continent, the Rata or Ironwood of New Zealand (*Metrosidéros robústa* and *M. lúcida*), and probably the Billian (*Eusideróxylon Zwágeri*) of Borneo, are but little inferior.

Lloyd's Register, classifying ship-building timbers in 17 lines, places Teak alone in the first; in the second, English Oak (*Quércus Róbur*), African Oak (*Oldfiéldia africána*), Live Oak (*Quércus vírens*), Adriatic, Italian, Spanish, Portuguese, and French Oak (*Q. Cérris, Ǽsculus, pyrenáica, Ílex, Súber*, and *Róbur*), Morung Saul (*Shórea robústa*), Greenheart (*Nectándra Rodiǽi*), Morra (*Móra excélsa*), Iron-bark (*Eucalýptus siderophlóia*, and probably *E. leucóxylon*, and *E. sideróxylon*), and White Ironbark (apparently *E. crébra, amygdalína*, and *paniculáta*); in the third, Cuba Sabicu (*Lysilóma Sábicu*), Pencil Cedar (*Juníperus Bermudiána*, or perhaps *Dysóxylon Muélleri*, and *D. Fraseriánum*), Angelly (*Artocárpus hirsúta*), Vanatica (*Pithecolóbium sp?*), Jarrah (*Eucalýptus marginída*), Karri (*E. diversícolor*), Blue Gum (*E. Glóbulus*), Red Gum (*E. rostráta*), Box (*E. hemiphlóia?*), Thingam (*Hópea odoráta*), Puhutukawa (*Metrosidéros tomentósa*),

Molave (*Vítex geniculáta* and *V. altíssima*), Dungon (*Stercúlia cymbifórmis*), Yacal (*Shórea reticuláta*), Mangachapuy (*Shórea Mangáchapoi*), Betis (*Payéna Bétis*), Ipil (*Afzélia bijúga*), Guijo (*Shórea robústa*), Narra (*Pterocárpus pállidus* and *P. santalínus*), Batitinan (?), and Palomaria de Playa (*Calophýllum Inophýllum* ?); in the fourth, those of the first and second line when second-hand; in the fifth, Stringy Bark (*Eucalýptus oblíqua*, etc.), Red Cedar (apparently *Cedréla Toóna*), Banaba, which is the Jarul of India, and Philippine Islands Cedar (probably chiefly *Cedréla Toóna*); in the sixth, Danish and other Continental White Oak, Mahogany (*Swieténia Mahógani*), Spanish Chestnut (*Castánea satíva*), Flooded Gum (*Eucalýptus salígna*), Spotted Gum (*E. maculáta*), Grey Gum (*E. viminális*), Turpentine (*E. Stuartiána*, chiefly), Black Butt (*E. piluláris*), Tulip-wood (*Harpúllia péndula* ?), Tallow-wood (*Eucalýptus microcórys*), and Mulberry (?); in the seventh, North American White Oak (*Quércus álba*); in the eighth, Pitch Pine (*Pínus rígida*), Oregon Pine (*Pseudotsúga Douglásii*) Huon Pine (*Dacrýdium Franklínii*), Kauri Pine (*Ágathis austrális*), Larch (*Lárix europǽa*), Hackmatack or Tamarac (*L. americána*), and Juniper (?); in the ninth, Dantzic, Memel, and Riga Pine (*Pínus sylvéstris*), and American Red Pine (*P. resinósa*); in the tenth, English Ash (*Fráxinus excélsior*); in the eleventh, foreign Ash (*F. sambucifólia, americána*, etc.), and Rock Maple (*Ácer barbátum*); in the twelfth, American Rock Elm (*Úlmus americána* and *racemósa*), and Hickory (*Hicória ováta, álba, glábra, mínima, Pécan*, etc.); in the thirteenth, European and American Grey Elm (*Úlmus campéstris* and others); in the fourteenth, Black Birch (*Bétula lénta*) and Black Walnut (*Júglans nígra*); in the fifteenth, Spruce Fir (*Pícea excélsa*), Swedish or Norway Red Pine, and Scotch Fir (*Pínus sylvéstris*); in the sixteenth, Beech (*Fágus sylvática*); and in the seventeenth, Yellow Pine (*Pínus Stróbus*).

The Turpentine-tree (*Syncárpia laurifólia*), White Box (*Tristánia conférta*), Box (*Eucalýptus hemiphloía*) and Spotted Gum (*E. maculáta*) of New South Wales are also generally useful. The Securipa and Guarabu of Brazil, the latter of which may be *Terminália acumináta* or *Peltogýné macrolóbium*, though little known, are employed locally; but the Stringy-bark of Tasmania (*Eucalýptus oblíqua*) and the Blue Gums (*E. Glóbulus* in Tasmania and *E. botryoídes* in Victoria) have been proved suitable both for beams and planks. Other dense timbers are employed mainly for beams and keelsons, such as the Mora of Demerara (*Dimorphándra Móra*, or *Móra excélsa*), Tewart (*Eucalýptus gomphocéphala*) of West Australia, Iron-bark (*E. siderophlóia*) of Queensland and New South Wales, and Sabicu (*Lysilóma Sábicu*) of Cuba. Chow, or Menkabang Penang (*Casuarína equisetifólia*) from Borneo, the "Cèdre" of the Seychelles, though a heavy wood, is mainly employed for masts, as are also the Poon, Tatamaka, or Alexandrian Laurel of India (*Calophýllum Inophýllum*), which is known as "Phung-nyet" in the Andaman Islands, as "Domba" in Sinhalese and as Penago, Panagah, Pingow, or Borneo Mahogany in Borneo, the Peroba branca (*Sapóta gonocárpa*) of Brazil; and, still more, such soft woods as Riga Fir (*Pínus sylvéstris*), Yellow Pine (*P. Stróbus*), Oregon or Douglas Fir (*Pseudotsúga Douglásii*), the unequalled Kauri Pine of New Zealand (*Ágathis austrális*), and the Huon Pine of Tasmania (*Dacrýdium Franklínii*). For this purpose a certain elasticity is requisite, resistance, that is, to wind. Other coniferous woods are of more general use, such as Dantzic Fir (*Pínus sylvéstris*), the Totara (*Podocárpus Tótara*) and Tanakaha (*P. asplenifólius*) of New Zealand, the Moreton Bay Pine (*Araucária Cunninghámi*), mostly for spars, Red Pine (*Pínus resinósa*) and Pitch Pine (*P. palústris*), which serve equally for spars and for planking. Other species, mainly on account of their dimensions, are chiefly employed in boat-building, such as the Black or Cypress Pine (*Cállitris robústa*), the Oyster Bay Pine (*C. rhomboídea*) and the Bermuda "Cedar" (*Juníperus bermudiána*) among conifers; and European and American Elm, Jarrah (*Eucalýptus margináta*) and Red Gum (*E. rostráta*), Pynkado or Pyengadu (*Xýlia dolabrifórmis*), which is the Ironwood of Pegu and the Acle of the Philippines, Anan (*Fagrǽa frágrans*), Gumbar (*Gmelína arbórea*), Sundri (*Heritiéra littorális*), and the Brazilian Camara (*Geissospérmum Vellósii*) among hard woods. Some timbers are most valuable for compass timbers, such as the Angelim vermelho (probably *Ándira fraxinifólia*) of Brazil and the Puriri (*Vítex littorális*) and Pohutukawa (*Metrosidéros tomentósa*) of New Zealand; whilst others are used almost exclusively for decks and planking, such as the Turpentine Tree or Stanthorpe Box (*Eucalýptus Stuártiana*) and White Beech (*Gmelína Leichhardtii*) of Eastern Australia, the Canella preta (*Nectándra átra*) of Brazil, and the Lauan (*Dipterocárpus thúrifer*) of the Philippines. Exceptionally hard and tough woods, such as Lignum Vitæ (*Guaíacum officinálé*) and the Ironwood of Tasmania (*Notelǽa ligustrína*) are required for blocks; whilst tough but flexible kinds, such as the Ash of Europe or America and the Silver Wattle (*Acácia dealbáta*), are employed for oars. For the internal fittings of ships almost any species can obviously be used which is employed in ordinary civil architecture or joinery.

Submerged structures.—Passing next to timbers used for piles or other submerged structures, such as locks and water-wheels, Elm, Larch, Chestnut (*Castánea*), Live Oak (*Quércus vírens*), Sal (*Shórea robústa*), Totara (*Podocárpus Tótara*), and Rassak (*Vática Rássak*) of Borneo, may be specially mentioned. Greenheart, Jarrah, Pynkado, Chow, Kapor (*Dryobálanops aromática*), another Bornean timber, Alder, and Beech are also used for these purposes. For the strouds of water-wheels and for paddle-boards Willow is employed.

Strength timbers.—For such engineering purposes as require considerable strength, and resistance to definitely calculable strain, for bridges, piers, or baulks of timber, Teak, Jarul, Sal, Sissoo (*Dalbérgia Síssoo*) and Anan (*Fagrǽa frágrans*) among Indian timbers, the Locust of Trinidad (*Hymenǽa Coúrbaril*), Oak, and the superior kinds of Pine may be mentioned.

Sleepers.—Railway-sleepers absorb enormous quantities of timber, which requires to be durable when in contact with the earth and with metal. Home-grown Douglas Fir seems to rival Larch for this purpose in Britain. Deodar (*Cédrus Deodára*), Sal, Blackwood (*Dalbérgia latifólia*), Poon (*Calophýllum Inophýllum*), Nagesar or Ironwood (*Mésua férrea*), and Chilauni (*Schíma Wallíchii*) among Indian timbers; the Box of New South Wales

(*Eucalýptus hemiphlóia*); Puriri (*Vítex littorális*), Hinau (*Elæocárpus dentátus*) and Totara (*Podocárpus Tótara*) in New Zealand; and, when creosoted, the Upright or Real Yellow-wood, Geel Hout, or Umceya (*Podocárpus latifólius* or *P. Thunbérgii*) in Cape Colony, are employed for this purpose; and one of the most important industries of the future in the colony last-mentioned is the cultivation of the European Cluster Pine (*Pínus Pináster*) for the same use.

Mining timber.—Less care is exercised in the selection of pit props for mines. Larch and pine, both home-grown and of Baltic origin, are largely used in English mines and *Pínus Pináster* is imported from Bordeaux to the Welsh collieries and Cornish tin-mines.

Telegraph poles.—For telegraph poles much the same characters are requisite as for masts, in addition to durability underground. Besides Larch and European Pine and Douglas Fir, the Black or Cypress Pine of New South Wales (*Cállitris robústa*), being proof against termites, is in request for this purpose, and, in the United States, Chestnut (*Castánea vulgáris*, var. *americána*) is used.

Building.—Less durability is essential is scaffold poles and ladders, for which Spruce (*Pícea excélsa*) is largely used. For joists, rafters, and flooring, no wood is so much used with us as Dantzic Fir (*Pínus sylvéstris*), though the somewhat shaky and cheaper Swedish Fir of the same species is also largely used, whilst that of Norway is imported in the form of ready-made flooring and match-boarding. In the west of England Baltic Pine is largely replaced by American White Pine (*Pínus Stróbus*). The Pitch-Pine of the United States (*Pínus palústris*) is now largely employed in match-boarding and other internal work in English buildings, and Larch is much used for flooring, as also are both Baltic and American Black Spruce (*Pícea excélsa* and *P. nígra*). Since the importation of these coniferous timbers from the Baltic and from America, which dates mainly from the beginning of the eighteenth century, Oak, till then the chief building timber in North-west Europe, has been but little used, though, of course, old oak beams, floors, and panellings are still abundant. From its not splintering, Willow is still occasionally used for flooring. In the United States, whilst White Oak (*Quércus álba*) is very largely employed for the main timbers of houses, the Pines, especially the soft White Pine (*Pínus Stróbus*), the Long-leaf Pine (*P. palústris*), the Loblolly Pine (*P. tǽda*) and the so-called Norway Pine (*P. resinósa*), with other species in the west, are (under a confusing jumble of popular names) the timbers most used. In Northern India, the Bhotan Pine (*Pínus excélsa*) and Himalayan Cypress (*Cupréssus torulósa*) are important coniferous timbers, and there are several valuable species of Oak, viz. *Quércus semecarpifólia, Q. dilatáta, Q. pachyphýlla, Q. lamellósa. Q. fenestráta, Q. spicáta and Q. Griffíthii.* Among the other hardwoods important in building are Champa (*Michélia Chámpaca*), Redwood (*Adenánthéra pavonína*) Sal, Ironwood (*Mésua férrea*), the Myrobalans, Babela, and Harra (*Terminália belérica* and *T. Chébula*), Shoondul (*Afzélia bíjuga*), Illupi (*Bássia longifólia*) and Ironwood or Pyengadu (*Xýlia dolabrifórmis*). In Australia, the Peppermint (*Eucalýptus amygdalína*) and the White Stringy Bark (*E. capitéllá*); in New Zealand, the Totara (*Podocárpus Tótara*) and Tanakaha (*Phyllocládus trichomanoídes*); the Yellow-wood (*Podocárpus elongatus* and *P. latifólius*) in South Africa; Mora and Angèlique in Guiana; Canella preta (*Nectándra átra* and *N. móllis*) in Brazil; and Cagüeyran (*Copaífera hymenæifólia*) in Cuba, are all timbers valuable to the builder.

Wood-paving.—The consumption of wood for paving in our large towns, already enormous, is rapidly increasing, although the comparative advantages of soft wood, in England mainly Pine, with its greater cheapness, and hard woods, with their greater durability and the chance of their becoming slippery, are not yet decided. The chief hard woods as yet used in England are Jarrah (*Eucalýptus marginátu*) and Karri (*E. diversícolor*) from South-western Australia. Black-butt (*E. piluláris*) and Crow's Ash (*Flindérsia austrális*), from Eastern Australia, were laid experimentally in Wellington Street, Strand, in 1895. Little can be said in favour of the Red Gum of the Eastern United States

(*Liquidámbar styracíflua*), a large quantity of which was ordered for use in Westminster in 1901. "Cedar," often spoken of in this connexion in Western American cities, is probably mostly the wood of *Thúya gigántea* (*T. plicáta*) and *Cupréssus lawsoniána.*

Shingles and fencing.—Wooden shingle roofs, for which Oak used to be employed, are of much less importance in England than in the United States, where White Cedar (*Thúya gigántea* and *T. occidentális, Cupréssus lawsoniána* and *C. Thyoídes* and *Libocédrus decúrrens*) is largely used for this purpose, which requires a straight-grained wood, easy to split. In all countries enormous quantities of split and sawn timber are consumed for fencing purposes; more especially Oak and Larch with us; "Cedar" in the United States; the so-called "Birch," really a Beech (*Fágus Solándri*), in New Zealand; and Beefwood or Swamp or Forest Oak (*Casuarína equisetifólia*) and allied species, together with various species of *Eucalýptus*, in Australia, of which, perhaps, *E. amygdalína, E. rostráta*, and *E. vimindális* are the chief.

Carpentry.—The work of the carpenter and joiner links that of the builder to that of the cabinetmaker. In Europe, in addition to much Baltic and American Pine, chiefly *Pínus sylvéstris, P. Stróbus*, and *P. palústris*, he uses much Spruce (*Pícea excelsa*), Bordeaux Pine (*Pínus Pináster*), and Swiss Pine (*Ábies pectináta*), besides Oak, Ash, and Chestnut. So also in the United States and Canada, the Hemlock Spruce (*Tsúga canadénsis*), White and Black Spruces (*Pícea álba* and *P. nígra*); and in the West Indies, Fiddlewood (various species of *Citharéxylum*) may be specially mentioned as carpenters' woods. In South Africa the Cedar Boom (*Widdringtónia juniperóides*), though not very durable, is a useful wood, as the allied species, *W. Whítei*, from the kloofs of the Shiré Highlands, may probably prove; and in Eastern Australia the Moreton Bay Pine (*Araucária Cunninghámi*) may be mentioned in this group. The carpenter requires cheap wood, easily worked, and of moderate strength.

Carriage-building.—We may class here the various woods employed in the many branches of the wheelwright's, waggon and carriage-builder's trade. Hornbeam (*Carpínus Bétulus*) and

Australian Blackwood (*Acácia melanóxylon*) are peculiarly fitted for the hubs; Ash and *Eucalýptus crebra* and *E. goniocályx* for spokes; Hickory (various species of *Hicória*) for axle-trees and shafts; Poplar, American White-wood (*Liriodéndron tulipífera*), Birch and Maple (*Ácer barbátum*) for panels; the dense Pyengadu (*Xýlia dolabrifórmis*) and Padouk (*Pterocárpus índicus*) of Burma, for gun-carriages or the frames of railway-waggons, and the Bastard Peppermint of New South Wales (*Tristánia suavéolens*) for somewhat similar purposes, in which tough hard wood is needed. About 1750, Satinwood, upon which Cipriani and Angelica Kauffmann executed their paintings, became fashionable for coach-panels; whilst for the humbler purposes of wheelbarrows Willow is useful from its freedom from splintering.

Furniture.—An immense variety of woods has been employed in the making of furniture, susceptibility to polish, beauty of colour or grain, and durability being their chief requisites, together with freedom from shrinkage, whilst they are variously employed either planed, carved, turned, or bent. Thus some wood known as "Cedar" seems to have been largely used in ancient Assyria and Egypt, forming the beams of the temple of Apollo at Utica, said by Pliny to have been sound 1200 years after their erection, employed alike in Solomon's temple, in Greek sculpture, and in carpentry, as for the chest in which Cypselus of Corinth is said to have been concealed about 550 B.C. As Vitruvius speaks of that of Crete, Africa, and Syria as the best, it is probable that then, as now, the wood of several species was confused under one name, probably the Lebanon Cedar (*Cédrus líbani*), that of Mount Atlas (*C. atlántica*) and the 'Arâr (*Tetraclínis articuláta*) of Morocco. This last sweet-scented wood, known also as Atlas Cypress, was the much-vaunted "Citrus" or "Citron" Wood of the Romans and probably the "Thyine Wood" of the Apocalypse. The roof of the cathedral at Cordova, originally a mosque, is built of it, it being there known as "Alerce." The true Cypress (*Cupréssus sempervírens*) was, no doubt, largely used, not only, as is related, for Alexander the Great's Babylonian fleet or Semiramis' bridge over the Euphrates, but owing to its durability and resistance to moth, for clothes-chests. An Italian chest of this wood of the 14th century is preserved at South Kensington, and John of Gaunt bequeaths one in his will in 1397. The Certosina work, or inlaying of this wood and walnut with ivory, so-called from the choir fittings of the Certosa between Milan and Pavia, an art practised at Florence in the 15th century, was perhaps brought by the Venetians from Persia, from which country it also reached Bombay. Sissoo (*Dalbérgia Síssoo*), possibly the Chittim of Holy Scripture, and other species of Rosewood, Ebony, Teak, and Walnut, may have reached Assyria, Syria, and even more western lands from India; but the Corsican Ebony used by the Romans for veneers was probably the Laburnum, the "Faux Ébénier" of the French. Lotos-wood, said to have been used in Greek sculpture, may have been that of the Nettle-tree (*Céltis austrális*), still much used in Southern Europe. We read of the Romans using Box and Beech for chairs and for veneers; Beech for chests; Olive, both wild and cultivated for veneers; Fig, Willow, Plane, Elm, Mulberry, Cherry, and Cork-Oak, as ground for veneers; Maple, especially Bird's-eye Maple (probably *Ácer campéstré*), for tables; and Syrian Terebinth (*Pistácia Terebínthus*), and Poplar for various other purposes. Though Norway Pine was imported by Henry III., in the 13th century, for panelling at Windsor, throughout the Middle Ages, Oak was the main furniture wood as it was the chief building material. As in the timber-frame houses of the Chester rows, the 14th century roof of Westminster Hall, or the marvellously carved one of the Palais de Justice at Rouen in the 16th; so in the great bed of Ware and other English and Flemish furniture during the Tudor period, Oak alone is employed. It was used as a bed wood for veneering by Boule under Louis XIV., and was painted white and gilt in the time of Louis XVI. Italian Walnut (*Júglans régia*) was much used in Italy for carving and gilding from the 15th century; and it was at Venice and Florence that the use of the soft white woods of Willow, Linden, and Sycamore for carved and gilt frames for mirrors originated in the 16th. The use of Ebony, especially for inlaying Walnut wardrobes became more general after the Dutch settlement in

Ceylon in 1695: Grinling Gibbons, who was partly of Dutch descent, employed Linden and other white woods for his inestimable carving; and the work of Thomas Chippendale in the 18th century gave Mahogany the popularity in England that Satinwood enjoyed at that time in France. Heppelwhite and Sheraton employed Mahogany not only for chairs, but for small articles such as tea-caddies, whilst in the inlaid work of the period it was used, not only with other dark woods, such as Rosewood, Laburnum, and Purple-heart (*Copaífera pubiflóra*), but also with Holly, Maple, and Pear. At the present day Mahogany is used for dining-room furniture and veneers, though much Oak, some of which is the Canadian Red Oak (*Quércus rúbra*), is used for the same purpose, whilst large quantities of Walnut (*Júglans nígra*), Ash (*Fráxinus americána*), Bass-wood (*Tília americána*), Maple (*Ácer barbátum*), and Birch (*Bétula lénta*) are imported from North America for library and bed room furniture, stained or painted Deals being employed for yet cheaper goods. Fifty years ago American Walnut was only used in England for inferior purposes, such as framing for veneers; but now it has much advanced in popularity with cabinet-makers and shopfitters with a doubling of its former price. Another American wood of increasing importance is the American Whitewood, or Canary Whitewood (*Liriodéndron tulipífera*), used for the seats of American Windsor chairs, and, from its suitability for staining or polishing, rapidly becoming a favourite with wood-workers. Beech and Yew are the staple woods of our Buckinghamshire chair factories, Ash being used in bent wood-work; whilst bamboo work and cane-seats are somewhat outside our present scope.

Among furniture woods in use in other countries we can only enumerate a few:

In India:—

Ebony (*Diospýros* spp.),
Rosewoods or Blackwoods (*Dalbérgia látifolia*, etc.),
Sissoo (*Dalbérgia Síssoo*),
Redwood (*Adenánthera pavonína*),
Padouk (*Pterocárpus índicus* and *P. dalbergióides*),

Bija Sâl or Bastard Teak (*Pterocárpus Marsúpium*),
Margosa or Neem (*Mélia* spp.),
Siris (*Albízzia* spp.),
Chittagong wood (*Chickrássia tabuláris*),
Chatwan (*Alstónia scholáris*), a soft wood, named from its use for blackboards in Indian schools,
Gumbar (*Gmelína arbórea*),
Toon, Moulmein Cedar or Indian Mahogany (*Cedrela Toóna*); and Jack or Ceylon Mahogany (*Artocárpus integrifólia*).

In Mauritius and other islands in the Indian Ocean:—
Tatamaka or Rosewood, under which name are confused *Thespésia popúlnea* and *Calophýllum Inophýllum*.

In South Africa:—
Sneezewood, Neishout, or Umtạti (*Pteróxylon útilé*),
Stinkwood (*Ocotéa bulláta*),
Cape Ebony (*Eúclea pseudébenus*, etc.),
Cape Ash, Essen Boom, or Umgwenyuizinja (*Eckebérgia capénsis*),
Saffron-wood, or Umbomoana (*Elæodéndron cróceum*),
Assegai-wood, or Umguna (*Curtísia fagínea*),
Salic-wood, or Unkaza (*Buddléia salviæfólia*), and
Red Cedar, or Rood Els (*Cunónia capénsis*).

In Yoruba Land, West Africa:—
Iroko (*Chloróphora excélsa*), resembling Satinwood.

In Borneo:—
Mirabow (*Afzélia palembánica*).

In Australia:—
Blackwood (*Acácia melanóxylon*, etc.),
Jarrah (*Eucalýptus margináta*),
Shingle Oak (*Casuarína strícta*),
Queenwood (*Daviésia arbórea*),
Rosewood (*Dysóxylon Fraseriánum*),
Beefwood (*Grevíllea striáta*),
Mulberry (*Hedyoárya angustifólia*),
Silky Oak (*Stenocárpus salígnus*),

Moreton Bay Pine (*Araucaria Cunninghámi*), and
Pencil Cedar (*Podocárpus eláta*).

In Tasmania :—

Honeysuckle (*Bánksia margináta*) and
Huon Pine (*Dacrýdium Franklínii*).

In New Zealand :—

Honeysuckle or Rewa-rewa (*Knightia excélsa*),
Kauri Pine (*Ágathis austrális*),
Rimu (*Dacrýdium cupressínum*),
Miro (*Podocárpus ferrugínea*), and
Totara (*P. tótara*).

And in Tropical America :—

Mahogany or Baywood (*Swieténia Mahógani*),
Sabicu (*Lysilóma Sabicu*),
Santa Maria, or Galba (*Calophýllum Cálaba*),
Green Ebony (*Brýa Ébenus*),
Zebra Wood (*Cónnarus guianénsis*, etc.),
Sapodilla (*Áchras Sapóta*), and
Braziletto (*Cæsalpínia brasiliénsis*, etc.).

Veneers.—Very choice ornamental woods are employed mainly as veneers. Such are, in addition to many of those just enumerated: Amboyna wood, the product, it is believed, of some species of *Pterocárpus*; the burrs of Yew, largely used for tea-caddies, etc., in the 18th century; those of Walnut; and the beautiful Lacewood or Honeysuckle wood of North America (*Plátanus occidentális*).

Turnery.—The turner requires a tough wood, which will often be also hard and susceptible of good polish. No wood is more generally useful to him than the Ash, as it does not splinter. Curiously enough, cankered Ash-wood, popularly known as "bee-sucken Ash," being apparently twisted in its grain, is extremely hard and tough, and, therefore, suitable for mallets. Beech is used for wedges, planes, and tool-handles; Hornbeam for the bearers of the cylinders of printing-machines; Pear for T-squares; and Elm, and in former times Maple, for bowls; whilst the record of the demand for Walnut for the manufacture of gunstocks reads like a romance. In 1806 France required 12,000 Walnut trees per annum; while in England, before the battle of Waterloo, £600 was paid for a single tree. For cheaper gunstocks American Walnut is now used, whilst the American species of Ash, Beech (*Fágus ferrugínea*), and Hornbeam (*Carpínus caroliniána*, known as "blue Beech"), are employed in the United States for purposes similar to those to which their European equivalents are put. The Hickories (*Hicória*), more especially for handles, the Persimmon (*Diospýros virginiána*) for shuttles, plane-stocks, etc., and the Cherry (*Prúnus serotína*) are also important to the American turner. In Japan, Kizi (*Paulównia imperiális*) is the main basis for lacquer-ware: the so-called Cherry (*Exocárpus cupressifórmis*) and the fragrant Musk-wood (*Oleária argophýlla*) of Australia, and the Violet-wood (*Copaífera bracteáta*) of Brazil may be specially mentioned; whilst in South Africa the various species of *Olea* known as Ironwood, the Silk-bark or Zybast (*Celástrus acuminátus*), Buffelsbal (*Gardénia Thunbérgii*), Ladle-wood (*Hartógia capénsis*), and Umzumbit (*Millétia Káfra*); and in India the Babul (*Acácia arábica*), Ironwood (*Mésua férrea*), Ebonies (*Diospýros* spp.), calamander (*D. quæsíta*), Anjan (*Hardwíckia bináta*), Tamarind (*Tamaríndus índica*), Dhaura (*Anogeíssus latifólia*), Bullet-wood (*Mímusops littorális*), Satin-wood (*Chloróxylon Swieténia*), and Sandal-wood (*Sántalum álbum*), are noteworthy.

Walking-sticks.—A great variety of woods are used in the manufacture of walking-sticks. Not to mention Jersey Cabbages and the leaf-stalks of the Date-palm and a great variety of Canes, imported specially from Singapore, these include English-grown Oak, Ash, Blackthorn, Holly and Hazel, Whitethorn, Aspen, Birch, Crab-apple, Furze, Maple, Hornbeam, and Rowan. Medlar (*Méspilus germánica*) and Chestnut (*Castánea satíva*) are imported from France; Cork Oak (*Quércus Súber*) from Spain; Carob (*Ceratónia Síliqua*) from Algeria; Guelder-rose (*Vibúrnum Ópulus*), under the names of "Teazle" or "Balkan-rose," from the Balkans; Olive and Orange from Southern Europe, while "Black Orange" is a trade name for the common Broom (*Cýtisus scopárius*); Box, from Persia; Ebony, from Ceylon; and, from the West Indies,

Cocus or "Flowered Ebony" (*Brýa Ebenus*), Partridge-wood (*Ándira inérmis*), Pimento (*Piménta officinális*), and Letter-wood or Leopard-wood (*Brósimum Aublétii*).

Engraving.—For wood-engraving, the Box (*Búxus sempervírens*) of Turkey is unequalled, and the use of metallic blocks has diminished the urgency of the search for a substitute for, as wasteful consumption threatened exhaustion of the supply of, this species. The Cape Box (*Búxus Macowánii*), introduced in 1885, is now considerably used: Ebony is nearly equal in texture to Box, but its colour militates against its use; Hawthorn is probably next best to Box of any known wood, but cannot readily be obtained of sufficient size: Pear (*Pýrus commúnis*), used for calico-printer's blocks, the Chinese T'eng li mu (*Pýrus betulæfólia*), and Pai'cha (*Euónymus europæus*, var. *Hamiltoniánus*), the American Box or Dogwood (*Córnus flórida*) and other species are suitable for coarse work; but Jamaica Box (*Tecóma pentaphýlla*) is on the whole the most likely successor to Box.

Musical instruments.—While any well-seasoned ornamental wood, such as Rosewood, Mahogany, or Walnut, is used for the cases of pianofortes, those parts of musical instruments in which resonance is produced must consist of wood of uniform texture, free from all knots or other defects or contrasts of grain. Ancient Etruscan flutes seem to have been made of Box; whilst at the present day the Green Ebony (*Brýa Ébenus*) of the West Indies is considered well fitted for this purpose. Evelyn writes that Cypress is a sonorous wood, and is employed in making harps, organ-pipes, and other musical instruments; but the Silver Fir (*Ábies pectináta*), known in the trade as "Swiss Pine," is now accounted the most resonant of all woods, and is used for the bellies of the violin and the sounding-boards of pianos, Sycamore (*Acer pseudo-plátanus*) or Hard Maple (*A. barbátum*) being employed for the back and sides of the former instrument.

Miscellaneous Uses.—Even tobacco-pipes consume large quantities of certain woods, such as the Bruyère, commonly known as Briar (*Eríca arbórea*), from Southern Europe, the Myall (*Acácia homalophýlla*) from Australia, and the Cherry (*Prúnus* *ávium, Máhaleb*, etc.), used for long pipe-stems and grown mainly in Austria. The light white woods of the Horse-chestnuts or Buckeyes (*Ǽsculus*) are used for artificial limbs, just as, judging by the writings of the comic dramatists, Linden-wood was employed in making corsets for male dandies in ancient Greece. Millions of cubic feet of Bermuda Cedar and of the Red or Pencil Cedar of Virginia (*Juníperus bermudiána* and *J. virginiána*) are cut annually for the manufacture of pencils alone. The quantities of Alder (*Álnus glutinósa*) Beech, Willow (*Sálix álba* more especially), Spruce or White Deal (*Pícea excélsa*), Birch (*Bétula álba*) and even Horse-chestnut (*Ǽsculus hippocástanum*) in Europe, and of Tupelo (*Nýssa sylvática*) and Canoe Birch (*Bétula papyrífera*) in North America, consumed for sabots must be immense, to say nothing of the quantities of these and other woods used for shoe-lasts, shoe-pegs, boot-trees, hat-blocks etc. Soft white woods, such as Willow, Alder, Linden, Poplar or "Cottonwood" that of the Tulip-tree (*Liriodéndron*) and the Cucumber-tree (*Magnólia acumináta*), confounded together as "Canary Whitewood," and the Spruces (*Pícea*) and Soft Pines (*Pínus Stróbus*, etc.), are those chiefly in demand by the toy-manufacturer.

Cooperage.—The requirements of the cooper are more varied than might be supposed, different woods being needed for staves, for hoops, or for head-pieces, and for dry, liquid, or volatile goods. Oak is largely used for staves, especially French Oak (*Quércus Róbur*), and American White Oak (*Q. álba*), but in Australia the Black Wattle (*Acácia mollíssima*) takes its place. Willow and Hickory are used for hoops and Ash for a great variety of purposes, but for dry goods the cooper employs cheap soft white woods such as those used for the manufacture of packing-cases.

Packing cases.—Packing-cases made of inferior Silver Fir (*Ábies pectináta*) are sent all over the world from Switzerland and the Tyrol: its cheapness causes Norway Spruce (*Pícea excelsa*) to be almost as universally employed; and on the continent of Europe the Black Austrian, Bordeaux Cluster, and Italian Stone Pines (*Pínus austríaca, Pináster*, and *Pínea*) are also largely used for this purpose. Their not splitting when nailed renders the

Poplars admirable for this purpose, and the White, Aspen, and Lombardy Poplars (*Pópulus canéscens, trémula,* and *fastigiáta*) are accordingly largely used in France, as are *Pópulus monilífera* and other "Cottonwoods," as they are there called, in the United States. *Pícea Smithiána*, the Himalayan Spruce, is in common use in India; but for tea-chests, though Chir (*Pínus longifólia*), Chatwan (*Alstónia scholáris*), Chaplash (*Artocárpus Chaplásha*), Toon (*Cedréla Toóna*), Shembal (*Bómbax malabáricum*), and Maples, such as *Ácer Campbéllii* in the north-east, and *A. píctum* in the north-west are employed, there is an inadequate supply of suitable native wood, which is being met by the importation of Birch from Russia.

Crates, paper-pulp, etc.—Ash, Alder, and Birch are largely used in the making of crates; and few persons probably, outside the trade, notice the variety of woods, in addition to Willow, which go to the making of our baskets. Enormous quantities of the Pine timber of Sweden (*Pínus sylvéstris*) are consumed in the form of lucifer matches; and, while wood-shavings and wood-wool, as it is called, much used in packing, are little more than bye-products in the conversion of timber for other purposes, the manufacture of wood-pulp for paper, an industry belonging almost entirely to the last twenty years, has grown to such dimensions as to seriously affect the question of our timber supplies. The Poplars, Alders, Buckeyes, and Spruces are the most suitable woods for this manufacture; but the coarser kinds of printing paper, packing paper, and pasteboard are made from Pine, even the branches and chips, formerly wasted, being utilized. The refuse of *Juníperus virginiána* from the pencil factories yields a paper useful for underlaying carpets or wrapping articles liable to be injured by moth. Two methods are followed, the mechanical, yielding a granular product called paper-pulp, and the chemical, yielding a more fibrous felt known as cellulose. Coniferous cellulose, prepared in Germany and the United States, is used for an infinity of purposes such as tubes, vases, wax-cloth, etc. As an illustration of the growth of the wood-pulp industry it may be stated that in 1891 the product of

Norway was valued at 8,600,000 kronor (about £430,000), and that of Sweden at 10,400,000 kronor (£520,000), whilst in 1900 they were 27,400,000 and 33,200,000 kronor respectively.

Fuel.—The heat-producing value of wood as fuel varies greatly, owing to the differing capacity that woods have for retaining moisture. Thus, while green wood may contain 50 per cent. of moisture, ordinary stack-wood may contain only 25 per cent., and kiln-dry wood only 2 per cent. With 25 lbs. of water, 100 lbs. of fire-wood will contain about 1 lb. of incombustible ash and 74 lbs. of the dry substance of wood. This last consists of 37 lbs. of carbon, 32 lbs. of oxygen and 4·4 lbs. of hydrogen; and in burning the whole of the oxygen combines with 4 lbs. of hydrogen to form water, so that only the 37 lbs. of carbon and 0·4 lb. of hydrogen, *i.e.* about half the weight of the dry substance of the wood, are available for heat-production. Every pound of water combined in the wood requires about 600 units of heat to evaporate it, the unit being the amount of heat necessary to raise 1 lb. of water 1° C.; so that 100 lbs. of stack-wood (25 per cent. moisture) only furnishes about 255,000 units, whilst if kiln-dry (2 per cent.) it would yield 350,000. The advantage of seasoning for firewood is, therefore, obvious. The resinous woods of the conifers produce most flame and are most useful accordingly in starting a fire; but the denser hard woods produce from 25 to 30 per cent. more heat.

Charcoal.—When wood is heated to 200° F. without access of air, it remains unaltered, at 220° it becomes brown, and at 270° to 300° it suffers decomposition, torrefied wood or red charcoal being formed. At 350° it is resolved into volatile products and true or black charcoal. If the temperature is raised gradually, so that 600° F. is not reached for several hours, the process is called dry distillation. The first product of distillation is almost entirely water; but at 500° pyroligneous (crude acetic) acid, or wood-vinegar, wood-spirit and uncondensable gases pass off, charcoal and some tar remaining. In the primitive method of the charcoal-burner, or *meiler*, in which billets of wood are stacked horizontally or inclined round a central chimney opening, most of the volatile products are lost; but for charcoal this process is still largely employed on the Continent. If the fire is steady and regular, the slower the process the better the yield. For gunpowder-charcoal, however, and acetic acid, iron or brick ovens are mostly employed. The best gunpowder-charcoal is produced from light woods, such as Willow, Buckthorn, or "Dogwood" (*Rhámnus Frángula*), and Alder. Charcoal is darker, heavier, a better conductor of heat and electricity, less easily ignited, and gives out greater heat in burning, the higher the temperature at which it has been made. The proportion of charcoal yielded is greater (24 to 30 per cent.) with a slow process, that of the volatile products with a rapid one. From experiments with Hornbeam, Alder, Birch, Rowan, Beech, Aspen, Oak, Buckthorn, Silver Fir, and Larch, we find the yield of charcoal to range from 20 per cent. with slow, to 34·6 per cent. with quick distillation; the total distillate from 43 to 53 per cent.; the pyroligneous acid from 47·5 in the hardwoods to 38 in the conifers; and the tar from 2·9 in Beech to 9·7 in conifers. In practice only about 18 to 20 per cent. by weight of charcoal is obtained, or about half the volume of the wood. Pyroligneous acid is in England largely manufactured from spent dye-woods, such as fustic, logwood, etc., the charcoal obtained being largely used for packing the meat refrigerators in ships. The products of distillation, under the most favourable circumstances, are stated as:

	Charcoal.	Tar.	Crude Pyroligneous Acid.	Pure Acetic Acid.
Birch, - - -	22·4	8·6	45·0	4·47
Beech, - - -	24·6	9·5	44·0	4·29
Oak, - - -	26·2	9·1	43·0	3·88
Juniper, - -	22·7	10·7	45·8	2·34
Silver Fir, - -	21·2	13·7	41·2	2·16
Scots Fir, - -	21·5	11·8	42·4	2·14

Purer acetic acid is obtained by re-distillation, and, when mixed with certain essences, constitutes aromatic vinegar. Among the acetates prepared on a large scale from pyroligneous acid are those of lime, the brown containing from 60 to 70, and the grey from 80 to 85 per cent. of acetate. In the preparation of these naphtha is recovered; and from this, by neutralizing with lime and re-distilling, wood spirit or methyl alcohol. Wood-tar, used for creosoting wood and in the manufacture of roofing-felts, is a thick, dark, viscous material, containing from 5 to 20 per cent. of acetic acid, from 30 to 65 per cent. of pitch, and from 20 to 45 per cent. of tar oils. From these last, creosote, a colourless, highly refracting oil, with a specific gravity of 1·04, boiling at 406° F., and paraffin, used for candle making, are obtained, by neutralizing with carbonate of soda and further distillation.

Dyeing and tanning.—Finally, somewhat apart from these other uses to which woods are applied, is the employment of certain species for dyeing and tanning. Of the former the most important are Logwood (*Hæmatóxylon campechiánum* L.), which dyes red or black, and of which we import over 50,000 tons annually from Central America; Fustic, a yellow dye, obtained from the wood of the large West Indian trees, *Chloróphora tinctória* Gaud. (= *Maclúra tinctória* D. Don) and its varieties, *xanthóxylon* and *áffinis*; Sappan or Yellow-wood, from *Cæsalpínia Sáppan* L.; the red dyes known as Brazil, Braziletto, Nicaragua, or Lima wood, from *Cæsalpínia crísta* L., *brasiliénsis* L., *echináta* Lam., *C. bíjuga*, and *C. tinctória*; Camwood, *Báphia nítida* Afz., from West Africa; and Red Sanders or Sandal-wood, *Pterocárpus santalínus* L. fil., and *Adenánthera pavonína* L., from India.

Barks are more used for tanning than are woods; but the Quebrachos, the produce of several South American species, have been a good deal employed of late years.

CHAPTER VI.

OUR SUPPLIES OF WOOD.

In spite of the substitution of iron or other substances for wood in shipbuilding and other industries, with the increasing numbers of civilized man the consumption of wood increases at such a rate as to demand serious attention.

The clearing of forest land for the purposes of agriculture has been most recklessly carried out, especially during the last century in the United States and in Canada, much of the wood being wasted. Where, too, the timber has been cut for use, this has in general been done so completely without any provision for the regeneration of the forest-lands as to lead to their extinction. The floods and famines of China, the waste of the agricultural soil in Ceylon, the barrenness of Mesopotamia, Syria, Asia Minor, and Cyprus, the drying up of the springs and deterioration of the climate in South Africa, Mauritius, Turkey, and Spain have been attributed mainly to wholesale destruction of forest. The felling of the woods on the Atlantic coast of Denmark has exposed the country to sharp sea winds and drifting sand, forming lagoons and bogs and causing a marked deterioration of the climate: the disafforesting of the Appennines during the last two centuries has much increased the violence of the mountain-torrents; and even in Russia, which has not only the largest area of forest of any European state, but the largest percentage of her whole area under forest, a decrease in the waters of the Volga has been attributed to the same cause.

Whilst all woodland has disappeared from some lands, special species are threatened with extinction in others. The pine forests of Tunis have disappeared during the last hundred years: some districts of Australia already experience a scarcity of fire-wood and of mine-props: until Government regulations put a stop to the felling of saplings to act as rollers in transporting the larger logs, the valuable Greenheart of Demerara was in imminent danger of extinction; and the enormous drain upon the supply of White Pine (*Pínus Stróbus*) is a grave danger in North America.

Great Britain.—In Great Britain the abundance of coal renders us independent of wood as fuel, and our geographical position so facilitates the importation of timber that we have to a great extent neglected our woodlands as a source of profit, while our mild insular climate has enabled us to overlook the hygienic importance of forests. There is accordingly little more than 2½ million acres of woods and forests in the United Kingdom, or only 3·8 per cent. of the entire area, a lower percentage than that of any other European state, while this country stands pre-eminent as the greatest importer of timber, exceeding 300 million cubic feet, or, including paper-pulp, gums, bark, and other forest produce, an annual value exceeding 35 millions sterling. No complete statistics are available as to our consumption of home-grown timber. Special local demand is to some extent met by local supply, as, for instance, in the case of the bobbin-wood in the cotton-mill districts, pit-props in the Scottish mining area, and the Beech of the Chilterns, from 12,000 to 15,000 loads of which are used annually in the Buckinghamshire chair-making industry, by which some 50,000 families are supported. Of our imports, nearly five millions sterling is the value of the timber received from Canada.

Sir J. F. L. Rolleston, M.P., in his presidential address to the Surveyors' Institution in November, 1901, said:

"Before leaving the subject of land and its future, I should like to say that of all its products the only one, the value of which appears to be in the ascending scale, is timber. In the midland counties I have been furnished with accounts of timber sales at which single Oak trees have realized up to £100, while other woods are commanding good prices, and poles and thinnings are readily sold. There is a reason for this. The great onslaught that has been made on the virgin forests of the world, from the time of the Phœnicians onwards, without artificial reafforestation, must at length be appreciably felt.

The increase of population and the advance of civilization must also point to an increased use of timber of all kinds for works of construction, for articles of use and ornamentation, and for fuel. A rise in the value of home-grown timber seems possible; in any case a ready sale may be anticipated.

With the decline in the value of cereals it can hardly be doubted that a considerable portion of the land of this country (some of which is derelict, and some let at a very low rental) might be planted to advantage."

The forest areas of the other countries of Europe are estimated as follows:

Russia, - - -	469,500,000 acres, *i.e.*	34 per cent.
Sweden, - - -	43,000,000 acres, *i.e.*	24 per cent.
Austria-Hungary, -	42,624,000 acres, *i.e.*	29 per cent.
France, - - -	20,642,000 acres, *i.e.*	19 per cent.
Spain, - - -	20,465,000 acres, *i.e.*	16·3 per cent.
Germany, - - -	20,047,000 acres, *i.e.*	25·6 per cent.
Norway, - - -	17,290,000 acres, *i.e.*	25 per cent.
Italy, - - - -	9,031,000 acres, *i.e.*	18 per cent.
Turkey, - - -	5,958,000 acres, *i.e.*	14 per cent.
Switzerland, - -	1,905,000 acres, *i.e.*	18·8 per cent.
Greece, - - -	1,886,000 acres, *i.e.*	11·8 per cent.
Wurtemberg, - -	1,494,000 acres, *i.e.*	31 per cent.
Baden, - - -	1,338,000 acres, *i.e.*	33 per cent.
Portugal, - - -	1,107,000 acres, *i.e.*	5 per cent.
Belgium, - - -	1,073,000 acres, *i.e.*	12 per cent.
Holland, - - -	486,000 acres, *i.e.*	6 per cent.
Denmark, - - -	364,000 acres, *i.e.*	4·6 per cent.

With civilization comes an increasing demand for timber for fencing, building, mine-props, railway-sleepers, and telegraph-poles, not to mention that for more valuable woods for furniture, etc., and the multitudinous other minor uses of timber. Thus American statisticians have estimated 3 million cords[1] of wood as used annually in brick-burning, a million cords of Birch for tool-

[1] A cord = $2\frac{1}{2}$ loads, $2\frac{1}{2}$ tons, or 125 cubic feet.

handles and boot-lasts, 100,000 cords of Soft Maple for shoe-pegs, and over 3000 cords of Pine for lucifer matches in the United States alone.

Russia.—In the well-managed forests of Germany the average yearly growth, and, therefore, the amount legitimately felled annually, is estimated at 2·3 cubic feet for every 100 cubic feet of standing timber, or 50 cubic feet per acre. But in spite of the enormous annual yield which this computation gives to the forests of Russia (viz. 23,475 million cubic feet), when we find nearly half that amount (10,000 million) now used within the country for fuel alone, and 30 million for house-building, it will be realized how little reliance can be placed in Russia as a permanent source of supply for Europe. Before reckoning for her increasing population we may recall the saying that Russia is burnt down every seven years. Of the total timber output from Russian government forests in 1880 of 2,900,000 cubic fathoms, Spruce (*Pícea excélsa*) constituted 37·5 per cent., Pine (mainly *Pínus sylvéstris*), 27·8, soft woods (Birch, Linden, Aspen, etc.), 19·5, and hard woods (Oak, Beech, etc.), 8·8 per cent. Besides paper-pulp from the Aspen, and a certain amount of Walnut, Russia exports Box from Odessa, and a large amount of Deal from the White Sea and Baltic ports. The supplies of timber at Archangel and the other White Sea ports is yearly drawn from a greater distance inland.

Scandinavia.—Sweden sends more than half of her exported timber to Great Britain. It consists largely of Pine, both as pit-props and in a manufactured form, as window and door-frames; Spruce or "White Deal," used for scaffolds, ladders, etc.; matches, of Pine and Aspen; and paper-pulp of Aspen, Spruce, and Pine. The exports of Norway are similar, a certain amount of Birch and Maple (*Ácer platanóides*) also coming from this country to England. Both Norway and Sweden are apparently reducing their forest areas by cutting more than the annual increment.

France.—Though a well-wooded country, with carefully managed forests in almost every department, exporting Oak and sending Bordeaux Pine (*Pínus Pináster*) as mine-props to our

Welsh collieries, France imports common building woods as well as the more costly kinds used for furniture, etc., her imports exceeding her exports to the value of over five million sterling per annum.

German Empire, etc.—Spain imports, but does not export timber. Prussia has 23 per cent. of its area under forest, over 6 million acres, or 30 per cent. of the whole, being under government administration. The yield is about 47 cubic feet per acre per annum, *i.e.* safely within the calculated annual increment of 50 cubic feet, the total expenditure about 1½ millions sterling, and the net surplus over a million, or about 3s. 6d. an acre for all ground in use. The chief species are Kiefer (*Pínus sylvéstris*), exported as Dantzic or Riga Fir or Prussian Deal, and Fichte or Roth Tanne (*Pícea excélsa*), forming between them three-fourths of the whole crop. Eiche (*Quércus Róbur*) is exported to England as Baltic or East Country Oak, and the Silver Fir, Edeltanne or Weissfichte (*Ábies pectináta*) abounds in the Vosges and occurs in Schleswig-Holstein and Silesia. More than a quarter of the area of Bavaria is under wood, and, though there is a large local demand for fuel, the careful foresight of the administration is evidenced by the fact that in 1885 a government forester was sent to study the timber-trees of the United States, who frankly explained his mission by saying, "In fifty years you will have to import your timber, and as you will probably have a preference for American kinds, we shall begin to grow them now, so as to be ready to send them to you at the proper time." Timber is the chief export of the country.

Saxony has over a million acres of forest, one-third of which belongs to the State, the annual cut, estimated at a million cubic feet, being much less than it might be. The Saxon forests include Oak, Beech, Ash, Birch, and Alder, as well as Pine, Spruce, Silver Fir, and Larch.

Wurtemberg has nearly 1½ million acres, or over 30 per cent. of its whole area, under forest, comprising the Pine-wood districts of the Black Forest and the hardwoods of the Swabian Alps. Pine, Spruce, Silver Fir, and Oak are floated down the Rhine to the Dutch shipbuilding yards, whilst Beech furnishes the chief fuel of the country, and is used for ships' keels, carriage-building, and chair-making, and Aspen is in demand for matches and paper-pulp.

Hesse-Darmstadt, the Fir-trees from which are in special demand in Holland, has no less than 595,000 acres, or nearly three-quarters of its area, under forest; whilst Baden has also over a million acres, or one-third of its area, so occupied.

Austria-Hungary.—The forests of the Austrian Empire occupy over 42½ million acres, those of Austria being 30 per cent., those of Hungary, 26·6 per cent. of the entire areas of the two countries. Spruce, Silver Fir, and Larch are the prevalent species, and the bulk of the timber is consumed, for building purposes or fuel, at home. Hungary has also some large forests of excellent Oak. **Switzerland.**—From the 1,900,000 acres of the forests of Switzerland it is estimated that over 89 million cubic feet of timber are cut annually, but, in addition to considerable clearing, the demands of a growing population for building purposes, and the use of much wood as fuel, there has been considerable waste, as, for instance, in cutting young trees for fencing, so that the total cut has been estimated as in excess of the yield, and the export has accordingly declined. Spruce, Silver Fir, and Pine are the predominant species.

Italy.—Italy exports a certain amount of Oak of various qualities, but of ill-ascertained origin. The best, the Tuscan, Neapolitan, and Sicilian, would seem to be *Quércus Róbur*, *Q. Ǽsculus*, and *Q. pyrenäica*. Modena, Roman, and Sardinian Oak and Adriatic Oak (*Q. Cérris*) are inferior. The country, is, however, deficient in timber, from the point of view both of climate and of demand. While with our moist climate we can manage with a far smaller proportion of forest, the countries bordering on the Mediterranean all suffer from the removal of their forests. Centuries ago the Karst region of Southern Austria was covered with magnificent Oak forests and furnished piles and shipbuilding timber to Venice in her palmy days. It was said that a squirrel could travel for miles along the Istrian coast from tree to tree.

Reckless felling by the Venetians led to the washing away of the surface soil, until the country for twenty miles north of Trieste was reduced to bare rock. Forty years ago the Austrian Government began a costly system of reafforestation.

Asia.—Turning from Europe to Asia, we find undoubtedly a large supply of Larch (*Lárix sibírica*), Pine, Spruce (*Pícea cephalónica*), Birch, and other species in Siberia; but, unless the Amoor can, to some extent, play the part of the St. Lawrence, the difficulty of transport will be insuperable. Neither China, the interior of which probably suffers much from the effects of disafforesting, nor Japan, hold out any prospect of any large export either of common or of choice woods, whilst, except perhaps in the remote future to western North America, cost of freight would put the former class of timber out of the question.

In Japan, where forest conservancy dates from the third century A.D., half the area of the country, or about 47,000,000 acres, are stated to be forest, yielding more than 120 species of valuable timbers, of which the Nikko Silver Fir (*Ábies homolépis* S. and Z.) and Saghalien Fir (*A. sachalinénsis* Masters) are the cheapest, and Hi-no-ki (*Cuprélssus obtúsa* Koch) and Ke-ya-ki (*Zelkówa acumináta* Planchon) are the most expensive.

India.—Taking British India as 480 million acres, 40 million, or one-twelfth of the whole area, are forest. In spite, however, of the enormous local consumption for fuel and the increasing demand for railway-sleepers, India produces such a variety of valuable ornamental and dense Hardwoods that conservation is likely to enable her long to continue her exportation. In 1899-1900 she exported Teak to the value of over £600,000, but her supply of cheap softwood for tea-chests, etc., is hardly equal to the demand. At the same time many of her ornamental furniture woods might well be more largely used in Europe.

Such woods as Pynkadoo or Miraboo (*Afzélia palembánica*), Kranji (*Diálium índicum*), and Tampinnis (*Sloétia sideróxylon*), in the Malay Peninsula, the Lauan (*Dipterocárpus thúrifer*) and Acle (*Xýlia dolabrifórmis*) of the Philippines, and the Rassak (*Vática Rássak*), Billian (*Eusideróxylon Zwágeri*), and Compass (*Kœmpássia *malaccénsis*) of Borneo, may well prove worthy of European attention, especially for density and durability, when they become better known, and the supply of them may be said to be as yet untapped.

Among 200 species thought worthy of trial in the arsenal at Manila, the essentially Malayan flora of the Philippines includes:

Acle (*Xýlia dolabrifórmis*),
Banaba (*Lagerstrœ́mia Flos-Regínæ*),
Betis (*Payéna Bétis*),
Bolongnita (*Diospýros pilosánthera*),
Cedar (*Cedréla Toóna*),
Dougon (*Stercúlia cymbifórmis*),
Guijo (*Shórea robústa*),
Ipel (*Afzélia bíjuga*),
Lauan (*Dipterocárpus thúrifer*),
Mangachapoi (*Shórea Mangáchapoi*),
Molave (*Vítex geniculátus* and *V. altíssima*),
Narra (*Pterocárpus pállidus* and *P. santalínus*)
Padouk (*Pterocárpus índicus*),
Palo Maria (*Calophýllum Inophýllum*), and
Yacal (*Shórea reticuláta*).

Australasia.—Australian timbers have, as we have already said, suffered in European repute by not being seasoned; and, as, in spite of a vast area of scrub, the area of timber-producing forest is comparatively small, wholesale clearing for the purposes of agriculture, the use of wood for fuel, and the great demand for building, fencing, railways, and telegraphs, have sensibly affected the supply. Conservation has begun; but mine-props and even firewood are locally scarce. Queensland exports Red Cedar (*Cedrela Toóna*), and Moreton Bay Kauri and Cypress Pines (*Araucária Cunninghámii*, *Ágathis robústa*, and *Cállitris robústa*). New South Wales sends Cedar (*Cedréla Toóna*) and Pine (*Araucária Cunninghámii*) to China and New Caledonia, and the area under the former species is now considerably reduced. Nearly half the area of the colony of Victoria (40,000 out of 88,198 square miles) was estimated as forest in 1878, most of it being in the hands of Government, and more than half of it consisting of *Eucalýptus*. Many Victorian timbers are extremely dense and hard, such as Red Gum, Blue Gum, White Gum or Peppermint, Messmate and Iron-bark (*Eucalýptus rostráta, glóbulus, amygdalína, oblíqua,* and *leucóxylon*), etc.; and accordingly, though some of them may well maintain a more than local value for sleepers, wood-paving, etc., timber at present appears among the imports rather than among the exports of the colony. The forest-area of South Australia, where *Eucalýptus* also forms the staple of the timber-supply, is not large. It is West Australia, however, and especially its south-western parts, from which we at present import the bulk of our Australian timber-supply. Besides Sandalwood (*Sántalum cygnórum* or *Fusánus spicátus*) to the value of nearly £30,000 annually, sent mainly to China, West Australia is exporting timber to the value of half-a-million sterling, the chief species being Jarrah (*Eucalýptus marginóta*), which is officially stated to be the predominant species over 14,000 square miles, Karri (*E. diversícolor*) occupying 2,300 square miles, Tewart (*E. gomphocéphala*) occupying some 500 square miles.

The timber areas in West Australia are stated as:

Jarrah (with Blackbutt and Red Gum),	8,000,000	acres.
Karri, - - - - - - -	1,000,000	,,
Tewart, - - - - - -	200,000	,,
Wandoo, - - - -	7,000,000	,,
York Gum, Yate, Raspberry-jam, and Sandalwood, - - - -	4,000,000	,,

This area is estimated to contain 62 million loads of mature timber worth £3 per load, a total value, deducting $\frac{1}{3}$ for waste in sawing, of £124,000,000.

Nearly one-half of the island of Tasmania (8,000,000 acres) is timbered, seven-eighths of the woodland being under Government, but the timber area is diminishing. The beautifully mottled, durable Huon Pine (*Dacrýdium Franklínii*) has become scarce and high-priced. The bulk of the timber exported consists of Stringybark (*Eucalýptus oblíqua*), sent in planks to Victoria, South

Australia, and New Zealand; but the most valuable timber of the colony is the Blue Gum (*E. glóbulus*), which is abundant in the south of the island.

The forest-area of New Zealand, estimated at over 20,000,000 acres in 1830, was only 12,000,000 acres in 1874, when clearing was proceeding at the rate of 4 per cent. per annum; but conservation was then inaugurated and the many valuable species of timber thereby saved from extermination.

Of these the most valuable is the Kauri Pine (*Ágathis austrális*), which is confined to the North Island. This fine durable timber is the soft wood of the country, and is extensively converted for export to Australia, the freight militating against it in competition with Baltic timber for the English market, though it is employed to some extent for the decks of yachts.

Africa.—Little can be said as to the timber resources of the African continent. Neither Atlas Cedar (*Cédrus atlántica*), resembling the Deodar, nor Atlas Cypress (*Tetraclínis articuláta*), the Citron-wood of the ancients, are at all known commercially, and the same must be said of Morocco Ironwood (*Argánia Sideróxylon*). Algeria, however, has nearly 5 million acres of forest, three-fifths of which are under State control, and its Evergreen Oaks (*Quércus Ilex, Súber, bállota,* etc.), its Kabyle Ash, said to be equal to English, and Maritime Pine (*Pínus Pináster*) should prove of value. From our West African colonies we import small quantities of African Oak or Teak (*Oldfiéldia africána*), a dense wood, shipped from Sierra Leone, and African Rosewood (*Pterocárpus erináceus*); but little is known of the timber-trees of tropical Africa; while the south of the continent is one of the districts of the world which suffers most in climate from the want of timber, partly from reckless destruction.

Little is known as yet as to the botanical nature or abundance of the undoubtedly valuable timbers of Rhodesia. It is estimated that there are about 2000 square miles of forest in Matabeleland, while Mashonaland is not so well timbered. Annual grass fires kill innumerable young trees: the natives are answerable for the destruction of many thousands; and the felling of large timber is attended with much unnecessary destruction of smaller trees. The Gwaai forest, which extends along the river of that name, fifty miles from Buluwayo, consists of Ikusi, or Native Teak, several kinds of Acacia, and Asenga Mopani. Large areas in Mashonaland also are covered with Ikusi, a handsome dark brown wood streaked with yellow, which is worked for building purposes. The Asenga Mopani, stated to grow 50-75 feet high with a straight stem, is a furniture wood. The Shangani river passes through a forest of Baobab, the largest tree of the country; whilst the Mahobohobo, valuable as a mine-timber because it is termite-proof, abounds in the Selukwe and Belingwe districts. Katope, resembling Pine; Mbawa and Malombwa, resembling Mahogany; and Muwowa, used for native canoes, and stated to reach an immense height, are also valuable species.

Cape Box (*Búxus Macowánii*) is far inferior to Turkey Box; but many of the cabinet-woods of Cape Colony, such as Stinkwood (*Ocotéa bulláta*) and Sneezewood (*Pteróxylon útilé*), deserve more than local repute. The remnants of the indigenous forests of "Pencil Cedar" (*Widdringtónia juniperóides*) will repay strict conservation, whilst one of the most important industries of the future will be the growth of the Maritime Pine (*Pínus Pináster*) for railway-sleepers. Natal has 165,000 acres of forest; but depends largely for firewood upon the rapid-growing *Eucalýptus* and *Casuarína* which have been introduced from Australia. Some of the indigenous timbers, such as Essenboom, or Cape Ash (*Eckebérgia capénsis*), Assegai-wood (*Curtísia fagínea*) and Umzimbit or White Ironwood (*Toddália lanceoláta*) may prove worthy of attention, especially by cart-builders. Like the as yet undetermined Pink Ivory, a singularly beautiful wood, they unfortunately grow mostly in kloofs or other somewhat inaccessible situations.

In 1898 Cape Colony imported over 3½ million cubic feet of rough timber, of which over 2,600,000 cubic feet came from Sweden, and 2 million cubic feet of planed timber, of which over 930,000 cubic feet came from Norway, and 691,000 from Sweden. In the same year Natal imported 1,687,000 cubic feet of rough timber, of which 1,292,000 were from Sweden, and 1,150,000 cubic feet in planks, 918,000 cubic feet of which were from the same country.

Three-quarters of the area of the island of Madagascar is stated to be forest, mainly as yet untouched. Its woods are as yet little known botanically. They include one or more Ebonies, a "Violet-wood" (perhaps an *Acácia*) and a "Rosewood," besides a valuable hard redwood suitable for joinery, known as "Lalona."

South America.—Timber does not form an article of export from the southern or western portion of South America; but Brazil resembles Australia in the extent and variety of its forests. At the Chicago Exhibition of 1893 no less than 440 different Brazilian timbers were exhibited; but unfortunately many of these have not yet been botanically identified. It is stated that some of the species vary much in durability according to the situation in which they are grown; that some of them are too hard and too heavy for many ordinary purposes; and that the absence of railway facilities for transporting the timber to the coast has much reduced the exports. These, however, exceed £100,000 annually, comprising Mahogany, Logwood, Rosewood, and Brazilwood. Rosewood is *Dalbérgia nígra*, shipped from Rio, whilst other species of the genus are known as Violet-wood and King-wood. Brazilwood, hard and heavy, but largely used as a dye, is *Cæsalpínia echináta.*

French Guiana produces many valuable timbers, including Angélique (*Dicorýnia paraénsis*), Cuamara or Tonka-bean (*Coumaroúna odoráta*), Courbaril or Locust (*Hymenǽa Coúrbaril*), Balata (*Mímusops Bálata*), Lancewood (*Duguétia quitarénsis*), and Crabwood (*Cárapa guianénsis*), several of which species grow also in Dutch and British Guiana. In all three colonies the forests cover almost the whole area. British Guiana, where forest conservation has been introduced, produces hundreds of species of timber, suitable for almost every purpose, growing, however, in a mixed virgin forest, though at present the exports amount only to about 170,000 cubic feet, valued at £11,000 a year. The most important species are Greenheart (*Nectándra Rodiǽi*), Mora

(*Dimorphándra Móra*), Crabwood (*Cárapa guianénsis*), Bullet (*Mímusops globósa*), and Locust. Trinidad grows Mora, Crabwood, Bullet, Locust, Lignum-Vitæ (*Guaiacum officinálé*), Galba, (*Calophýllum Cálaba*), the dye-wood Fustic (*Chloróphora tinctória*), and other valuable species; but its export is insignificant. Ecuador, Colombia, and Venezuela have extensive forest resources, but export little or no timber. Honduras, however, exports Mora, Mahogany, Fustic, and Zebra-wood (*Guettárda speciósa*), whilst British Honduras now only exports Cedar (*Cedréla odoráta*), Mahogany, and Logwood. The annual export of Mahogany is 6½ million cubic feet, costing £8 to £10 per thousand feet; and that of Logwood 24,000 tons at £2 to £3 a ton.

West Indies.—Though exporting little timber save Mahogany, and even employing Pine imported from the United States in its sea-ports, Cuba possesses extensive and valuable forests, yielding Cedar, Logwood, Fustic, Lignum-Vitæ, Ocuje (*Calophýllum Cálaba*), Roble Blanco or Jamaica Box (*Tecóma pentaphýlla*) an Ebony (*Diospýros tetraspérma*), Cocus-wood or Granadillo (*Brýa Ébenus*), and the valuable Sabicu (*Lysilóma Sábicu*).

In 1873 Jamaica was estimated to contain 800,000 acres of timber, of which 20,000 were in the hands of Government; but clearing was then proceeding at the rate of 30,000 acres a year. Dye-woods, such as Logwood, Fustic, etc., now form over 8 per cent. of the exports of the island, which exceed 1½ million sterling; but the forests contain many valuable cabinet-woods, the Mahogany being harder and richer in grain than that of Honduras.

While Barbadoes and some others of the Windward Islands are wholly dependent, even for fuel, upon imported timber, Grenada, Tobago, St. Lucia, and Dominica produce Cedar, Galba (*Calophýllum Cálaba*), Angelin (*Ándira inérmis*), Bullet-wood (*Mímusops globósa*), and Bois Riviere or Water wood (*Chimárrhis cymósa*), and have a small export.

North America.—In the United States and Canada during the last twenty years timely, if somewhat alarmist, warnings have been put forward against the reckless waste of the timber resources of the continent. Mr. B. E. Fernow, Chief of the Forestry Division of the United States Department of Agriculture, in 1886 expressed the opinion that the reason why the prophecies of a dearth of timber made for more than a century by alarmists in Europe have not been realized is that their clamour has induced more careful husbanding of forest-resources. He then estimated the forest area of the United States, exclusive of Alaska, as less than 500 million acres, much of this being only brushwood or thinly stocked with trees. The amount of wood then used he quotes as 20,000 million cubic feet, made up as follows:

Lumber-market and manufacture, - -	2,500 million.
Railroad construction, - -	360 ,,
Charcoal, - - - -	250 ,,
Fences, - -	500 ,,
Fuel, - - - - - - - -	17,500 ,,

"There is also to be added," he writes, "an item requiring yearly a considerable amount of wood for a use to which no other civilized nation puts its forests. I refer to the 10,000,000 acres or so of woodland burnt over every year, intentionally or unintentionally, by which a large amount of timber is killed or made useless; and, what is worse . . . the capacity of the soil for tree growth is diminished." Reckoning 50 cubic feet as the yearly accretion per acre, the 20,000 million cubic feet consumption here indicated would require an area of not less than 400 million acres to be kept well stocked.

Some day, no doubt, the development of the coal-fields of the United States will considerably lessen the consumption under the largest of the above-mentioned items, and there is certainly room for economizing in other directions. It is computed, for instance, that, in the Californian Redwood (*Sequóia sempervírens*) forests, to produce a railroad-tie worth 35 cents, timber to the value of 1·87 dollars is wasted. In 1894 there were in the United States 156,497 miles of railroad; there were in 1899 189,294 miles. Reckoning 2640 as the average number of sleepers per mile would make the number used by 1884 413,152,080. The young sound trees employed will not commonly make more than two sleepers each, *i.e.* not more than 100 sleepers could be cut from an acre of such timber-land as prevails in the States, so that the lines existing in 1884 had required all available timber from 4,131,520 acres. The average "life" of a sleeper is seven years, so that 59,021,700 ties, or the product of 590,217 acres, would be requisite to keep the existing lines in repair. The average length of new line built every year was then about 5000 miles, requiring 13,200,000 ties, or the timber of 132,000 acres. If we allow twenty-five years as the time necessary for trees to attain a size suitable for making ties, then it would require the annual growth of 14,755,425 acres to keep good the existing lines, and 3,300,000 to supply the annual demand for new lines, to say nothing of keeping the latter in repair. Not less than 18,000,000 acres of woodland need, therefore, to be kept in reserve for the sole maintenance of the permanent way of the railroads of the United States. The annual fuel consumption is reckoned as the produce of 6¼ million acres annually, and the entire consumption as 25 million acres. Not only have too many Redwood trees been used for fuel, but of late ordinary building has absorbed a great many, panels of Redwood having become very popular in San Francisco as a substitute for plastered walls, whilst there has also been considerable exportation to China, Hawaii, and the Philippines. Some lumbermen predict that within a few years the Redwood tree will be as scarce as the buffalo, and that a shortage has already begun is evident from the fact that the price of Redwood has risen rapidly from 25 to 45 dollars per 1000 square feet. Another serious factor in the question of timber supply in the United States is the extravagant manner in which the turpentine industry is conducted. Instead of any care being taken not to destroy the timber (as is done in the south of France) it has been said that there is no business connected with the products of the soil which yields so little return in proportion to the destruction of the material involved. The turpentine is chiefly obtained in Georgia from the Long-leafed and Loblolly Pines (*Pínus palústris* and *Tǽda*), and the forests of this State were once unsurpassed, and, if properly husbanded, might have continued indefinitely to yield a handsome return. The turpentine farmers, however, aim only at obtaining the maximum amount of crde-resin with the smallest expenditure of labour, caring nothing for the fate of the trees they attack.

If, however, 500 million acres of true timber-forest were maintained in the United States, an annual cut of 20,000 million cubic feet, or 40 cubic feet per acre, would not at first sight appear excessive. It is, however, important to bear in mind that the White Pine (*Pínus Stróbus*) requires 90 years to reach the dimensions attained by the Northern Pine of Europe (*Pínus sylvéstris*) in 70, whilst the Long-leafed Pine (*P. palústris*) requires 200 years for the same growth.

The White Pine has for half a century been the most important timber of the United States, furnishing, as it does, the best quality of soft Pine. Of the home consumption of this wood some idea may be formed from the fact that the city of Chicago alone received in one year over 2000 million feet, principally of this species, or an amount equal to the entire produce of Canada during that year. Speaking of this species, in 1882 Professor C. S. Sargent of Harvard wrote, "It has been wantonly and stupidly cut, as if its resources were endless: what has not been sacrificed to the axe has been allowed to perish by fire. The Pine of New England and New York has already disappeared. Pennsylvania is nearly stripped of her Pine, which only a few years ago appeared inexhaustible." . . . "In Michigan there remained of standing White Pine timber, suitable for market, but 35,000 million feet, board measure," whilst in 1880 there had been cut in the State over 4000 million feet, "requiring only eight years at this rate to exhaust the supply." In Wisconsin there were standing 41,000 million feet, with a cut of over 2000 million for that year, "leaving a supply that would last but fourteen years." In Minnesota there were remaining 8170 million feet, and 541 million were cut in 1880, leaving a supply for fifteen years; so that the supply in the three States would be exhausted in twelve years. There was in fact little more than 80,000 million feet in the United States, whilst consumption was at the rate of 10,000 million per annum and the demand constantly and rapidly increasing. Already by 1885 the United States were importing timber from Canada to the value of nearly two million sterling, or about 75 million cubic feet, more than the entire cut of the province of Ontario. That the extreme forecasts of the alarmists have not been wholly realized throughout the United States may be owing to the fact that it has been cheaper for the more densely populated north-eastern States to supplement their own dwindling resources from Canada rather than from the southern States. Thus Mississippi, with 18,200 million feet board measure of Long-leaf Pine standing in 1880 and with an annual cut of 102 million feet, can supply timber at the same rate for 150 years, a period sufficient, with proper conservation, to enable the supply to renew itself.

Throughout Newfoundland and the Dominion of Canada reckless waste has prevailed in the past. Forest fires and the absence of replanting has reduced the forest area of Newfoundland to about $\frac{1}{55}$ of the whole area of the country, or some 464,000 acres, bearing White Pine, Spruce (*Pícea álba* and *nígra*), Tamarack or Red Larch (*Lárix microcárpa*), said to be better than that of the mainland, Yellow Birch (*Bétula excelsa*), and Poplar. Prince Edward's Island produces the same species, together with Rock Maple (*Ácer barbátum*), Hemlock Spruce (*Tsúga canadénsis*), and the valuable Cedar (*Juníperus virginiána*), which has been largely sacrificed for railway purposes; but fires and clearings have largely diminished the supply, the annual cut being more than 17 times the increment. Nova Scotia had 9 million acres of timber land in 1875, but the annual cut was for years 25 per cent. more than the increment. Hackmatack or Larch (*Lárix americána*), White Pine and Hemlock Spruce, are the chief species. New Brunswick had but 6 million ares of timber land in 1874, mainly covered with hardwoods. Sleepers of Cedar (*Cuprèssus thyóides*) and Hemlock bark-extract for tanning are important articles of export, besides deals, consisting mostly of Black Spruce (*Pícea nígra*). In the province of Quebec the lumber industry is still by far the most important trade; but, whereas in 1874 there were 74 million acres, there are now only 62 million, 32 million of which are under license to cut timber. The species are mostly the same as those of New Brunswick, including White Pine and a scarce but valuable Oak (*Quércus álba*). The wood-pulp industry has grown from an annual value of £160,000 sterling in 1890 to nearly ten times that amount; and a service of rangers has been organized to prevent forest-fires. In Ontario lumbering has ceased to be the sole industry that it once was; but almost the whole amount felled is exported, and the demand of the adjoining States of the Union keeps the annual consumption far in excess of the increment. Though two-thirds of British Columbia, or about 110 million acres, were under timber in 1874, and almost all was under Government control, destructive fires and wholesale clearing have very much lessened the supply. There is, however, a very extensive timber reserve on the coast, consisting of Douglas Fir (*Pseudotsúga Douglásii*), Spruce, Red Cedar (*Juníperus virginiána*), Yellow Cedar (*Cupréssus nootkaténsis*), and Hemlock (*Tsúga Mertensiána*), the available supply of which is from 40,000 to 100,000 million feet. British Columbia has now a wooded area estimated at 285,000 square miles, extending along the coast, river-valleys, and foot-hills as far north as Alaska, and producing many useful species besides the Douglas Spruce. There are, however, sixty saw-mills in operation, with an annual capacity of 550 million feet.

In the early days of its occupation by the French, the forests of Eastern Canada, which then stretched unbroken from the Atlantic to the head of the St. Lawrence basin, a distance of over 2000 miles, engaged the attention of the Government, who drew from them large numbers of masts and spars for their navy and issued stringent regulations for the preservation of the Oak. On the conquest of the country by Great Britain, which then had almost the entire trade with the Baltic, Canadian lumber was neglected; but the continental blockade during the war with Napoleon directed the attention of our timber importers to the resources of Canada, and an import of 2600 loads in 1800 grew to one of 125,300 loads in 1810, and over 300,000 loads in

1820, whilst for the last fifty years it has exceeded a million loads annually. Whilst during the first half of the last century Canada only exported wood to the United Kingdom and the West Indies, for the last twenty years she has experienced a steadily increasing demand from the United States, which now take about half her annual export, or some 13 million dollars worth annually. For many years past the Pine logs floated down to Ottawa have numbered nearly four million a year; and now the demand for paper-pulp has given the Spruce, owing to the far greater area of its distribution, a value in the aggregate much greater than that of the Pines.

In addition to the southern forest belt, now so largely cleared or depleted in the eastern half, there is the great northern forest which stretches from the Straits of Belle-Isle round by the southern end of James Bay to Alaska, a distance of about 4000 miles, with a breadth of some 700 miles. "This vast forest," says Dr. Robert Bell of the Canadian Geological Survey, "has everywhere the same characteristics. The tree, as a rule, are not large, and they consist essentially of the following nine species: Black and White Spruce, Banksian Pine, Larch, Balsam Fir, Aspen, Balsam Poplar, Canoe Birch, Bird-Cherry, White Cedar, White and Red Pine: Black Ash and Rowan occur sparingly in the southern part of this belt."

With nearly 38 per cent. of the whole area of the Dominion under forest, Canadians have in the past given little heed to conservation, believing in the power of natural reproduction to balance the forces of destruction, a belief which, when not substantiated by careful statistical investigation, is a dangerous fool's paradise.

Conclusions.—A most valuable practical test of the increased consumption and the growing scarcity of timber is the advance in prices. It has been estimated that in Germany from about 1550 to 1750 wood quadrupled in price, from 1750 to 1830 the progressive increase of price was at the same rate, but from 1830 to 1880 the rate was much higher, reaching in some cases 300 per cent. within the half century. What was worth 100 francs in 1840 was worth 150 francs in 1850, 260 francs in 1860, 360 francs in 1865 and 400 francs by 1877. In the United States prices rose 100 per cent. between 1874 and 1882; and an equal rise took place in Russia; whilst in Sweden and Norway between 1847 and 1882 (35 years) a rise of from 150 to 200 per cent. according to species occurred.

The obvious conclusions to be drawn from this necessarily incomplete survey of the world's resources and consumption of timber are that, in spite of substitutes, the use of wood increases with advancing population and civilization; that there is still in many lands much waste, much over-felling and but little conservation or forethought; that no country can safely declare its supply inexhaustible; and that, though an absolute dearth of timber may be far distant, some valuable species are in danger of extermination, and we may expect a considerable enhancement of the price of the commoner kinds as the supply has to be drawn from more and more remote sources.

It is undoubtedly, from the magnitude of the interests at stake, a question which demands the attention of the economists, landowners and legislature of every country. If, as Bernard Palissy wrote in the sixteenth century, "after all the trees have been cut down it will be necessary for all the arts to cease"; and if even Colbert could prophesy that "France will perish for want of wood," the danger, in our own time and in many lands besides France, is far more serious.

CHAPTER VII.

TESTING WOOD.

THE very general substitution of iron or steel for wood in permanent structures renders the exact investigation of the strength of timber less important now than formerly. Nevertheless, in merely temporary structures, such as scaffoldings or centerings, its power of withstanding different strains is of very serious concern. Practically, although not scientifically, every joist, rafter, window-sash or door-frame, the chair we sit on, the floor we walk on, the wood of the cart or boat we ride in, are all tested as to their strength, their elasticity, their hardness and their toughness. In the workshop it is recognized that the fitness of a wood for a given purpose invariably depends upon a combination of several qualities. A spoke, for instance, must not only be strong, it must be stiff to keep its shape, tough and hard; and accordingly it must be made from wood split with the grain, and not from sawn or cut material. The experienced wood-worker judges the suitability of any particular piece of wood for his purpose by rule-of-thumb. This rule-of-thumb guess is largely based on the general rule that, in timber, weight, hardness, and power of resistance to most strains, vary together. To this rule, however, there are many important exceptions, where testing would prove what no rule-of-thumb is likely to perceive; and it was in reference to this that Tredgold remarked that actual testing may take the place of a life-time of practical experience in carpentry.

In the scientific testing of timber each property is examined separately. A beam resists bending, and is accordingly termed *stiff*; wicker bends readily, or is *flexible*; while the rod or beam that straightens itself again on the removal of a load that has been applied to it is termed *elastic*. Resistance to a pull in the direction of the grain is known as *tensile strength*; whilst a force applied in an opposite but parallel direction is a *crushing force*. The pressure of a hammer-head across the grain of the handle tends to *shear* the fibres, and a nail entering a board tests its *cleavability* or tendency to split.

The results of the many tests that have been published are often widely discordant. This arises from various causes, *e.g.* incorrect identification of the species, nature of the locality where the tree was grown, the age of the tree when felled, the part of the tree from which the test specimen was taken, the extent to which it was seasoned, the size of the piece tested, and the method of stating the experimental results. The use of popular names, such as Ironbark or Blue Gum, each applicable to half-a dozen different species, is an obvious source of error. As we have already seen, the same species grown under different conditions of heat, moisture, etc., varies widely in rate of growth, and accordingly in strength also. Timber is at its best when the tree is at its maturity, an age which depends upon the species, the climate and the soil. Before that age not only does the less durable sapwood predominate, but the heartwood has not yet reached its full strength; whilst after maturity the heartwood is the first to show symptoms of weakness. As we have already seen, the centre, with its many knots, is generally the weakest part of the heartwood, and a scantling will have greater transverse strength, or resistance to bending stress, and tensile strength in proportion to the number of rings that occur both at its butt and its top. Seasoning, as we have seen, may double the strength of timber. Early experiments on the strength of timber were generally made with very small pieces owing to the difficulty of holding and bringing strains to bear upon large scantlings. Pieces less than a quarter of an inch square were often used. Such pieces might give an unduly unfavourable result from the cutting across of individual fibres; or, on the other hand, being freer from knots or other defects, more readily seasoned throughout and more homogeneous, they are rather picked than average samples, and may give an unduly favourable result.

Density.—We can only give here a bare outline of the principles, methods, and results of testing. Much, as we have seen, depends upon density; and, admitting that, owing to air or moisture in the wood, the results are not as satisfactory as could be wished, we have two simple methods, described by Professor Unwin, for determining this character, viz. (i) by measuring and weighing planed rectangular blocks, and (ii) by weighing the block and the water it displaces. In the former method, if b = the breadth, t = the thickness, and h = the height of the block in inches, its volume will $= \frac{bth}{1728}$ cubic feet; and if W = its weight in pounds, the heaviness of the wood per cubic foot will $= \frac{1728W}{bth}$ lbs.

In the second method, if W = the weight of the block in pounds, W′ = the weight in pounds of the water it displaces, and G = the weight of a cubic foot of water, *i.e.* 62·4 lbs. at the normal temperature, then $\frac{W'}{G}$ = the volume of the block in cubic feet, and the heaviness of the wood per cubic foot will be $\frac{GW}{W'}$.

Mr. Stephen P. Sharples, who made the examinations of North American timbers incorporated by Prof. Sargent in the *Ninth Census of the United States*, vol. ix. (1880), making at least two determinations for each species, calculated the specific gravity by measurement with micrometer calipers and weighing. The specimens tested were 100 millimetres long and about 35 millimetres square, and were dried at 100° C. until they ceased to lose in weight. Of the 429 species experimented upon the specific gravity ranged from 0·2616 in the Small-fruited Fig (*Fícus aúrea*) to 1·3020 in Black Ironwood (*Condália férrea*).

Bauschinger found, by experiments made at Munich in 1883 and 1887, that the density and strength of timber is greatly affected by the amount of moisture it contains. To determine the percentage of moisture Prof. Unwin gives this method. Drill a hole through the test block and weigh the shavings at once. Dry them in an oven at a temperature of from 200° to 212° F. for 8 or 10 hours, and, when they cease to lose weight, re-weigh them. If then $W =$ their first or wet weight, and $D =$ their second or dry weight, $W - D =$ the weight of moisture they contain, and $\frac{100(W - D)}{D} =$ the percentage of moisture.

Bauschinger decided on 15 per cent. of moisture as the standard dryness of air-dried wood.

Ash percentage and fuel value.—From the specific gravity Mr. Sharples deducted the percentage of ash (determined by burning small dried blocks at a low temperature in a muffle furnace), in order to obtain the relative approximate fuel value. This calculation is based on the assumption that the real value of the combustible or volatile substance of all woods is the same. Though resinous woods give off more than 12 per cent. more heat on burning than do non-resinous woods, at least this amount is lost in the case of the former in the form of unconsumed carbon in the smoke. The amount of heat obtained is, in fact, very nearly in direct proportion to the specific gravity, *i.e.* the heavier the wood the greater the amount of heat obtained. Taking as the unit of fuel value an imaginary wood with no ash and a specific gravity of 1, the relative fuel value of 430 woods examined varied from 0·248 in *Yucca* to 1·194 in Black Ironwood (*Condália férrea*). Taking as a unit of heat the amount necessary to raise 1 cubic decimetre or 1 kilogram of water 1° C., 4000 units will be produced by burning a kilogram of dry wood, *i.e.* the relative fuel value of any wood multiplied by 4000 will give approximately the amount of heat obtained by burning a cubic decimetre of it.

Strength.—All measurements of the strengths of timbers are determinations of their powers of resisting certain stresses, or forces tending to produce strains, or changes of shape. It must always be remembered that, unlike metals or many artificial products, wood is not, and cannot be, considered as uniform in structure and composition: it is not homogeneous or isotropic. Stresses applied to it, and the resultant strains must, therefore, be considered separately. Those stresses which are exerted in a direction normal, or at right angles, to a cross-section or imaginary surface of division are termed pushes or pulls, and being continuous, or in parallel though opposite directions, may be considered as identical, or rather as differing only in mathematical sign ($+$ or $-$). Those which are exerted at a tangent to such a cross-section are termed shearing stresses. The intensity of a stress is its amount per unit of surface, and, may, therefore, be expressed in pounds or tons per square inch, or in kilograms per square millimetre, or per square centimetre.[1]

In 1676 Robert Hooke enunciated the law that (using modern terminology) within the limits of elasticity, or recovery from strain when the stress is removed, strain is proportional to stress. In accordance with Hooke's law, Thomas Young postulated the modulus of longitudinal extensibility that bears his name. This is generally called the modulus of elasticity, but incorrectly, since there are others. It is constant for any material, being represented by the letter E, and is, in fact, the ratio of the intensity of push or pull to longitudinal strain. Thus, if l = length and δl = change of length under a stress p, then $\delta l : l :: p : E$, or

$$E = \frac{lp}{\delta l}.$$

Obviously, a stress applied to a transverse section of wood will have to break the fibres across, while one applied to a longitudinal section tends to separate the fibres from each other. Thus the

[1] To facilitate the conversion of results thus variously stated, it may be mentioned that 1 ton, or 2240 lbs. per sq. inch = 1·511 kilos. per sq. mm., or 151·1 kilos. per sq. centim.; whilst 1 kilo. per sq. centim. = 14·22 lbs. per sq. inch.

strength of wood along the grain depends upon the strength of the fibres; that across the grain, upon their cohesion. This latter or lateral strength is, in broad-leaved trees, from $\frac{1}{6}$ to $\frac{1}{4}$ of the longitudinal strength; but in coniferous woods it is only from $\frac{1}{20}$ to $\frac{1}{10}$.

One of the simplest and most instructive tests of timber is that of transverse strength or breaking weight. Laslett, in his Woolwich experiments, took pieces 84 inches long, 2 inches wide, and 2 inches thick, placed upon supports 72 inches apart, and then poured water gradually into a scale suspended from the middle, noting the deflection with 390 lbs. weight and at the breaking point. The transverse strength (p) is calculated from the formula $p = \frac{w'l}{\frac{2}{3}bd^2}$, where w' = the breaking-weight in pounds, l = the length between supports, b = the breadth, and d = the thickness of the sample, or with the dimensions employed,

$$p = \frac{w' \times 72}{\frac{2}{3} \times 2 \times 4} = 13\tfrac{1}{2}w'.$$

Mr. Gamble uses the formula $\frac{w'L}{bd^2}$, where L = the length between supports in feet, b = the breadth of the bar in inches, and d = its thickness in inches. Bauschinger employed for bending tests beams 20 inches square and 9 feet long, with 98·4 inches between their supports; and Professor Lanza of the Massachusetts Institute of Technology employed beams varying from 4 to 20 feet in length, from 2 to 6 inches in width, and from 2 to 12 inches in thickness. Then, W being the load at the centre in tons, l the length in inches of the beam between supports, b its breadth, and h its thickness, also in inches, f, the greatest direct stress on the fibres, or coefficient of bending strength, is obtained in tons per square inch from the formula $f = \frac{3}{2}\frac{Wl}{bh^2}$. If δ = the deflection at the centre in bending in inches, the coefficient of elasticity (E) in tons per square inch is obtained from the formula $E = \frac{1}{4}\frac{Wl^3}{\delta bh^3}$. Sir John Anderson has reduced the results of many experimenters to a simple comparative table of mean breaking weight for beams 1 foot long and 1 inch square in timbers employed in England, which, with some slight modifications, is as follows :—

Ash (*Fráxinus exclésior*), - -	690 lbs.
Beech (*Fágus sylvática*), - -	625 ,,
Elm (*Úlmus campéstris*), - - -	405 ,,
Larch (*Lárix europǽa*), - - - -	440 ,,
Memel Fir (*Pínus sylvéstris*), - - -	561 ,,
Riga Fir ,, ,, - - - -	457 ,,
Scots Fir ,, ,, - - -	381 ,,
Christiania Fir ,, ,, - - -	574 ,,
American Red Pine, - - - -	501 ,,
,, White Spruce, - - - - -	570 ,,
Oak, English (*Quércus Róbur*), - -	591 ,,
,, Dantzic, - - - - - - -	513 ,,
,, Adriatic, - - - - - -	460 ,,
,, Canadian, - - - - - - -	580 ,,
,, or Teak, African (*Oldfiéldia africána*), - -	855 ,,
Mahogany (*Swieténia Mahágoni*), - - - - -	531 ,,
Teak (*Tectóna grándis*), - - - - -	814 ,,

The ultimate strength of a material is that stress which is required to produce rupture, and this may be either tensile stress or that exerted longitudinally or parallel to the axis of a beam, crushing stress, or resistance to compression in the direction of the fibres, or shearing stress, *i.e.* tangential.

Professor Unwin figures details of various instruments employed for testing timbers, more especially for tensile strength, including Bauschinger's roller and mirror extensometer, and several shackles for holding the test-specimens. The principle of most modern instruments for these purposes is the same, the weight being applied gradually, either by small weights or by hydraulic action, to a system of levers, the force exerted being shown by a delicately adjusted steelyard. Thus the comparatively simple instrument of American design, introduced at Woolwich in 1854 by Sir John Anderson, and figured in his work,[1] consists of a combination of two levers which together give a purchase of

[1] *The Strength of Materials and Structures*, London, 1872, p. 16.

200 to 1, that is to say, 1 lb applied to the end of the long arm of the upper lever will exert a stress of 200 lbs. on the specimen attached by shackles to the lower one.

The dimensions of the specimens tested by different experimenters, whether for breaking weights, tensile strength, or other measurements, have unfortunately varied greatly. In contradistinction to the long beams just mentioned as used by Bauschinger and Lanza, Captain Fowke, in testing the New South Wales timbers at the Paris Exhibition of 1855 for breaking weight, etc., used samples 2 inches square and 12 inches between supports. Mr. Laslett used samples of the same sectional area but 72 inches between supports; whilst Mr. F. A. Campbell, experimenting on Australian timbers in 1879, employed a sectional area of only $\frac{1}{16}$ of an inch.

The term *strength*, when used absolutely, generally means the breaking weight under a bending test, and in English books is expressed in pounds. It is found by the formula $\frac{b \times d^2 \times E}{l}$, where b = breadth in inches, d = depth in inches, l = length in feet, and E = the constant or modulus. This constant, in England, means the number of pounds' weight applied in the middle of a bar one inch square and twelve inches between supports required to break the bar.

When a beam is supported at each end in such experiments as these, the distance to which the middle of the beam is forced down below its original position by the load, is termed its deflection. In solid rectangular beams the deflection varies directly as the load and the cube of the length, and inversely as the breadth and the cube of the depth. The resistance to deflection is known as *stiffness* or *rigidity*. If then we require two beams of the same breadth, but of different lengths, to be equal in stiffness, then their respective depths must be in proportion to their lengths. Thus, if the beams are 24 and 12 feet long respectively, and the latter is 12 inches deep, the former will have, in order to be equally stiff or rigid, to be 24 inches deep. Strength, on the other hand, in solid rectangular beams, varies inversely as the length, directly as the breadth, and directly as the square of the depth, so that, in the example given above, the longer beam will only require to be 17 inches deep in order to be as strong as the shorter. If the beams are equal in breadth, but of different length, and are required to be equal in stiffness, their breadths must be as the cubes of the lengths. In two beams 24 and 12 feet long, for example, the breadths must be in the ratio of 24^3 to 12^3, *i.e.* 13,824 to 1728, or as 8 is to 1. In other words, the long beam would have to be eight times as broad as the shorter one to be equally rigid, whereas it only requires to be twice as broad to be equally strong. So, too, in cylinders, the strength varies as the cube, the stiffness as the fourth power of the diameter.

The constants or values of deflection were deduced by Barlow from the formula $D = \frac{l^3 \times W}{b \times d^3 \times \delta}$, where l = length in feet, W = the greatest weight in pounds which the beam can bear without losing its elasticity or acquiring a permanent set, b = breadth in inches, d = depth in inches, and δ = deflection in inches. From this it obviously follows that

$$\delta = \frac{l^3 \times W}{b \times d^3 \times D}.$$

It is found in practical engineering that the deflection of timber beams (δ) should not exceed $\frac{1}{480}$th of their length.

Bauschinger employed, for testing tensile strength, rods 18 inches long and 1 or $2\frac{7}{8}$ inches square for $5\frac{1}{4}$ inches at each end, reducing to $\frac{1}{4}$ or $1\frac{5}{8}$ inch in the middle. He does not, however, consider these, or his experiments on bending (in which the individual variation of the large beams employed, as to knots, etc., produces wide differences in the results) so instructive as to the relative values of timbers as are crushing experiments. For such experiments he used blocks 6 inches high and $3\frac{1}{2}$ inches square, protected at the ends with metal plates.

Results will be affected by so many circumstances that it is most important that the history of logs experimented with should be known. The nature of the locality in which the timber is grown, the age of the tree, the part of the tree from which the timber is taken, and the extent to which it has been seasoned, will all modify the results. Thus Bauschinger showed that strength varies according to the proportion of summer to spring wood, and that the centre of a tree is therefore weaker; whilst the following table of the range of variation in 26 trees of *Pínus palústris*, quoted by Professor Unwin from a Report of the U.S. Department of Agriculture, shows how butt, middle, and top logs differ in strength, largely no doubt for the reason, which we explained in a previous chapter, that the annual increments of wood forming cones do not extend uniformly from end to end of a log. [See pp. 79-80 and Fig. 41.]

	Heaviness in lbs. per cubic foot.	Coefficient of Elasticity from Bending Test.	Tensile Strength.	Crushing Strength.	Coefficient of Bending Strength.	Shearing Strength.
		All in tons per square inch.				
Butt,	28–64·8	500–1380	3·84–14·4	2·13–4·40	2·12–7·25	·21–·58
Middle,	36–53·5	510–1369	2·82–13·4	2·25–4·15	3·40–7·65	·24–·55
Top,	32–56·5	375–1200	1·85–10·8	2·04–4·06	1·90–7·00	·22–·52

As to seasoning, since timber loses from $\frac{1}{7}$ to $\frac{1}{5}$ or, when perfectly dried, $\frac{1}{3}$ of its weight in the process, and strength and the co-efficient of elasticity vary directly with density, its effect is obvious.

Unfortunately the systems employed for stating the results of experiments vary almost as much as the dimensions of the specimens tested, so that it is a matter of considerable calculation to compare the records of different experimenters. Mr. Sharples, for instance, defines the co-efficient of elasticity, or rather of longitudinal extensibility, as the weight in kilograms sufficient to elongate a stick one centimetre square to double its original length, were that possible, and states results ranging from 25,699 in *Fícus aúrea* to 165,810 in the Western Tamarack (*Lárix occidentális*). To translate his results into the tons per square inch usual in England it is necessary to divide them by 151·1. (See footnote on p. 133.)

So too while Professor Thurston defines the *modulus of rupture* as "the quantity which represents the stress upon a unit of area of cross-section . . . at the instant of breaking under the transverse stress," and Mr. Sharples expresses this *breaking-weight*, as it is generally termed, in kilograms per square centimetre, English writers here also use tons or pounds per square inch. So too Mr. Sharples gives the *resistance to longitudinal presssure*, or ultimate weight which a stick will support, in the number of kilograms required to crush a stick one centimetre square by such pressure, while Mr. Laslett terms this, *vertical strength*, and states it in the number of pounds of vertical force required to crush 1 square inch of base. Mr. Sharples also gives the *resistance to indentation* or number of kilograms required to sink a punch one centimetre square to the depth of 1·27 millimetres perpendicularly to the fibres.

It is well nigh impossible to reduce all the results of different experimenters. They will, therefore, be here stated mainly in the form and with the terminology of their respective authors.

The following symbols will be employed:

S.G. = Specific gravity, compared to water as 1000.

W = Weight of a cubic foot in pounds.

E = Co-efficient of elasticity, stated in tons or pounds per square inch, or in kilograms per square centimetre.

e' = Elasticity compared to Oak as 1·00.

p = Transverse strength in pounds per square inch.

p' = Transverse strength compared to Oak as 1·00.

f = Co-efficient of bending strength in tons per square inch.

ft = Tensile strength or tenacity along the fibre, in tons per square inch.

c = Direct cohesion, in pounds per square inch.

c' = Cohesion compared with Oak as 1·00.

fc = Crushing strength along the fibre, in tons per square inch.

v' = Crushing strain as compared to Oak as 1·00.

fs = Shearing resistance along the fibre, in tons per square inch.

R = Modulus of rupture for transverse strain, stated either in kilograms per square centimetre, or in pounds per square inch.

PART II.

WOODS OF COMMERCE, THEIR SOURCES, CHARACTERS, AND USES.

Acacia, in England, *Robínia Pseudacácia* L. See **Locust.**

Acacia (*Eucrýphia Moórei* F. *v.* M.: Order *Rosáceæ*). New South Wales and Victoria. Known also as "Plum, Acacia Plum," or "White Sally." Warm, light brown, moderately hard, of considerable dimensions, easily worked. Used for the bodies of buggies.

Acajou, a general name in the French timber-trade for *Swieténia Mahágoni.* See **Mahogany.** In French Guiana it is applied also to *Cedréla guianénsis* A. Juss.: (Order *Meliáceæ*). S.G. 577. Reaching large dimensions, soft, not very flexible, very homogeneous and free from flaws, working well, without splitting, durable, owing to a bitter principle obnoxious to insects, and termite-proof. Fairly common and in much request as a furniture-wood. Used in Europe for cigar-boxes.

Acle (*Xýlia dolabrifórmis* Benth.: Order *Leguminósæ*). India, the Malay Peninsula, and the Philippines. "Ironwood" of Pegu and Arracan. *Hindi* "Jambu," *Burm.* "Pyengadu," *Philipp.* "Acle." Formerly named *Mimósa Aclé* and *Ínga xylocárpa.* Height 70—80 ft., yielding timber 1—2½ ft. square; S.G. 934—1225, W 63, *e′* 2·19, *p* 17,200, *p′* 1·58, *c* 8960—10,360, *c′* 1·275, *v′* 1·527, *fc* 5·2. Heartwood dark brown or reddish-brown, often beautifully mottled with a waved and twisting grain, heavier than water, hard, tough, strong, rigid, its pores filled with a thick, oily resin, which renders it clammy until completely seasoned, difficult to cut, causing sneezing in working, taking a good polish, shrinking $\frac{1}{8}$ in. per foot in seasoning, "more indestructible than iron," being both termite and teredo-proof, but having sometimes extensive heart-shakes which unfit it for constructive purposes. The Burmese wood contains more resin than that from the Deccan. It is used for piles and beams of bridges; in Bengal and Burma for telegraph-posts; in Southern India for posts, railway-sleepers (for which purpose it ranks next to Teak), carts, etc.; in Burma for agricultural implements; and for house and boat-building in the Philippines, and is probably the best hardwood in India for paving.

African (*Oldfiéldia africána* Benth. and Hook.: Order *Euphorbiáceæ*). Western Tropical Africa. Known also as "African Oak, African Teak," though not connected with Oak or Teak. Height upwards of 30 or 40 feet, girth 7—8 feet, S.G. 934—1086, *c* 7052, *fc* 4·9, *c'* ·931, *t'* 1·341, *p* 15,000, R 855 lbs. Dark red, very hard, strong, rigid, fine close and straight in grain, free from shakes, shrinking and warping little, very durable; but difficult to work, and shipped from Sierra Leone in logs so badly hewn as to yield little more than 50 per cent. of well-squared timber. Used in ship-building for keelsons, beams, etc., and classed in the second line in Lloyd's Register.

Ah-pill (*Erythrophlœum Labouchérii* F. v. M.: Order *Leguminósæ*). Northern Queensland and North Australia. Probably the "Leguminous Ironbark" of Leichhardt, and also named *Labouchéria chlorostáchys* F. v. M. A medium-sized tree. Wood red, close-grained, very durable, and the hardest in Australia. Used by natives for spear-heads.

Ailantus (*Ailánthus glandulósa* Desf.: Order *Simarúbeæ*). *Molucca* "Ailanto," *French* "Ailante," *Germ.* "Götterbaum," *Ital.* "Albero di paradiso," *Russ.* "Pajasan," *Span.* "Barniz falso de Japan." Height 50—60 feet; diam. 1—2 ft.

Sapwood broad, yellowish. Heartwood not dissimilar, greyish-orange Rings wide and distinct. Springwood very broad with numerous large vessels towards its inner margin and few small ones, scattered, or in segments of circles, four or more together, towards its outer part. Medullary rays distinctly visible to the naked eye, with a satiny lustre. Pith-mass very large. The wood contains vessels, tracheids, wood-fibres, fibre-cells, and parenchyma. It is moderately heavy, tolerably hard, somewhat difficult to split, and of a beautiful satin-like lustre. It is durable, and is appreciated by cabinet-makers; but the tree is mainly grown for shade. A native of Japan and Northern China, it is grown for ornament in England and the Eastern United States. It is used for charcoal in Europe.

Akagashi (*Quércus acúta* Thunb.: Order *Cupulíferæ*). South Japan. The dark red-brown, very hard and heavy wood of an Evergreen Oak, which with that of some allied species, such as the grey-white **Shiragashi** (*Q. vibrayeana* Tr. and Tav.) is used in boat and waggon building.

Akashide (*Carpínus laxiflóra* Bl.; Order *Cupulíferæ*). Japan. Height 40 ft.; diam. 1 ft. It is used for furniture, waggon-building, agricultural implements and firewood.

Alder, Common or Black (*Álnus glutinósa* Medic.: Order *Betuláceæ*). *French* "aune," *Ital.* "alno" or "ontano," *Span.* "alano" or "aliso," *Germ.* "schwarz Erle" or "Else," *Russ.* "olse." S.G. fresh 901, dry 551. W 50—62 when green, 50—34 when dry. Strength, compared to Oak, 80; stiffness, 63. Height 20—40, very rarely 70—80 feet; diam. 1—2 feet. No heart-wood. Wood white when alive, red when cut, becoming pinkish-brown. Rings rather broad, not very distinct, waving inwards where they cross the few, lighter-coloured, medullary rays. Brown pith-flecks are frequent. Pith-mass triangular with rounded angles, from which the medullary rays radiate in curves. The wood contains vessels, tracheids, wood-fibres, fibre-cells, and parenchyma; but the vessels are small, few, and uniformly distributed. It is soft, easily split, rather light, with a smooth, fine grain, and lustrous. It does not warp or splinter. When kept wholly submerged it is very durable, but not at all so otherwise. To preserve the finer pieces from insect attack they are sometimes, in Scotland, immersed for some months in peat-water, to which some lime is added, which gives the wood some resemblance to Mahogany. It has then been used for tables. Alder was formerly used for piles, water-pipes, sluices, etc, but Elm, being far more durable when alternately wet and dry, is much better for such purposes. The piles of Ravenna, according to Vitruvius, and those of the Rialto at Venice, and those of

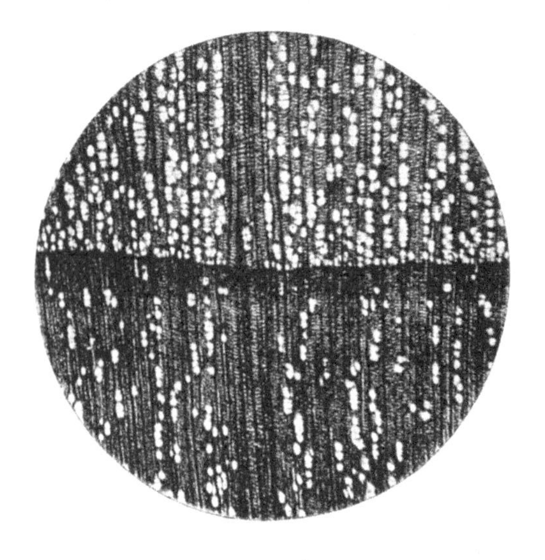

FIG. 45.—Transverse section of Alder (Álnus glutinósa) highly magnified.

Amsterdam, according to Evelyn, were largely of Alder, and Pliny speaks of it as "eternal" when so used. Alder is employed for packing-cases, the staves of herring-barrels, shovels, clogs and sabots, bobbins, barrows, kneading-troughs, etc. The roots and knots, being often handsomely veined, are used in small articles of turnery and cabinet-making. Alder is practically the best wood for gunpowder-charcoal. It is imported from the Baltic ports of North Germany, where there are extensive pure forests of this species, sometimes mixed with Birch; and it is mainly bought by the Lancashire clog-makers. (Fig. 45.)

Alder, American or **Hoary** (*Álnus incana* Willd : Order *Betulácecæ*). *Germ.* "Weisserle." A similar but inferior wood, with more lustre, fewer pith-flecks, very few, wide, but indistinct medullary rays, has a wide range in North America.

Alder, Red, *Álnus rúbra* Bong., a native of the Pacific slope, is sometimes employed for charcoal.

Alder, White, the name in the western United States for *Álnus rhombifólia* Nutt.; but applied in Cape Colony to *Platylóphus trifoliátus* Don, a yellowish-white, hard, tough, durable wood, from a tree 20—40 ft. high and 2—4 ft. in diam., much used in the Colony for furniture and boat-keels. It takes a good polish and in the lower part of the tree has generally a fine twisted grain. W 38. Boer name, "White Els."

Alerce (*Tetraclínis articuláta*). See **'Arar.**

Almond, Indian (*Terminália Catáppa* L.: Order *Combretáceæ*). India, Queensland, Fiji, etc. "Tavola" of Fiji. A large deciduous tree. Wood reddish-brown, waved, rather close-grained, taking good polish, durable. Sapwood light-coloured. W 32—41.

Almond, Wild (*Brabéjum stellatifólium* L.: Order *Proteáceæ*). South Africa. Known also as "Red Stinkwood," "Caffre" or "Wild Chestnut." Usually small but reaching 60 ft. in height and 1½—6 ft. in diam. Durable. Used for waggons and furniture.

Amboyna-wood (*Pterocárpus índicus* Willd: Order *Leguminósæ*). Moluccas, especially Ceram and Amboyna, Arru and New Guinea. Malay name, "Kaya Boka," known also as "Lingoa wood." The tree throws out burrs which are sawn off in slabs 2—8 in. thick and up to 9 ft. in diam. Light reddish-brown to orange, fragrant, somewhat resembling Yew, very hard, beautifully mottled and curled, taking a good polish and very durable; but of uncertain botanical origin. Sold in Singapore by weight. Used in inlaying and for snuff-boxes, etc. S.G. 634, W 39.

Amaranthe. See **Purple-heart.**

Amarello Pao. See **Fustic.**

Ameixera. See **Sanders, Yellow.**

Amla ka (*Phyllánthus Emblica* Gaertn: Order *Euphorbiáceæ*). India, China, and Japan. Red, hard, elastic, durable, especially under water; but seldom straight. Its fruit is known as Emblic Myrobalans.

Anan (*Fagræ̃a frágrans* Roxb.: Order *Loganiáceæ*). Burma and Indian Archipelago. Red-brown, hard, close-grained, beautifully mottled, very durable, teredo-proof. Used for bridges and boat-building. W 52·5, S.G. 840.

Andiroba. See **Crabwood.**

Angelim vermelho (*Ándira fraxinifólia* Benth.?: Order *Leguminósæ*). Brazil. Of crooked growth, reddish-brown, moderately heavy. Used for ship-timbers.

Angelin (*Ándira inérmis* Kunth: Order *Leguminósæ*). West Indies. "Cabbage" or "Bastard Cabbage Tree." *Cuba* "Yaba." Height 20—30 ft. S.G. 688. W 56·8-60·4. E 563 639. R 300 lbs. *f* 2·01—5·44. *fc* 1·98—2·99. *fs* ·32—·45. Brown, veined, hard, very durable underground or under water. Used for mill-rollers and in house and ship-building. Some "Partridge-wood" may be the timber of this species.

Angelique (*Dicorýnia paraénsis* Benth.: Order *Leguminósæ*). Brazil and Guiana. Height 20—54 ft., yielding timber 12—22 in. square. S.G. 746—916. R 215 kilos. Reddish-brown, moderately hard, tough, strong, elastic, straight and even-grained, easily worked, durable in sea-water, insect-proof, sometimes with an ornamental waviness of grain, with but little sapwood; but with slight shakes and an unpleasant odour, and said to rust nails. Used in French dockyards, as a substitute for Teak, for backing armour plates, etc., far more durable than Oak.

Angelly (*Artocárpus hirsúta* Lam.: Order *Artocarpáceæ*). India, chiefly in the south-west, Ceylon, Burma. Known also as "Jungle Jack" and "Aini." Height 50—60 ft.; diam. 2½—3 ft. S.G. 590. W 36—51. Very tough, bears exposure to water well, and, if kept oiled, is very durable. Used for planks in house-building, canoes, fishing-boats, etc. Classed in the third line in Lloyd's Register.

Anjan (*Hardwíckia bináta* Roxb. : Order *Leguminósæ*). Central and Southern India. *Telugu* "Epe," *Tamil* "Acha maram." Height 50—60 or 120 ft., straight. Dark red, often purplish, close-grained, "perhaps the hardest and heaviest wood in India," very durable, not warping, but liable to split, very strong. W 84—85. Used for bridges, sleepers, beams, rafters, house posts, and ornamental work.

Apple (*Pýrus Málus* L.: Order *Rosáceæ*). *German* "Apfelbaum," *French* "pommier," *Span.* "manzana," *Russ.* "jablon."

Height seldom over 30 ft., or diam. over 2½ ft. Dark brown, generally strongly tinted with red. Sapwood, dull white. Rings distinct, broad, often over ½ in. across; with no pith-flecks; medullary rays not visible to the naked eye; vessels small, uniformly distributed, 1—4 together. Heavy, very hard, close-grained, brittle, taking a high polish, but warping badly on drying. Used for mallets, tool-handles, and other turnery, and traditionally preferred in Cornwall for "poling" tin-ore.

Apple, Black or **Brush** (*Sideróxylon austrálé*, Benth. and Hook. fil.: Order *Sapotáceæ*). New South Wales and Queensland. Also known as "Native" or "Wild Plum." Height 80—100 ft., diam. 1—3 ft., W 55—58. Pale yellow, close-grained, prettily veined, but requires careful seasoning. Used for staves, laths, and building, but suitable for cabinet work.

Apple, Emu. See **Plum, Sour.**

Apple, Oregon Crab (*Pýrus rivuláris* Dougl. : Order *Rosáceæ*). Alaska—California. Height 15—25—40 ft., diam. 1—1½ ft. S.G. 832, W 51·8. Light reddish-brown, heavy, very hard, close-grained, taking a fine polish. Used in mill work, and for handles.

Apple-tree (*Angóphora lanceoláta* Cav. : Order *Myrtáceæ*). New South Wales and Queensland. Also known as "Mountain Apple Tree, Orange, Red" or "Rusty Gum." *Aborig.* "Toolookar." Height 70—80 ft., diam. 2—3 ft., S.G. 893. Hard, used for rough work and fuel.

Apple-tree, Broad-leaved (*Angóphora subvelutína* F. v. M.). New South Wales and Queensland. *Aborig.* "Illarega."

Reddish, tough, polishes well, durable, but requires careful seasoning. W 52—53. Used by wheelwrights and for fencing.

Apple-tree, Narrow-leaved (*Angóphora intermédia* DC.). Eastern Australia. Formerly *Metrosidéros floribúnda* Sm. Height 40—100 ft., diam. 1—3 ft. Hard, tough, bears damp well, but is subject to gum-veins. Much used by wheelwrights.

'Arar (*Tetraclínis articuláta* Masters: Order *Cupressíneæ*). Morocco and Algeria. The "alerce" of the roof of Cordova Cathedral, and probably the "citron-wood" of the Romans and the "thyine wood" of the Apocalypse (xviii, 12). Also named *Thúja articuláta* Vahl. and *Cállitris quadriválvis* Vent. Height 30 ft., fragrant.

Arbor-vitæ. See **Cedar, White.**

Argan (*Argánia Sideróxylon* Rom. and Schult.: Order *Sapotáceæ*). Morocco. "Morocco Ironwood." Height seldom more than 20—30 ft., diam. 8 ft. or more. Heavier than water, very hard.

Arjun (*Terminália Arjúna* Bedd.: Order *Combretáceæ*). India, Burmah, Ceylon. Apparently known also as "Kahua" or "Kowah." Height 50—80 ft.; diam. 2—8 ft. Dark brown or brown-red, very heavy, strong, but sometimes rotten at the heart, splits freely when exposed to the sun, durable, but liable to termite attack. Recommended for beams, rafters, and masts.

Asada. See **Hornbeam, Hop.**

Ash (*Fráxinus excelsior* L.: Order *Oleáceæ*). *Germ.* "Esche." *Dutch* "aesche," *Dan.* and *Swed.* "ask," *French* "frêne," *Ital.* "frassino," *Span.* "fresno," *Russ.* "jasan." S.G. fresh 852, dry 750—692, W 43—53, E 750 tons, e' 1·28, p 11,600, p' 1·05, ft 4—7, c 3780, c' ·499, fc 2—4, v' ·912, fs ·5, R 15,120 lbs. Stress required to indent $\frac{1}{20}$ in. transversely to the fibre, 2300 lbs. per square inch. (Fig. 31.)

Height 30—50, or even 80 ft.; diam. $1\frac{1}{2}$—2 ft.

The sapwood, very broad, about 40 rings, yellowish or greyish-white; the heart, light brown, or greyish-white, not very different from the sapwood, or, in colour, from light Oak. Rings very distinct, with a broad zone of spring-wood with numerous larger vessels, sharply marked off from the autumn wood, in which the few small vessels stand singly or from 2—4 together. Pith-rays, scarcely recognizable. Pith-mass, ovoid, very large. The wood contains vessels, wood-fibres, fibre-cells, and parenchyma. It is of moderate weight and hardness, very even and close in grain, lustrous and susceptible, of a good polish, the toughest of European woods, and very pliable. It warps but little; and, if felled in winter and properly seasoned, is extremely durable, though few woods are more perishable if these precautions are neglected.

"Very great advantage will be found in reducing the Ash logs soon after they are felled into plank or board for seasoning, since, if left for only a short time in the round state, deep shakes open from the surface, which involve a very heavy loss when brought on later for conversion" (*Laslett*).

The compression or contortion of its fibre produces a lateral grain or figure in Ash known as "ram's-horn," or, from its resemblance to the figured Maple used for the backs of violins, as "fiddle-back." This is best shown in billets imported from Austria and Hungary. Though the Ash grows in almost any soil, it produces the best light coloured wood when grown quickly in rich loam and a moist climate, as in the valleys of Britain and Central Europe. The slower-grown wood of poorer soils, mountains, and northern Europe is apt to become "black-hearted," as also does that of pollard trees. This is sometimes attributed to incipient decay, and is held to lessen the strength of the wood, but produces the figured veneers imported from the Pyrenees, as "Pyrenean Ash." Wounds or cankers also occur in the heart-wood, which are believed, in the north-east of England, to be caused by bees; but this "Bee-sucken Ash," as it is termed, is extremely hard and tough, so as to be suitable for mallets, etc.

The utility of the Ash has long been recognized, and few woods have a greater variety of uses, so that the poet Spenser terms it "The ash, for nothing ill." Greeks, Romans, and Teutons alike, used its tough saplings for lances, the Romans preferring the wood obtained from Gaul, and the Teutons also employing it for bows, arrows, shields, and boat-building.

Roman agricultural writers recommended it for implements, and from its varied uses in this respect it has been called "the husbandman's tree." It is frequently coppiced, this young, or "Maiden Ash," and the "stooled" shoots, or second growth from the original roots, which are very tough, being fit for walking-sticks or whip-handles when four or five years' growth, for lance-poles or hop-poles a year or two later, for spade-handles at nine years, and when 3 inches in diameter as valuable as the timber of the largest tree. These growths are frequently termed "Ground Ash." In the Potteries it is largely used for crate-making, for which purpose it is cut every five or six years, though for other purposes only every seven or eight. Since, when steamed or heated, it can be easily bent, without injury, into any curve, it is invaluable for hoops. Larger wood is largely used by the wheel-wright, for both spokes and felloes, and by the carriage-builder, and for oars. As it does not splinter it is also useful for chopping blocks and shop-boards. For furniture it is chiefly used where softer, as in Central Russia. Its flexibility unfits it for use in architectural work. When seasoned, the sapwood is as valuable as the heart. The roots and knotty parts of the stem were formerly valued by cabinetmakers, being known, according to Evelyn, as "Green Ebony." There being no bitter principle in the heartwood, Ash is very liable to the attacks of the larvæ of the furniture-beetle, though painting renders it more durable.

Ash is valuable as fuel, and its residue is rich in potash.

The tree is a native of Europe and Northern Africa.

Ash (*Elæocárpus obovátus* Don: Order *Tiliáceæ*). North-eastern Australia. "Chereen" of natives in New South Wales, "Woolal" in Queensland. Height 80—90 feet, diam. 2—2½ ft. White, hard, tough, easily worked. Used for oars.

Ash (*Flindérsia*). See **Flindosa**.

Ash, American, Quebec, or **White** (*Fráxinus americána* L.: Order *Oleáceæ*.) S.G. 654, W 30—40, Co-efficient of elasticity 101,668, R 861 kilos, Resistance to longitudinal pressure 463, Resistance to indentation 171.

Height 70—100 ft, diam. 1—3 ft. Imported in partly

squared logs, 18—35 ft. long, and 10—18 in. square, in planks and partly manufactured, as oars, etc.

Wood generally much whiter, and with narrower rings than Common Ash (*F. excelsior*); the sapwood, when well seasoned, nearly white; the heart, light reddish, contrasting with the sapwood more than in Common Ash, but less than in other American kinds, in the best quality, lightest and most uniform; in second quality, slightly stained alternately red and yellow; and in the third quality, mottled red. It is of much slower growth than Common Ash, the rings being only about half the width of those in that species, very distinct, with a narrower zone of spring-wood and fewer larger vessels. Though it may be termed rather heavy, it is less so than the European species, moderately hard, but very tough and elastic, except in the oldest timber, clean and straight in grain, very easy to work, and standing well after seasoning.

In America it is used for all purposes to which Common Ash is applied in England. The small wood of young trees or stools, which is mostly sapwood and white, is the best material for oars. Larger logs, when white, are much sought after for bedroom and other furniture, and for coach-panels; but the more coloured logs are universally considered by the trade inferior in strength and durability, though, being more easily worked, they are used by cabinet-makers for drawers and carcass work, for which European Ash is never employed.

It occurs generally throughout Canada and the Eastern United States, chiefly on river-banks.

Ash, Black, Hoop, or **Ground** (*Fráxinus nígra* Marshall = *F. sambucifólia* Lam.: Order *Oleáceæ*). *Germ.* "schwarze Esche," *French* "frêne noire," *Span.* "fresno negro." S.G. 632,. W 39·37—30, Co-efficient of elasticity 87,185, R 806 kilos, Resistance to longitudinal pressure 423, Resistance to indentation 194.

Height 80 ft. or more, diam. 2—2½ ft.

Slow-grown, trees 22 in. in diameter having 234 annual rings. Sapwood thin, light brown, or nearly white, sharply contrasting with the dark brown heart. Numerous thin medullary rays. Spring-wood with crowded ducts forming a narrow sharply-defined zone. Moderately heavy, rather soft, not strong, but tough, elastic, coarse grained, separating easily between the rings, not durable, except under water.

Less valuable than White Ash, but much used in America for furniture and interior finishing, and for fencing and hoops. The Indians use it for making chair-bottoms and "splint" baskets, working it "into sticks as wide along the rings as the splints are to be, and perhaps two inches thick. These are then bent sharply in the plane of the radius of the rings when they separate into thin strips nearly or quite as many as the rings of growth" (Romeyn Hough). Large wart-like swellings, or "burls," on the trunk, with much contorted grain crossed by innumerable radiating "pins," or abortive branches, form, when cut tangentially, very valuable veneers.

Swampy situations from Newfoundland and Winnipeg southward, the most northern American Ash.

Ash, Black (*Litsœa dealbáta* Nees : Order *Lauríneæ*), New South Wales and Queensland. Height, 100—150 ft., diam. 2—3 ft. Yellowish, streaked with brown longitudinally, fragrant, close-grained, tough. Used for indoor work. The name is also applied to the smaller white wood of *Cupánia semigláuca* F. v. Muell., the "Tyal-dyal" of New South Wales, which is not used.

Ash, Blue (*Fráxinus quadranguláta* Michx : Order *Oleáceæ*).

Dry woods in the central United States, Michigan to Tennessee, best in the basin of the lower Wabash river. Height, 70 feet, diam., 2 ft. Heavy, hard, and more durable, especially when alternately wet and dry, than any other Ash. Valuable for tool-handles, used also in carriage-making, flooring, and other purposes, as is White Ash.

Ash, Blue (*Elœodéndron austrálé* Vent. : Order *Celastráceæ*), north-east Australia. Known also as "White Cedar" or "Couraivo." Height, 20—30 ft.; diam. 4—12 in. W 49·5. Pinkish, close-grained, prettily marked, but liable to shakes. Valuable for oars, staves, or shingles.

Ash, Blueberry (*Elæocárpus cyáneus* Ait. and *E. holopétalus* F. v. Muell.: Order *Tiliáceæ*), the former in Tasmania and throughout Eastern Australia, the latter in Victoria and New South Wales. The former is known also as "White Boree" or "Native Olive," the latter as "Maddagowrie" or "Prickly Fig." *E. cyáneus* grows 40—50 ft. high, with a diameter of 12—15 in., has a dark heart and white sapwood and is very tough, and is useful for tool-handles. *E. holopétalus* reaches 60—80 ft. with a diam. 1—2 ft., and W 37·5, is white, close-grained and suitable for cabinet work.

Ash, Brush (*Acronýchia Baúeri* Schott.: Order *Rutáceæ*). New South Wales and Queensland. Height 40—70 ft., diam. 1—2 ft. Very hard, close-grained, and strong. Excellent for tool-handles and might be used for cabinet work.

Ash, Cape (*Ekebérgia capénsis* Sparrm.: Order *Meliaceæ*). Natal and Cape Colony. *Boer* "Essen-boom," *Zulu* "Umgwenyuizinja." Height 20—30 ft.; diam. 2—3 ft., close-grained, tough. Used for furniture, the sides of waggons, etc.

Ash, Crow's. See **Flindosa.**

Ash, Elderberry (*Pánax sambucifólius* Sieber: Order *Araliáceæ*). Northern and Eastern Australia. A small tree. Wood very tough, prettily marked, and used locally for axe-handles.

Ash, Green (*Fráxinus lanceoláta* Borkh. = *F. víridis* Michx. = *F. Pennsylvánica*, var. *lanceoláta* Sargent: Order *Oleáceæ*). Vermont and Saskatchewan to Texas, Arizona, and Florida, most common and best in the Mississippi valley. Height 50—60 ft.; diam. 1½—2 ft. S.G. 718. W 44·35. Brown, with obscure pith-rays and several rows of open ducts in each ring. Heavy, hard, and strong. Somewhat inferior to White Ash; but often employed for the same purposes.

Ash, Grey. See **Ash, Red.**

Ash, Moreton Bay (*Eucalyptus tesseláris* F. v. Muell.: Order *Myrtáceæ*). Central and North-eastern Australia. "Ilumba," "Corang," and "Carbeen" of natives. Height 30—60 ft.; diam. 1—2 ft. Dark brown, tough, not hard nor durable. Used for building purposes, being of better quality in the warmer parts of its range.

Ash, Mountain (*Pyrus Aucupária* L. Order *Rosáceæ*). See **Rowan.**

Ash, Mountain (*Alphitónia excelsa* Reisseek: Order *Rhamnáceæ*). North-eastern Australia. Also known as "Red Ash," "Leather-jacket," "Humbug," "Murr-rung." Height 45—50 ft.; diam. 1—2 ft. W 53. Heart dark-brown; sapwood pinkish, darkening with age, hard, close-grained, very tough, polishing well, durable, but warps in drying. Valuable for staves and, perhaps, for gun-stocks.

The name is also applied in the Illawarra district of New South Wales to *Elæocárpus longifólia* C. Moore and *E. Kírtoni* Bailey, close and fine-grained, light-brown woods, easily worked and employed by wheelwrights and for oars; and in other parts of Australia to *Eucalýptus Stuartiána* F. v. M., the Apple-scented Gum, to *E. Sieberiána* F. v. M., the Cabbage Gum, to *E. piluláris* Sm., the Blackbutt, to *E. pauciflóra* Sieb., to *E. hæmástoma* Sm., the Scribbly Gum, in East Gippsland to *E. goniocályx* F. v. M. (See **Box, Bastard**), and in Victoria to *E. amygdalína* Labill., and especially the tall variety *régnans*. This last-named species is a native of Southern Australia and Tasmania, and is probably the tallest tree on the globe, often reaching 100—150 ft., with diam. from 3—8 ft., but in some cases 400, 410 or even 420 or 471 ft. It is, therefore, appropriately called "Giant Gum," the name "Peppermint tree" belonging apparently to smaller specimens. The Gippsland aboriginal name "Wangara" is the equivalent of "Stringybark," applied to many other species. S.G. air-dried 1045—1076, dry 908—703. W 47·54. R. 778—1152 lbs. Light yellowish brown, with a neat striped figuring, straight in grain, easily worked, sometimes proving durable under water, not twisting in drying, suitable for fencing, shingles, ships' planks, keelsons, carpentry, or railway-carriage building.

Ash, Black Mountain (*Eucalýptus lencóxylon* F. v. M.). See **Iron-bark.**

Ash, Oregon (*Fráxinus Oregóna* Nutt. Order *Oleáceæ*). North-western United States. Height 30—40 or 60 ft.; diam. 1—1½ ft. Resembling White Ash, rather heavy, sometimes brittle, not strong. Used for furniture, waggon and carriage-frames and cooperage.

Ash, Prickly (*Zanthóxylum Clava-Hérculis* L.: Order *Rutáceæ*). West Indies. The prickly young stems are imported under the name of "Briar" for walking-sticks. In Jamaica it is known as "Prickly Yellow-wood." W 60·66. E 499. *f* 2·7. *fc* 1·77. *fs* ·418.

Ash, Red, Grey, or **Brown-barked** (*Fráxinus Pennsylvánica* Marshall = *F. pubéscens* Lam. = *F. tomentósa* Michx.: Order *Oleáceæ*). New Brunswick and South Ontario to Louisiana and Florida, along streams, chiefly in the north. *French* "Frêne rouge," *Germ.* "Rothesche," *Span.* "Fresno colorado." Height 50 ft.; diam. 2 ft. S.G. 625. W 38—96. E 8122 lbs. R 869 kilos.

Sapwood light brown or nearly white, sharply defined; sometimes streaked with yellow; heart rich or light reddish brown, moderately heavy, hard, rather strong, coarse-grained, brittle; pith-rays numerous, thin. Used locally for agricultural implements, fence-rails, interior finishing or furniture, as a substitue for Black Ash.

In Australia the name "Red Ash" is applied to *Alphitónia excélsa* [See **Ash, Mountain,**] and to *Orítes excélsa* [See **Oak, Silky.**]

Ash, Rock, of Cape Colony. See **Els, Klip.**

Ash, Water or **Swamp** (*Fráxinus Caroliniána* Miller = *F. platycárpa* Michaux.: Order *Oleáceæ*). Swamps: Virginia—Loui-

Other less important species, used locally, are *Fráxinus anómala* Watson. S.G. 660. W 41—11. Light brown with thick sapwood (30—40 rings), thin medullary rays, many large scattered ducts and several rows of small ducts. Heavy, hard, close-grained. Colorado, Utah, Nevada.

F. Berlandieriána D.C. Used for tool-handles in Mexico.

F. velutína Torrey (= *F. pistaciæfólia* Torrey). Used, for axe-handles and waggons, in New Mexico, Texas, and Arizona.

Aspen (*Pópulus trémula* L. : Order *Salicineæ*). *German* "Aspe," "Espe," "Zitterpappel," *French* "peuplier tremble," *Span.* "alamo tremblon," *Russ.* "osyka."

Dingy white, looking reddish brown in transverse section, with no heartwood. Rings circular, broad, distinct; medullary rays not visible to the naked eye; vessels small, uniformly distributed, dendritic, 2—7 together; generally with white pith-flecks near the centre. Soft, light, elastic, easily split, warping and cracking but little. Used as blindwood, for cooper's ware, milk-pails, herring casks, butchers' trays, clogs, pack-saddles and paper-pulp; and, in France, for sabots and for flooring. Imported in small quantities from the southern Baltic ports, mainly for turnery.

Aspen, American or **Quaking Asp** (*Pópulus tremuloídes* Michaux : Order *Salicíneæ*). Throughout North America, coming up after fires and replacing destroyed fir-forests. *Germ.* "Amerikanishe Zitter-Espe," *French* "tremble d'Amerique," *Span.* "Alamo tremblon." S.G. 403. W 25—13. Co efficient of elasticity 81441. R 677 kilos. Resistance to longitudinal pressure 330 kilos. Resistance to indentation 80 kilos. Height seldom 60 ft. or 2 ft. in diam. Light brown with very thick, very white sapwood. Resembling Aspen in texture, close-grained, cottony in fibre, light, soft, not strong, soon decaying in contact with damp. Used in turnery, and occasionally for flooring, chiefly in the Western United States; but chiefly for paper-pulp, for which purpose, though very white, in strength it is inferior to Spruce.

Aspen, Large-tooth (*Pópulus grandidentáta* Michaux). [See **Poplar, Large-toothed.**]

Assegai-wood (*Curtísia fagínea* Ait. : Order *Cornáceæ*). *Zulu* "Umguna," "Umnoiso." Cape Colony and Natal. Height 40—80 ft.; diam. 3—4 ft. Bright red, becoming dull on exposure, close-grained, very strong, tough, elastic and durable even in damp situations. Used for furniture, shafts of assegais, tool-handles, spokes and felloes, and is one of the best woods for waggon-building.

Avocado Pear (*Pérsea gratíssima* Gærtn.: Order *Lauraceæ*). *Cuba* "Aguacate." West Indies. S.G. 661. Grown chiefly for its fruit.

Axe-breaker (*Notelǽa longifólia* Vent.: Order *Jasmináceæ*). "Mock Olive." *Aborig.* "Coobagum." New South Wales. Height 48—50 ft.; diam. 1—1½ ft. Hard, close-grained, and firm.

Babela (*Terminália belérica* Roxb.). See **Myrobalan wood.**

Babul (*Acácia arábica* Willd.: Order *Leguminósæ*). *Hindi* "Babul." *Bengali* "Babla." *Panjabi* "Kikar." Height 50—60 ft.; diam. 2—2½ ft. W 45. Heartwood, pinkish to brown, mottled, hard, and, if well seasoned, very durable. Used extensively in Northern India for wheels, sugar and oil presses, rice-pounders, agricultural implements, and tool-handles; and in Sind for boat-building, rafters, occasionally for railway-sleepers, and for fuel.

Bagasse, the name of two similar but distinct timbers of French Guiana: (i) *Bagássa guianénsis* Aubl. (Order *Artocarpáceæ*), a large, straight-growing tree. S.G. 1130—730. R 215 kilos. Very durable and excellent for flooring; and (ii) *Icíca altíssima* Aubl. (Order *Burseráceæ*), known also as "Cèdre blanc," "Kurana" or "Carana," also of large size. S.G. 1036—842. R 226 kilos. Soft, but of excellent quality. Used for canoes and for cabinetmaking.

Bakula (*Mímusops Eléngi* L.: Order *Sapotáceæ*). *Sinhalese* "Munamal." *Tamil* "Makulai." India, Ceylon, Moluccas. S.G. 736. W 46—61. Red, hard, strong, fairly durable. Used in house-building in Southern India and Ceylon, and for furniture in the latter country.

Balata, Bully, Bullet-tree, or **Buruch** (*Mímusops globósa* Gaertner, *Sapóta Mulléri* Miquel: Order *Sapotáceæ*). Apparently identical with *Mímusops Bálata* Crueg. *Surinam* "Balata rouge," "Horse-flesh wood." S.G. 1232—1032. W 80·97. E 1097. *f* 8·58. *fc* 4·77. *fs* ·494. R 353 kilos. Height 100 ft.; diam. 2—5 ft. In logs 20—50 ft. long, squared up to 36 in. or 42 in. Dark red-brown, fine, straight and close-grained, very heavy, hard, strong, easily worked, taking a fine polish, very durable; but subject to serious heart-shake, unfitting it for use in large scantlings. Much used in house-building for beams, floors, and mill-work, being said to have more than three times the resistance of Oak, and nearly twice that of the best Teak. This tree yields the elastic substance Balata. British Guiana.

The name is also applied to the wood of the allied **Sapodilla** (*Áchras Sapóta* L.), (q.v.).

Ballow, as yet undetermined. W 61. Resembling Oak, hard, strong, durable, easily worked. Extensively used for railway-sleepers, piles, beams, and planks. British North Borneo.

Banaba. See **Jarul.**

Bancoulier. See **Walnut, Belgaum.**

Bandara (*Lagerstrœ́mia parviflóra* Roxb. var. *majúscula* C.B. Clarke: Order *Lythráceæ*). India. A large tree. Diam. 2 ft., W 40. Red or light brown, compact, moderately hard, tough, elastic, seasons well and is easily worked, and durable in water. It is used for beams, rafters, and boat-building, and is recommended for sleepers.

Banksia (*Bánksia littorális* R. Br.: Order *Proteáceæ*). West Australia. Height 20—40 ft.; diam. 1 ft. Rich brown, beautifully grained. Coming into use for furniture.

Barranduna (*Trochocárpa laurina* R. Br.: Order *Epacridáceæ*). New South Wales and Queensland. "Beech, Brush Cherry," or "Myrtle." Height 20—30 ft.; diam. 6—12 in. W. 48. Warm brown, prettily grained, hard, close-grained, tough, requiring careful seasoning. Useful for turnery.

Barberry (*Bérberis vulgáris* L.: Order *Berberidáceæ*). *Germ.* "Sauerdorn." Too small to be of use. Sapwood lemon-yellow; heart bluish-red. Pith-rays widening outwards, all broad: pore circle very narrow, with very large vessels in radial lines. Europe.

Barwood (*Báphia nítida* Afz.: Order *Leguminósæ*) See **Camwood.**

Basswood (*Tília americána* L.: Order *Tiliáceæ*). "American Linden" or "Lime," "Bee tree." *Germ.* "Amerikanische Linde," *French* "Tilleul d'Amerique," *Span.* "Tilio Americano." Eastern

United States and Canada. Height 80—100 ft.; diam. 3—4 ft. S.G. 452. W 28·2. R 589 kilos. Ash percentage ·55. Relative fuel value ·45. Co-efficiency of elasticity 84010 kilos. Resistance to longitudinal pressure 348 kilos. Resistance to identation 63 kilos. White to light brown, light, soft, tough, close-grained, easily worked, but not strong, shrinking considerably in drying; but durable. Extensively used for cheap furniture, toys, carriage-panels, chair-seats, carpentry, turnery, cooperage, and to some extent for paper-pulp and charcoal. It is sometimes worked up by a spirally-cutting saw, sawing round the log, so as to make a thin board as long as the log and as much as 100 ft. broad.

Basswood, White (*Tília heterophýlla* Vent.). Middle and Southern States. Not distinguished commercially from the preceding.

Bay-wood. See **Mahogany.**

Bead-tree (*Mélia compósita* Willd. = *M. Azedarách* L.: Order *Meliáceæ*). "Persian Lilac," "Pride of India." In Australia "White Cedar" or "Cape Lilac." *Hind.* "Nim." *French* "Lilas des Indes," "Sykomore," "Laurier grec." *Port.* "Margosa." *Tamil* "Vem-pu." Syria, India, China, Australia. Height 40—50 ft.; diam. 1—2 ft. W 30—38. Sapwood yellowish-white; heart yellowish to reddish-brown, handsomely marked, especially vertically, soft and rather loose-textured, easily worked, but taking a good polish and becoming hard and durable, but warping and splitting. It is used in India for furniture, being known in the south as "Bastard Cedar."

Beati (*Cássia siámea* Lamk.: Order *Leguminósæ*). *Sinhalese* "Wa." Southern India, Ceylon, Burma. Heart nearly black, often beautifully mottled longitudinally, very hard. Used in Burma for mallets and walking-sticks, and in Ceylon for fuel for railway-engines. Exported to Bombay and thence to England as "Bombay black-wood" or "Rosewood."

Beech, Common (*Fágus sylvática* L.: Order *Cupulíferæ*). *Germ.* "Gemeiner" or "Roth Buche," *Dutch* "rood beuke," *Danish* "bög," *Swedish* "bok," *Russian* and *Polish* "buk," *Ital.* "faggio," *Portug.* "faya," *Span.* "haya," *French* "hétre." S.G. 700—720, average 705. W 43—53. E 603 tons. *ft* 4—6. *fc* 3—4. *f* 4·7. *fs* ·4. Breaking weight (tensile) 4853 lbs. per sq. in. Crushing strain on sq. in. 9363—7733 lbs. Lighter than Oak, with less tensile but almost equal crushing strength. Height 70—100 ft.; diam. 3—4½ ft. Wood varying in colour from red to yellow or white, the red being the better, grown on richer soil. Rings

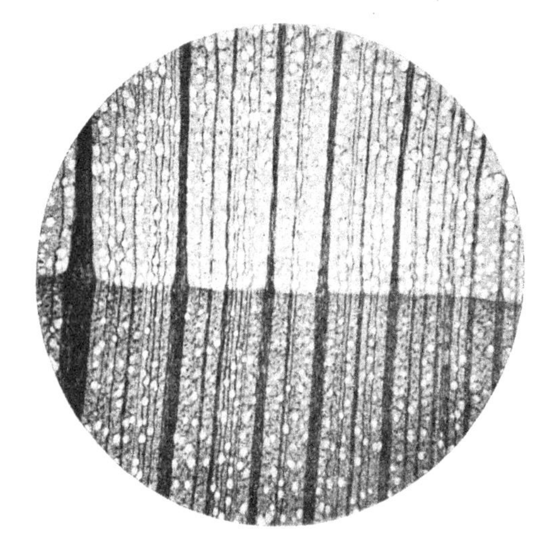

FIG. 46.—Transverse section of Beech (*Fagus sylvatica* L.) highly magnified.

distinct, bulging between the medullary rays. Vessels small, much more numerous in the spring wood, 1—5 together, so that autumn wood appears darker. Medullary rays large, very distinct, with a satiny lustre, occupying nearly a tenth of the transverse section. Pith-mass triangular, small. The wood contains vessels, tracheids, wood-fibres and parenchyma. It is hard, heavy, as strong as Oak and tougher, but 25 per cent. less stiff, close and even in texture, with a fine silky grain, easily cleaved along the rays, very durable under water, and, when well seasoned, not liable to split. (Figs. 26 and 46, 47.) It must, however, be kept either wet or dry. Beech is largely used for chair-making in Buckinghamshire and in Vienna, in the latter district being often stained. Burning rapidly with a bright flame, it is the chief fuel on the Continent of Europe. It also yields one of the best gunpowder-charcoals. It is in great request among turners for tool-handles, wooden screws, wheel cogs (for which it ranks next to Hornbeam), shoe-lasts, printers' rollers, wood type, knife-handles and bobbins, and makes excellent wedges. In France and Germany it is considered the best of all woods, except Walnut, for sabots and wooden soles, for which purpose it is "smoked" over branches and chips

FIG. 47.—Tangential section of Beech (*Fagus sylvatica* L.), highly magnified.

of beech, so as to become charged with pyroligneous acid, when it is extremely impermeable. It is imported from Holland and Germany to our eastern ports.

The name is applied in Australia to *Trochocárpa laurína* R. Br. [See **Barranduna**], *Flindérsia austrális* R. Br. [See **Flindosa**], *F. Oxleyána* F. v. M. [See **Jack, Long**], *Tristánia laurína* R. Br. [See **Box, Bastard**], *Schizoméria ováta* D. Don [See **Coach-wood**], *Monótoca ellíptica* R. Br. [See **Wallang-unda**],

Cryptocárya glaucéscens R. Br. [see **Beech, She**], and to *Gmelína Leichhárdtii* F. v. M. (Order *Verbenáceæ*). This last is also termed "White Beech," and by the aborigines "Binburra" and "Cullonen." It is a native of Queensland and New South Wales, 80—150 ft. in height; 2—4 ft. in diam. W 36. Light-coloured, with a fine, bright, silvery grain, strong, not warping, if moderately seasoned, durable, not easily attacked by termites, and easily worked. It is one of the most valuable of Australian timbers, being useful for turnery or for floats of mill-wheels, but specially valued for the flooring of verandahs and the decks of coasting vessels.

Beech, American (*Fágus ferrugínea* Ait.: Order *Cupulíferæ*). *Germ.* "Amerikanische Buche," *French* "Hêtre d'Amerique," *Span.* "Haya Americana." Eastern North America. Height up to 100 ft.; diam. up to 4 ft. S.G. 688. W 42·9. Ash percentage ·51. Relative fuel value ·685. E 120996 kilos. R 1148 kilos. Resistance to longitudinal pressure 478, to indentation 196 kilos. White to light brown; heart reddish: pith-rays large and conspicuous. Heavy, hard, stiff, strong, tough, rather coarse in texture, warping in drying, but taking a very smooth and beautiful polish, liable to insect attack, and not tolerant of contact with the ground, but otherwise durable. Used for plane-stocks, shoe-lasts, chairs, tool-handles, furniture, ships' timbers, and fuel. It is exported to England, but is considered inferior to English Beech.

Beech, Black. See **Beech, She.**

Beech, Blue. See **Hornbeam.**

Beech, Evergreen (*Fágus Cunninghámii* Hook: Order *Cupulíferæ*). "Myrtle." "Negro-head Beech." Tasmania and Victoria. Height 100—200 ft.; diam. sometimes 8 ft. or more. S.G. 972—593. W 37—53·69. R 548—692 lbs. E 842. *f* 4·06. *fc* 2·58. *fs* ·557. Brownish, satiny, with beautiful feathery cross veins, especially in the warty protuberances on the trunk, hard, and susceptible of an excellent polish. Used for cogs, doors, furniture, and carpentry.

Beech, Indian (*Pongámia glábra* Vent.: Order *Leguminósæ*). *Bengal.* "Kurunja," *Burm.* "Karung," *Tamil* "Poonga," *Fiji*

"Vesivesi." India, Tropical Australia, and the Pacific. Height 40—50 ft.; diam. 1—2 ft. W 40—42. White, turning yellow on exposure, moderately hard, close-grained, tough, prettily marked, but much attacked by insects, and not durable, though improved by seasoning in water. No distinct annual rings, but marked wavy "false rings"; vessels few, scattered; pith-rays distinct. In India used mainly for fuel.

Beech, She (*Cryptocárya glaucéscens* R. Br.: Order *Lauríneæ*). "Black Beech, Sassafras, White Laurel." North-Eastern Australia. Height 70—80 ft.; diam. 1½—2 ft. Soft, ornamental, not durable. Used in cooperage. The name has also been applied to *C. obováta* R. Br. [See **Sycamore, White.**]

Beech, White, in Canada is *Fágus sylvática* L. [See **Beech, Common**]. In Australia the name is applied to *Elæocárpus Kírtoni* F. v. M. [See **Ash, Mountain**], *Gmelína Leichhárdtii* F. v. M. [See **Beech, Common**], and to *Phyllánthus Ferdinándi* Müll. Arg. (Order *Euphorbiáceæ*). This tree, also called "Pencil Cedar," "Lignum-vitæ," and by the aborigines "Chow-way" and "Tow-war," a native of the north-east, reaches a height of 70—80 ft. and a diameter of 1—1½ ft. Its wood is grey, close-grained, and easy to work, but warps. It is used in building and for staves.

Beech, Water. See **Hornbeam.**

Beef-wood in Trinidad is *Rhópala montána* (Order *Proteáceæ*), a valuable timber; but in Australia the name is hopelessly vague, being applied to members of four genera of *Proteáceæ*, viz.: *Bánksia*, *Grevíllea*, *Hákea* and *Stenocárpus*, and to several species of the widely differing genus *Casuarína*. For *Bánksia* see **Honeysuckle**; for *Hákea*, **Pinbush**; for *Stenocárpus*, **Oak, Silky**; for *Casuarína equisetifólia*, **Oak, Swamp**; for *C. suberósa*, **She-oak, Erect**; and for *C. torulósa*, **Oak, Forest.**

Grevíllea striáta R. Br. (Order *Proteáceæ*), also known as "Silvery Honeysuckle," and by the aborigines as "Turraic," reaches a height of 40—50 ft., with a diam. of 1½ ft. Its timber is red and prettily marked, though named from its resemblance to raw beef, hard, close-grained, and susceptible of a good polish. It is used for fencing, cabinet work, and furniture.

Bendy-tree. See **Umbrella-tree.**

Betis (*Payéna Bétis* Villar: Order *Sapotáceæ*). Philippines. Used in ship-building, and classed in the third line of Lloyd's Register.

Big-tree (*Sequoía gigántea* Decaisne: Order *Coníferæ*). "Mammoth tree of California." "Wellingtonia." *French* "Sequoia gigantesque," *Germ.* "Riesen Sequoia," *Ital.* "Gigante

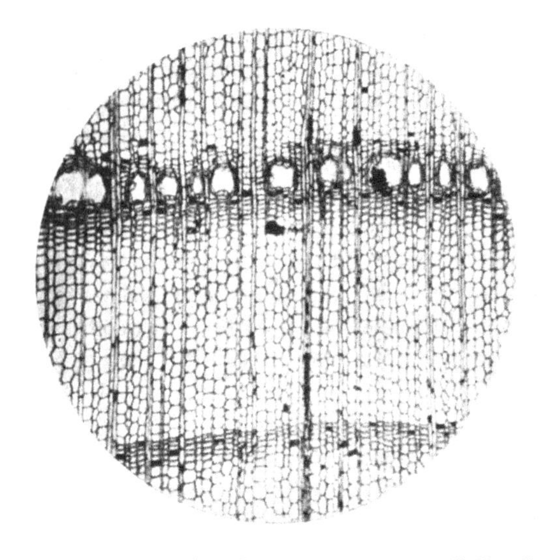

FIG. 48.—Transverse section of Wellingtonia or Mammoth Tree (*Sequoia gigantea* Decaisne), highly magnified.

della California." Western slopes of Sierra Nevada, California, 5000—8000 ft. above sea-level. Height 250—400 ft.; diam. sometimes exceeding 35 ft. The specimen, 62 ft. in girth, of which a section is exhibited in the Natural History Museum, South Kensington, has 1335 annual rings. Wood red-brown, light, soft, brittle, weak, cross-grained, durable in contact with soil. Formerly used locally for lumber, fencing, shingles, etc., but now only of historical interest. Introduced into England as an ornamental tree by William Lobb in 1853 (Figs. 48 and 49).

Bija. See **Teak, Bastard.**

Billian or **Borneo Iron Wood** (*Eusideróxylon Zwágeri* T. and B.: Order *Lauráceæ*). British North Borneo. A very large tree. W 67. Resembling Oak when newly cut, but with age or exposure becoming black as Ebony. Very heavy and hard, strong and durable, bearing exposure, resisting termites and the

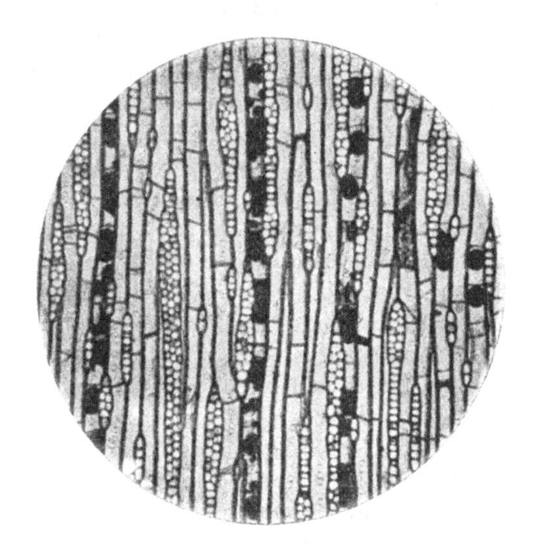

FIG. 49.—Tangential section of Wellingtonia or Mammoth Tree (*Sequoia gigantea* Decaisne), highly magnified.

ship-worm. The best wood in the Bornean and Chinese area for piles, beams, or planks.

Billy Web. See **Ebony, American.**

Bilsted (*Liquidámbar styracíflua* L.). See **Gum, Sweet.**

Birch, Common, European, White or **Silver** (*Bétula álba* L.: Order *Betuláceæ*). *French* "Bouleau commun," *Germ.* "Gemeine Birke," *Dutch* "Berk," *Danish* "Birk," *Swedish* "Bjork," *Russian* "Bereza," *Port.* "Bettula," *Span.* "Abedul."

Northern Europe and Asia. Height 40—50 ft.; diam. 1½ ft. S.G. fresh 919, dried 664. W 45—49. Yellowish or reddish-white to light brown, the vessels so minute as to be almost imperceptible to the naked eye, a smooth transverse section appearing as though sprinkled with flour. Rings and pith-rays distinctly marked: pith-flecks numerous near the centre: wood consisting of tracheæ, tracheids, fibres, fibre-cells, and parenchyma. Moderately hard and heavy, even-grained, difficult to split, but easily worked, neither strong nor durable, and liable to the attacks of worms. Burrs are occasionally produced on the stem, with solid marbled wood, valued by turners, and made into cups and bowls in Lapland. In many countries on the Continent, Birch, as the cheapest native hardwood, is largely used for furniture and turnery: in France it is largely used for felloes of wheels, cooperage, and sabots; and in the Scottish Highlands for an infinity of purposes, including spoons and plates, as in Russia. It is a valuable fuel in Northern Europe, comparing with Beech as a heat-producer as 13 to 15. It also produces excellent crayon charcoal, and its coppice-wood is largely used for brooms, hoops, and crate-making; for tanning leather; for a yellow-brown, or, with alum, a brownish-red, dye; and, when burnt, for distilling Scotch whisky and smoking herrings and hams. Birch timber is imported, mostly with the bark on, from Prussia and the south of Sweden, to Grimsby, Hull, and Ireland; but that from Sweden is often crooked; and the sapwood, especially if felled in the spring, left on the ground, kept too long on the voyage, or stored without ventilation, will become "doated" or foxed, undergoing, that is, a fungoid fermentation. A new industry has been recently started in Russia in the manufacture of Birch planks for export to India for tea-chests. The logs are sawn spirally, as we have already mentioned in the case of Basswood, and two thicknesses are then glued back to back with their grains crossing so as to correct warping. We are not here concerned with the manifold uses of Birch bark—especially in Russia, Sweden, and Scotland—for boxes, baskets, boots, boats, cordage, dyeing, tanning, and even bread-making. The two forms known as *Bétula verrucósa* Ehrh., the "Raubbirke" of Germany, and *B. pubéscens* Ehrh., the "Haarbirke," are here treated as one species. In Jamaica the name Birch is applied to *Búrsera gummífera*, a tree in no way similar.

Birch, American, a trade name for the imported timber, which is the product of more than one species, though chiefly of *Bétula lénta*, the Cherry Birch, and, to a much less extent, *B. lútea*, the Yellow Birch.

Birch, Black (*Bétula lénta* L.). See **Birch, Cherry.**

Birch, Black or **White,** of New Zealand (*Fágus Solándri* Hook. fil.: Order *Cupulíferæ*). *Maori* "Tawhai." Really a Beech, but known in Nelson as "White," and in Wellington as "Black Birch." Height 100 ft.; diam. 4—5 ft. Hard and very durable, and suited for fencing or fresh-water piles, but liable to attack by the ship-worm.

Birch, Canoe (*Bétula papyrífera* Marshall: Order *Betuláceæ*). "White" or "Paper Birch." *Germ.* "Nachen Birke." Canada and the Northern United States. Height 60 ft. or more; diam. 3 ft. or more. Sapwood white, heart reddish-brown, with a fine glossy grain. Rather heavy, hard, tough, and strong, not durable where exposed to alternations of moisture and heat. Used for bobbins, shoe-lasts and pegs, turnery, and extensively for paper-pulp and fuel. A curl in the grain where the branches are given off is sought after by Boston cabinet-makers for veneers.

Birch, Cherry (*Bétula lénta* L.: Order *Betuláceæ*). "Black, Sweet," or "Mahogany Birch," "Mountain Mahogany." *French* "Bouleau doux," *Germ.* "Kirsche-Birke," *Span.* "Abedul dulce." Canada and Eastern United States. Height 60—80 ft.; diam. 3—4 ft. S.G. 762. W 47·47. Ash percentage ·26. Relative fuel value ·759. Co-efficient of elasticity 141398. R 1216 kilos. Resistance to longitudinal pressure 619, to indentation 226 kilos. Sapwood yellowish-white; heart reddish-brown or rose-coloured, heavy, hard, very strong, close-grained, taking a beautiful satiny polish, not attacked by worms, and fairly durable, but becoming duller after conversion. Largely used, especially in the Northern States, for furniture, sometimes stained to imitate Mahogany or

Cherry, and also for turnery, in ship-building, and for fuel. It is exported to England in sawn planks and in slightly waney logs 6—20 ft. long and 1—2½ ft. square, the best coming from Quebec. It is here chiefly used for bedroom furniture, and fetches from 11d. to 2s. 7d. per foot.

Birch, Grey. See **Birch, Yellow** and **Birch, Old-Field.**

Birch, Mahogany. See **Birch, Cherry.**

Birch, Old-Field (*Bétula populifólia* Marshall: Order *Betuláceæ*). "White, Grey," or "Poplar-leaved Birch." From the St. Lawrence to the Delaware. Height 25—30 ft.; diam. 1 foot. S.G. 576. W 35·9. R 778 kilos. Sapwood brownish-white; heart light-brown, light, soft, fine-grained, taking a fine satiny polish, not strong, nor durable. Used for bobbins, shoe-pegs, paper-pulp, and fuel.

Birch, Paper. See **Birch, Canoe.**

Birch, Indian Paper (*Bétula Bhojpátra* Wall.). *Sanskrit* "Bhurjama," *Telugu* "Bharjapatri," *Japan* "Onoore." Northern India. Yellow to reddish-white, hard. Used for furniture and exported.

Birch, Red or **River** (*Bétula nígra* L.). Eastern United States. *French* "Merisier rouge." Smaller, lighter, and less valuable than the other American species, nearly white. Used like poplar, but not exported.

Birch, Sweet. See **Birch, Cherry.**

Birch, White. See **Birch, Common, Canoe, Black,** and **Old-Field.**

Birch, Yellow (*Bétula lútea* Michaux fil.). "Grey" or "Tall Birch." *Germ.* "Gelbe Birke," *French* "Bouleau jaune," *Span.* "Abedul amarillo." Eastern Canada and United States. Height 60—80 ft.; diam. 2—3 ft. S.G. 655. W 40·8. R 1248 kilos. Sapwood nearly white; heart light reddish-brown, heavy, very hard, close-grained, tough, very strong, taking a beautiful satiny polish. Burrs occur, which are used for mallets. The saplings are split for hoops and the older wood for very small woodware, such as button-moulds, for chair-seats, wheel-hubs, and, in Canada, for frames of sledges. It is also largely used for fuel.

Biwa. See **Loquat.**

Blackbutt (*Eucalýptus piluláris* Sm.: Order *Myrtáceæ*). "Flintwood, White-top," or sometimes "Mountain Ash, Willow," or "Stringy-bark." *Aborig.* "Toi, Tcheergun, Benaroon." South-eastern Australia. Height 50—150 ft.; diam. 2—4 ft. S.G. 990. W 59·36—62. E 1152. *f* 5·79. *fc* 3·75. *fs* ·36. Warm-brown or light yellowish, close and straight-grained, moderately heavy, very strong, but occasionally liable to gum-veins and shakes, working fairly well, but warping and requiring careful seasoning. Suitable for sleepers, paving, telegraph-poles, planks, or house-carpentry. It fetches 1s. 3d. a cubic ft. in London. The name is applied in New South Wales to *E. hæmástoma* [See **Gum, Spotted**]; in Tasmania to *E. Sieberiána* [See **Gum, Cabbage**]; occasionally to *E. piperíta* [See **Peppermint**]; and in South-west Australia to *E. patens* Benth., which reaches a height of 100 ft., with a diam. of 6 ft., is tough and durable, and is used for wheelwrights' work.

Blackthorn (*Prúnus spinósa* L.: Order *Rosáceæ*). "Sloe." Hard, tough. Sapwood reddish; heart blackish-brown, often with pith-flecks, pith-rays distinct, vessels small, numerous, equally distributed. Used for walking-sticks.

Blackwood, in Australasia (*Acácia melanóxylon* R. B.: Order *Leguminósæ*). In Tasmania "Lightwood," in New South Wales "Hickory, Silver Wattle," or "Black Sally," *Aborig.* "Moot-chong," "Mooeyang." Tasmania, and South-east Australia, and naturalised in India. Height 60—100 ft.; diam. 1½—3 ft. S.G. 854—529. W 36—63·5. E 1064. *f* 5·45. *fc* 3·24. *fs* ·687. Dark brown, the older growth beautifully figured, with about an inch of nearly white sapwood, hard, close and very even in grain, easily worked and taking an excellent polish, but warping unless very carefully seasoned. One of the most valuable of Australian timbers, an excellent substitute for American walnut. Largely used for oil-casks, in staves three inches thick, for furniture, gun-stocks, tool-handles, crutches, the sounding-boards of pianos, picture-frames, etc., the figured wood being cut into veneers.

Blackwood, in Cape Colony (*Royéna lúcida* L.: Order *Ebenáceæ*). *Boer* "Zwartbast," *Zulu* "Umcaza." Height 40—50 ft.; diam. 1—2 ft. Yellowish with brown stripes, hard, tough, taking a good polish. Used in waggon-building, but adapted for furniture or turnery. *R. nítida* Thunb., from Natal, is a smaller but similar wood, known by the same vernacular names.

Blackwood, Indian (*Dalbérgia latifólia* Roxb.: Order *Leguminósæ*). "Malabar Blackwood," "Rose-wood." *Beng.* "Sit Sal," *Tam.* "Iti." India. A large tree reaching 5 feet diam. S.G. 1064—818. W 66·5. R 522—602 lbs. Sapwood whitish; heart greenish-black, often mottled with lighter purplish streaks, heavy, hard, tough, close but cross-grained, and, therefore, difficult to work, taking a fine polish. Used for sleepers, agricultural implements, gun-carriages, cart-wheels, tool-handles, carving, and especially furniture, for which purpose it is exported, *via* Bombay, and has fetched £13 10s. per ton in London. With it is confused the wood of its variety *D. latifólia*, var. *sissoídes*, known in Tamil as "Biti," a smaller tree common in the extreme south of India, very strong and tough, but with much heart-shake and so much oil as to be unfit to receive paint; and that of *D. cultráta* in Burma. [See **Yen-dike.**].

Bloodwood (*Eucalýptus corymbósa* Sm.: Order *Myrtáceæ*). *Aborig.* "Boona." Southern Queensland and New South Wales. Height 30—100 ft.; diam. 1—4 ft. S.G. 983—853. W 72·6. E 1023. *f* 7·57. *fc* 4·48. *fs* ·615. Dark red-brown, moderately heavy, easily dressed, straight and close in grain, but full of gum-veins and not, therefore, a favourite with sawyers, becoming hard on drying, very strong and durable, little attacked by termites. Used chiefly for posts and rails, but also for piles and sleepers. The name is also applied to the allied *E. terminális* F. v. M. of the interior, the "Arang mill" of the aborigines, a very red wood, forming the chief large timber of the area, but not otherwise valuable; and also to *E. paniculáta* Sm. [See **Ironbark, White** and **Ironwood** xxi.].

Bloodwood, Brush, or **Scrub** (*Balóghia lúcida* Endl.: Order *Euphorbiáceæ*). "Roger Gough." New South Wales, Queensland,

Norfolk Island, New Caledonia. Height 70—80 ft.; diam. 2—2½ ft. W 44—45. Buff or light reddish, fine and close-grained, but rather soft and not used. [See also **Rosewood** (*Synóum glandulósum*)].

Bloodwood, Mountain, Smooth-barked, or **Yellow** (*Eucalýptus exímia* Schauer: Order *Myrtáceæ*), "Rusty Gum." Blue Mountains. Height 80 ft. Light-coloured, soft, not durable, except under water. Useful for work under water and as fuel.

Boco or **Coco Wood** (*Bocóa prouacénsis* Aubl.: Order *Leguminósæ*). Guiana. *Demerara* "Eta-balli." S.G. 1234—1208. R 402 kilos. Tall, yielding logs 29 ft. long and 18 in. in diam. Sapwood very hard and compact, clear yellow; heart dark-brown black, working well. Suitable for walking-sticks, carving and cabinet-work.

Bois Chaire (*Tecóma leucóxylon* Mart.: Order *Bignoniáceæ*). Brazil, Guiana, Trinidad. *Brazil* "Quirapaïba," *Cayenne* "Ebène verte." A large tree yielding logs 14 ft. long, squaring 14—16 in. S.G. 1220—1211. R 480 kilos. Hard, even-grained, durable, dark-green, or, when polished and varnished, a dark brownish-black. Used in chair-making, whence its name; in building and for the sounding-boards of pianos.

Bois d'arc. See **Osage-Orange.**

Bois de feroles. See **Satine'.**

Bois de natte (*Labourdonnaísia calophyllóides* Boj., L. *glaúca*, *Mimusops Imbricaria* Willd., etc.: Order *Sapotáceæ*). Mauritius and neighbouring islands. Extensively used for ship-building, cabinet-work, and furniture.

Bois graine bleu (*Symplócos martinicénsis* Jacq.: Order *Styráceæ*). "Kakarat." Dominica, Martinique, etc. A small tree. Used for planks in internal work.

Bois Lezard (*Vitex divaricáta* Swartz: Order *Verbenáceæ*). Brazil and West Indies. Strong and durable.

Bois Mulatre (*Pentaclēthra filamentósa* Benth.: Order *Leguminósæ*). "Palo Mulato." Trinidad. Height 30—40 ft.; diam. 1—2 ft. Dark, even-grained, and said to be durable under ground.

Bois Rivière. See **Water-wood.**

Bois Violet. See **Purple-heart.**

Bolongnita (*Diospýros pilosánthera* Bl.: Order *Ebenáceæ*). Philippines.

Bottle-brush, Red (*Callistémon lanceolátus* DC.: Order *Myrtáceæ*). "Water Gum." *Aborig.* "Marum." Eastern Australia. Height 30—40 ft.; diam. 1—1½ ft. Hard, heavy. Used for mallets, in ship-building and wheelwrights' work.

Bottle-brush, White (*C. salígnus* DC.). "River or Broad-leaved Tea-tree," "Stone-wood," "River Oak," *Aborig.* "Unoyie, Humbah." Eastern Australia. Height 40—50 ft.; diam. 1½—2 ft. S.G. 983. W 57—61. Drab to dark red, sometimes prettily-grained, very hard, close-grained, fairly easy to work, and said to be very durable underground.

Bow-wood. See **Osage-Orange.**

Box (*Búxus sempervírens* L.: Order *Euphorbiáceæ*). *French* "Bois commun," "Bois beni," *Germ.* "Buchsbaum." Northern and Western Asia, North Africa, and Central and Southern Europe. Height 8—30 ft.; diam. small. S.G. 950—980. W 80·5—68·75. Light yellow, very homogeneous, almost horn-like, neither rings, pith-rays, nor vessels being distinct, hard, heavy, firm, free from heart-shake, difficult to split, but works up smoothly, with a slight silky lustre and is durable, when thoroughly seasoned. "Boxwood is very apt to split in drying; and to prevent this, the French turners put the wood designed for their finest work into a dark cellar as soon as it is cut, where they keep it from three to five years. . . . They strike off the sapwood with a hatchet, and place the hardwood again in the cellar till it is wanted for the lathe. For the most delicate articles, the wood is soaked for twenty-four hours in fresh very clear water, and then boiled for some time. When taken out of the boiling water, it is wiped perfectly dry and buried, till wanted for use, in sand or bran." Compared for closeness of grain to Ebony by Theophrastus, used by the Romans for veneers and flutes, Virgil mentions the Box as used by the turner:

"Smooth-grained and proper for the turner's trade,
Which curious hands may carve, and steel with ease invade."
(Dryden's translation.)

Invaluable for mathematical instruments, the chief use of Box since the 15th century has been for wood-engraving, for which purpose it is chiefly imported from Abasia in Circassia and from Odessa in billets 3—8 ft. long and 3—12 in. across. In spite of the advance of other methods of engraving, Box is so unequalled for this purpose that careful search is being made for any wood likely to approach it in suitability. (See p. 103 *supra*). The largest number of box-trees in Europe are in the mixed forests of Ligny and of St. Claude on the Jura. At the latter place the wood, which is not of large dimensions, is turned into small boxes, beads, spoons, forks, etc.

In Australia the name Box is applied to a great number of Eucalypti, such as *E. hemiphlóia* [See **Canary Wood**], *E. largiflórens* [See **Gum, Slaty**], *E. leucóxylon* [See **Ironbark**], *E. odoráta* [See **Peppermint**], *E. stelluláta*, *E. Stuartiána*, and *E. viminális* [See **Gum, Manna**]. In America the name is applied to *Cornus flórida* L., the American Dogwood, an abundant wood fitted only for coarse engraving; in the Bahamas to *Vítex umbrósa* Sw.

Box, Bastard is also a name of wide application in Australia, sometimes referring to *Eucalýptus largiflórens* [See **Gum, Slaty**], *E. longifólia* [See **Woolly Butt**], *E. microthéca* [See **Box, Dwarf**], *E. polyánthema* [See **Box, Red**], *E. punctáta* [See **Leather-jacket**], or *E. tereticórnis* [See **Gum, Slaty**]; but chiefly to *Tristánia conférta*, *T. laurína*, and *Eucalýptus goniocályx.*

Tristánia conférta R. Br. (Order *Myrtáceæ*), otherwise "Brisbane, Brush, Red," or "White Box," or "Brisbane Mahogany." *Aborig.* "Tubbil-pulla." North-eastern Australia. Height 80—120 ft.; diam. 1—3 ft. W 59—64. Sometimes prettily grained, strong and durable, but warping very much unless carefully seasoned, dressing well and not attacked by termites. Used in ship-building, bridges, etc.

T. laurína R. Br., of Eastern Australia, known also as "Water Gum, Beech," and "Swamp Mahogany," is a smaller tree. Height

50—60 ft.; diam. 1—2 ft. Dark-coloured, hard, tough, close-grained, difficult to season. Used for tool-handles and cogs.

Eucalýptus goniocályx F. v. M.: (Order *Myrtáceæ*), of Tasmania and South-eastern Australia. "Grey Box, Mountain Ash, Spotted, Grey, White," or "Blue Gum." Height up to 300 ft.; diam. to 6 or 10 ft. S.G. 1152—798. W 72—74. R 799 lbs. Pale yellow to light brown, very hard, tough, usually free from gum-veins, straight-grained, difficult to split, not warping, very durable, especially underground. Used for joists, beams, rafters, sleepers, spokes, staves, boat-building, and fuel.

Box, Black (*Eucalýptus largiflórens*). See **Gum, Slaty.**

" " (*E. oblíqua*). See **Stringybark.**

" " (*E. microthéca*). See **Box, Dwarf.**

Box, Brisbane or **Brush** (*Tristánia conférta*). See **Box, Bastard.**

Box, Brown. See **Box, Red.**

Box, Cape (*Buxus Macówani* Oliv.: Order *Euphorbiáceæ*). Eastern Cape Colony. Height 40—80 ft.; diam. up to 4 ft. Yellow, hard, close and even-grained, resembling European (Turkish) Box, but not equally valuable for engraving. Used for turnery. The name is also applied to *Gonióma Kamássi* E. Mey. [See **Kamassi.**]

Box, China (*Múrraya exótica* L.: Order *Rutáceæ*). Queensland. Small, resembling Box, but apt to crack, requiring careful seasoning. W 61—63. Used for tool-handles.

Box, Cooburn (*Eucalýptus largiflórens*). See **Gum, Slaty.**

Box, Dwarf, Flooded, or **Narrow-leaved** (*E. microthéca* F. v. M.: Order *Myrtáceæ*). Also known as "Bastard" or "Black Box." Australia, in most of the colonies. Reddish, hard, heavy, elastic, sometimes figured like walnut, but darker, heavier, closer-grained, and too hard for ordinary cabinet-work. Used in building.

Box, Grey (*Eucalýptus goniocályx*). See **Box, Bastard.**

" " (*E. hemiphloia*). See **Canarywood.**

" " (*E. largiflórens*). See **Gum, Slaty.**

" " (*E. polyánthema*). See **Box, Red.**

Box, Grey (*E. salígna*). See **Gum, Blue.**

,, **Ironbark** (*E. oblíqua*). See **Stringybark.**

,, **Jamaica** (*Tecóma pentaphýlla* Juss.: Order *Bignoniáceæ*). Brazil, Venezuela, West Indies, etc. "White Cedar," "Cog-wood," "Roble blanco." S.G. 876—834. Used for piles, house and boat-building, and suggested for engraving.

Box, Knysna. See **Kamassi.**

,, **Native.** See **Boxthorn.**

,, **Poplar** (*Eucalýptus populifólia* Hook.: Order *Myrtáceæ*). "Red" or "White Box, Nankeen," or "White Gum." *Aborig.* "Egolla." North-east Australia. Height 50—60 ft.; diam. 2 ft. Grey or light-brown, hard, heavy, close-grained, tough, strong, hard to work, often unsound, handsome when polished, durable. Used for sleepers, posts, building, etc.

Box, Red (*Tristánia conférta*). See **Box, Bastard.**

,, ,, (*Eucalýptus populifólia*). See **Box, Poplar.**

,, ,, (*Eucalýptus polyánthema* Schau.: Order *Myrtáceæ*). South-eastern Australia. Also known as "Bastard, Brown" or "Grey Box, Lignum-vitæ," or "Poplar-leaved Gum." *Aborig.* "Den." Height sometimes 250 ft. S.G. 1010—1248. R 749—803 lbs. Brownish-red. very hard, fine-grained, very tough, often hollow, but very durable. Used for pit-props, cogs, naves, felloes and fuel.

Box, Scrub (*Tristánia conférta*). See **Box, Bastard.**

Box, White (*Eucalýptus hemiphlóia*). See **Canarywood.**

,, ,, (*E. odoráta*) See **Peppermint.**

,, ,, (*E. populifólia*). See **Box, Poplar.**

,, ,, (*Tristánia conférta*). See **Box, Bastard.**

,, **Yellow** (*Eucalýptus melliodóra* A. Cunn.: Order *Myrtáceæ*). Eastern Australia. Also known as "Yellow-Jacket, Honey-scented" or "Red Gum." *Aborig.* "Dargan." Height 40—80 ft.; diam. $1\frac{1}{2}$—$3\frac{1}{2}$ ft. S.G. 876—1125. W 60—70. R 725—695 lbs. Light yellow or pale brown, very hard, heavy, close-grained, tough, durable, with a wavy figure, but with some gum-veins and cup-shake. Used for spokes, naves, cogs, treenails, posts, and, to some extent, engraving. The name Yellow Box is also applied to

Eucalýptus hemiphlóia [See **Canarywood**], and to *E. largiflórens* [See **Gum, Slaty**].

Box-Elder. See **Maple, Ash-leaved.**

Boxthorn (*Bursária spinósa* Cav.: Order *Pittosporáceæ*). Australia and Tasmania. "Native Box" or "Olive." Height 20—30 ft.; diam. less than a foot. White, close-grained, and taking a fine polish. Used in turnery.

Braziletto or **Brazil-wood** (*Cæsalpínia brasiliénsis* Sw.: Order *Leguminósæ*), now almost extinct, *C. crísta* L., *C. bijúga* Sw., and *C. tinctória* H.B.K. Tropical America. Hard, heavy woods, taking a polish, and employed in cabinet work, but mainly as a red dye. They contain a red colouring-matter known as *Braziline*, soluble in water, and giving, with lime, baryta and tin chloride, a red precipitate, whilst Logwood gives a blue one. [See also **Nicaragua wood, Peach wood, Pernambuco wood** and **Sappan wood.**]

Break-axe. See **Ironwood** xxii.

Briar. See **Ash, Prickly.**

Briar-root (*Erica arbórea* L.: Order *Ericáceæ*). *French* "Bruyère." Southern Europe. Dark brown, dense, mottled. Small wood, used exclusively for tobacco pipes, imported from France.

Broom (*Cýtisus scopárius* Link.: Order *Leguminósæ*). Western Europe. Cultivated in Algeria, its stems being imported under the trade name of "Black Orange" as walking-sticks.

Buckeye, Ohio (*Æsculus glábra* Willd.) and **Sweet Buckeye** *Æ. fláva* Ait.: Order *Sapindáceæ*). Eastern United States. Small trees with creamy-white, light, soft, fine, and even-grained wood, not strong, but often tough, easily worked. Used locally for building, but more for turnery, artificial limbs, and especially paper-pulp.

Buckthorn (*Rhámnus cathárticus* L.: Order *Rhamnáceæ*). Europe, Siberia, North Africa. *Germ.* "Kreuzdorn." A shrub with narrow, greenish-yellow sapwood and orange-red heart, with a narrow but distinct zone of pores in the spring wood, and remarkable flamboyant groups of vessels in the autumn wood

(Fig. 28); pith-rays indistinguishable. The wood is hard and heavy, and suitable for turnery, but small. The other British species, *R. Frángula*, the Alder or Berry-bearing Buckthorn, yields the wood known as **Dogwood.**

Buffelsbal (*Gardénia Thunbérgii* L.: Order *Rubiáceæ*). Cape Colony. Hard and heavy. Used for clubs, tool-handles, axles, etc.

Bullet-wood, Andaman (*Mímusops littorális* Kurz.: Order *Sapotáceæ*). Tenasserim and Andaman Islands. A large tree, yielding a very hard, red-brown, close-grained, durable wood, which is, however, apt to split. Used for bridges and house-posts, and recommended for sleepers. [See also **Balata** and **Sapodilla.**]

Bully, Naseberry (*Sapóta Sideróxylon* Gr.: Order *Sapotáceæ*). Jamaica. W 74. E 1080. *f* 9·16. *fc* 4·31. *fs* ·50. One of the most valuable woods of the Colony.

Bunya-bunya (*Araucária Bidwílli* Hook: Order *Coníferæ*). Queensland. Height 100—150 ft.; diam. 2½—4 ft. Light-coloured, straight-grained, beautifully veined, very strong, easily worked, susceptible of polish, not warping, durable. Suitable for cheap furniture, but seldom felled, as its seeds are eaten by the aborigines.

Butternut (*Júglans cinérea* L.: Order *Juglandáceæ*). Eastern North America. "White Walnut." "Oil nut." *French* "Noyer cendre," *Germ.* "Graue Walnuss." Height 50—60 ft.; diam. 2—3 ft. Reddish-brown, light, soft, not strong, free from the attacks of worms, not easily split. Used for sleepers, for internal work in building, coach-panels, boats and canoes, wooden dishes, shovels, and cabinet-work. Seldom imported into England.

Button-wood. See **Plane.**

Cabbage-bark. See **Angelin.**

Cagueyran (*Copaífera hymenæifólia* Moric.: Order *Leguminósæ*). Cuba. A large tree, the wood of which is used in building.

Cajeput (*Melaléuca leucadéndron* L.: Order *Myrtáceæ*). Malaysia and tropical Australia. "Milkwood, White" or "Swamp Tea-tree." *Aborig.* "Atchoourgo." *Malay.* "Kayu puti." *New Caledonia* "Niaouli." Height 40—50 ft.; diam. 1—2 ft. Hard, heavy, with a ripple-like mottling, close-grained, very durable underground. Used for posts and in ship-building, turns well, and is suitable for carriage-work.

Calabash (*Crescéntia Cujété* L.: Order *Bignoniáceæ*). *Cuba* "Guira." West Indies. S.G. 580. W 54·69. E ·230. R 243 lbs. *f* 2·62. *fc* 1·69. *fs* ·358. Light brown, soft, elastic. Pith irregular, rather large; rings and pith-rays indistinct; vessels equally but sporadically distributed, small. Used for lasts, saddle-bows, etc.

Calamander-wood (*Diospýros quæsíta* Thwaites, *D. oocárpa* Thw. and *D. hirsúta* Linn. fil.: Order *Ebenáceæ*). *Tamil* "Calamander maram," *Sinh.* "Kalu-médiríya." W 57. One of the most valuable ornamental woods of Ceylon, but now scarce, red-hazel-brown or chocolate brown, with handsome black stripes, intermediate between Rosewood and Zebra-wood, hard, close-grained. Used for turning and veneers.

Camara (*Geissospérmum Vellósii* Allem.: Order *Apocynáceæ*). Brazil. S.G. 746. Light-coloured, moderately heavy, strong, but small. Used in boat-building.

Camphor (*Cinnamómum Cámphora* Nees: Order *Lauráceæ*). China, Japan, etc. Height 30—60 ft.; diam. 2—3 ft. Light-coloured, fragrant, soft. Used for entomological and ornamental cabinets.

Camphor, Borneo (*Dryobálanops aromática* Gaert.: Order *Dipterocarpáceæ*). *Malay* "Kayu Kapor Barus." Borneo and Sumatra. A large tree yielding timber 25—45 ft. long and 1—2 ft. square. S.G. 895—1053. E 3770,000. *p* 16000. *e*′ 3·36. *p*′ 1·47. *c* 6790. *c*′ ·896. *v*′ 1·561. Light red, resembling Honduras Mahogany, not fragrant, plain, close-, straight-grained, moderately hard and tough, durable, but liable to star-shake. Used for planks, beams, piles, etc.

Camphor, Nepal (*Cinnamómum glandulíferum* Meissn.: Order *Laurı́neæ*). India. Also known as "Nepal" or "Assam Sassafras," and apparently identical with "Martaban Camphor-wood" and "Burmese Sassafras." A large tree yielding timber 20—30 ft. long and 1½ ft. in diam., floating when seasoned, tough, strong, durable, with the odour of Sassafras. Used for house-carpentry, cabinet-work, tool-handles, etc.

Camphor-wood (Australian). See **Cypress-Pine.**

Camwood (*Báphia nítida* Afz.: Order *Leguminósæ*). West Africa. "Barwood." Height 8—10 ft. Red. Imported from Sierra Leone in logs 4 ft. long and 1 ft. in diam., or powdered. Used by native women for painting their bodies, and in England, with iron-sulphate, as a dye for red handkerchiefs.

Canary-wood (*Eucalýptus hemiphloía* F. v. M.: Order *Myrtáceæ*). South-east Australia. Known also as "Box, Grey, White" or "Yellow Box," "White Gum." *Aborig.* "Narulgun." Height 50—60 ft. or more; diam. 2—4 ft. or more. S.G. 1230—773. W 48·27. Yellow-white or light buff, heavy, very hard, strong, tough, cross-grained, not readily split, durable, but often hollow and subject to dry-rot and termite-attack. Used for sleepers, piles, planks, pit-props, fencing, handles, cogs, naves, felloes, screws, tree-nails, etc.

Canary-wood (*Moríndа citrifólia* L.: Order *Rubiáceæ*), (= *Naúclea unduláta* Roxb., *Sarcocéphalus cordátus* Miq.). "Indian Mulberry," "Leichhardt's Tree." *Hind.* "Al," *Tam.* "Nonna maram," *Fiji* "Kura." India and the Eastern Tropics to Queensland. Height 50—70 ft.; diam. 2—2½ ft. W 30—41. Yellow, soft, cross-grained, sometimes with a beautiful wavy figure, easily worked and taking a good polish. Suitable for turning or cabinet-work; but little used, save as a dye. (See also **Tulip-tree.**)

Canella or **Canelle,** a name applied to various Brazilian woods belonging to the *Lauráceæ*, including *Dicypéllium caryophyllátum* [See **Rosewood, Cayenne**], *Aydéndron canella* Meissn., *Nectándra atra*, and *N. móllis.* The latter is brown, light, easily worked, but not durable. S.G. 744. It is procurable of a considerable size, and is used for decks, house-building, and carpentry.

Cannon-ball tree (*Xylocárpus Granátum* Kœn. = *Cárapa moluccénsis* Lam.: Order *Meliáceæ*). Ceylon, Burma, Fiji Islands, etc. "Sea Cocoa-nut." *Burm.* "Pen-lay," *Fiji* "Dabi." Height

20 ft.; diam. 4 ft. W 47. Whitish, turning red, hard. Used for house-building, handles, spokes, furniture, etc. In South America the same name is applied to *Couroúpita guianénsis*, belonging to the Brazil nut group (*Lecythidáceæ*).

Carapa or **Carapo**. See **Crab-wood**.

Carob (*Ceratónia Siliqua* L.: Order *Leguminósæ*). *French* "Caroubier." Chiefly known for its fruit, the "Locust bean" or "St. John's-bread"; but imported from Algeria as walking-sticks.

Cashew-nut (*Anacárdium occidentálé* L.: Order *Anacardiáceæ*). A native of South America, cultivated for its fruit throughout the Tropics. *French* "Acajou à fruits, à pommes" or "de Guadeloupe," *Germ.* "Acajoubaum," *Hind.* "Kaju," *Tam.* "Mundiri." Red to brown, moderately hard. Used in boat-building, for packing-cases and for charcoal, especially for iron-smelting.

Catalpa (*Catálpa speciósa* Warder: Order *Bignoniáceæ*). South Central United States. Height 80 ft.; diam. 4 ft. Brown, light, soft, not strong, brittle, coarse-grained, durable, especially in contact with the soil. Used for posts and fencing; but suited for internal fittings, and being now much planted.

Cedar, a name extended from the Lebanon Cedar (*Cedrus Líbani* Loud.: Order *Coníferæ*), to other species of the genus, to various Junipers and other coniferous woods, and to many other woods of broad-leaved trees, especially the Meliaceous genus *Cedréla*, most of which resemble the true Cedar in being brown, even-grained woods of moderate hardness and often fragrant. The true Cedar is a native of the Lebanon, Taurus, and neighbouring ranges of South-west Asia, and was introduced into England as an ornamental tree after the middle of the 17th century, that at Enfield being perhaps the oldest existing English tree. *French* "Cedre du Liban," *Germ.* "Libanon Ceder," *Ital.* "Cedro del Libano." Height 50—80 ft.; diam. 3—4 ft. or more. Reddish-brown, light, straight and open-grained, very porous, soft and spongy in the centre, easily worked, but rather brittle, liable to extensive heart- and cup-shakes, not strong. Mountain-grown Cedar is harder, stronger, less liable to warp and more durable. The wood has a pleasant odour, which is obnoxious to insects. It is, therefore, suitable for cabinets, internal work, carving, etc., for which purposes it seems to have been mainly employed by the ancients, with whom it had so great a repute for durability. In the Cilician Taurus it is used for the best household furniture and for church-fittings. The "Cedar" of the English timber trade is the West Indian *Cedrela odoráta*,

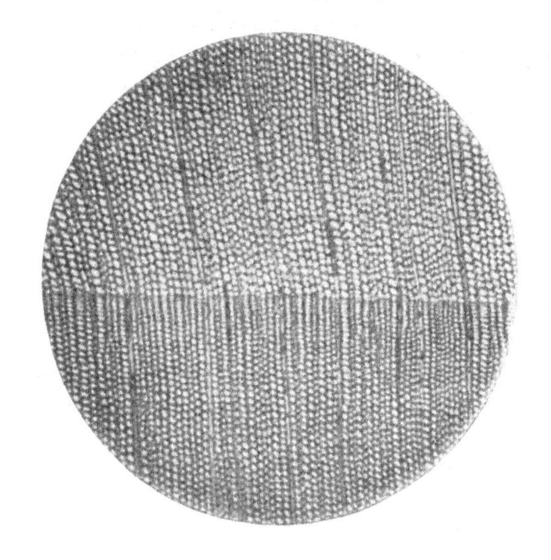

FIG. 50.—Transverse section of Cedar of Lebanon (*Cedrus Libani*).

shipped from Cuba, Trinidad, Honduras, and Tabasco, and fetching 3d. to 4½d. per foot. [See **Cedar, West Indian.**]

Cedar, Atlas (*Cédrus atlántica* Manetti). Mount Atlas. Height 80—100 ft.; diam. up to 5 ft. The outer part of the heartwood beautifully veined, resembling in quality and value that of the Deodar of India. Used in North Africa for sleepers, paving, cabinet-making, and carpentry.

Cedar, Barbadoes (*Juníperus barbadénsis* L.: Order *Cupressíneæ*). "Juniper Cedar." Barbadoes, Jamaica, etc. Closely allied to, if not identical with, the Red Cedar of the United States (*Juníperus virginiána* L.).

Cedar, Bastard, in India. See **Bead-tree** and **Mahogany, East Indian.**

Cedar, Bastard, in Jamaica (*Guazúma tomentósa* H.B.: Order *Sterculiáceæ*). *French* "Orme d'Amerique," *Telugu* "Rudraksha chettu." West Indies. Introduced into Ceylon and Southern India more than a century ago and now common. Light, splitting easily. Used in Jamaica for staves for sugar hogsheads.

Cedar, Bastard Pencil (*Dysóxylon rúfum* Benth.: Order *Meliáceæ*). Queensland and Northern New South Wales. Height 40—50 ft.; diam. 1½—2 ft. Red, nicely grained, easily worked. Used for cabinet-work.

Cedar, Bermuda (*Juníperus bermudiána* L.: Order *Cupressíneæ*). Bermudas. Closely allied to the Red Cedar, *Juníperus virginiána*, of which it is supposed to be a geographical variety. Height 50—60 ft.; diam. 2—3 ft. Used for boat-building and formerly exported for pencils, for which purpose the Red Cedar has superseded it.

Cedar, Black (*Nectándra pisi* Miq.: Order *Lauráceæ*). French Guiana. Of large dimensions. S.G. 818—531. R 130—159 kilos. Dark-brown, moderately heavy, firm, easily worked, teredo-proof. Useful for building and for ships' planking, but corroding iron.

Cedar, Borneo. See **Serayah.**

Cedar, Canoe (*Thúya gigántea* Nutt.: Order *Cupressíneæ*). "Yellow Cedar." "Red Cedar" (of the Western States). One of the "White Cedars" of the American trade. "Lobb's Arborvitæ." *French* "Thuia géant de Californie," *Germ.* "Riesens-Lebensbaum," *Ital.* "Albero de la vita di Lobb." Western North America, from Alaska to California. Height 130—200 ft.; diam. 9 ft. or more. Heartwood light greyish-brown, light, soft, brittle, not strong, free from knots, easily split or worked, not warping, and very durable in contact with soil. An important lumber tree, the only wood used by the Red Indians of the North-west for canoes. Used by white settlers for fencing, shingles, cooperage, doors, window-sills, indoor fittings, and the coarser kinds of furniture.

Cedar, Cuba. See **Cedar, West Indian.**

Cedar, Deodar (*Cédrus Deodára* Loud.: Order *Coníferæ*). Afghanistan to the Western Himalayas. "Indian Cedar." *French* "Cedre de l'Himalaya," *Hind.* "Devaderu." Height 150—250 ft.; diam. 5—7 ft. at base, tapering to one-third at 80 ft. up. Heartwood light yellowish-brown, compact, even-grained, moderately hard, not readily splitting or warping, fragrant, and exceedingly durable: annual rings uniform, with well-marked autumn zones: resin-canals absent. The pillars in the Shah Hamaden Mosque at Srinagar of this wood are probably over 400 years old, and some of the bridges in the same city, though their piers are alternately wet and dry, are said to have lasted even longer. This species was introduced into England as an ornamental tree in 1831. It is the chief timber of North-west India, where it is used for sleepers, for all purposes of construction and even for furniture.

Cedar, Florida. See **Cedar, Red.**

Cedar, Guiana (*Icíca altíssima* Aubl.: Order *Burserácecæ*). "Kurana" or "Carana-gum." *French* "Cèdre blanc, Cèdre bagasse." A large and very valuable wood of French and British Guiana, easily worked, fragrant, durable. Height up to 100 ft. S.G. 1036—842. R 226 kilos. Used for canoes, book-cases, internal house-fittings, etc.

Cedar, Honduras. See **Cedar, West Indian.**

Cedar, Incense (*Libocédrus decúrrens* Torr.: Order *Cupressíneæ*). "Bastard, White or Post Cedar." *French* "Cèdre blanc de Californie," *Germ.* "Californische Flussceder." Pacific slope of the United States. Height 100 ft. or more; diam. 6 ft. or more. Light greyish-brown, light, soft, fine- and close-grained, but brittle and not strong, very durable in contact with soil. Extensively used for posts, fencing, shingles, laths, internal work and furniture.

Cedar, Indian. See **Cedar, Deodar.**

Cedar, Japanese. See **Sugi.**

Cedar, Mexican. See **Cedar, West Indian.**

Cedar, Moulmein (*Cedréla Toóna* Roxb.: Order *Meliáceæ*). India, Java, Australia. "Bastard Cedar," "Bastard" or "Indian Mahogany," or "Chittagong wood" of India, "Cedar" or "Red Cedar" of Australia. *Hind.* "Toon," "Toona." *Burm.* "Thitkado." *French* "Cèdre de Singapore." Height 70—180 ft.; diam. 2—10 ft. S.G. 508—576, W 28—36. Timber generally from 14 to 40 ft. long, and from 11 to 26 in. square. Pale brick-red, resembling mahogany, often beautifully curled near the root or branches, very fragrant, clean and straight, but open in grain, moderately hard, easily worked, does not warp, but splits somewhat in seasoning, and is liable to heart- and star-shakes, durable, termite-proof. A very valuable wood, formerly hollowed out for canoes in North-east India and largely used for tea-chests, but not now sufficiently abundant. Much employed for furniture, door-panels and carving, and yielding beautiful veneers. The best of the woods known as Chittagong woods.

Cedar, New Zealand (*Libocédrus Bidwíllii* Hook. fil. or *L. Doniana* Endl.: Order *Cupressíneæ*). *Maori* "Pahautea," "Kawaka." Soft, close-grained woods, resembling the Incense Cedar of the Western United States. Suitable for planks and spars.

Cedar, Oregon (*Cupréssus Lawsoniána* Murr. = *Chamæcýparis Lawsoniána* Sargent: Order *Cupressíneæ*). "Port Orford Cedar, Lawson's Cypress, Ginger Pine." South Oregon and North California. Height up to 200 ft.; diam. up to 12 ft., but generally 120—150 ft. high. S.G. 460. Light yellowish-brown, light, but heavier than other "white Cedars," hard, strong, very close-grained, abounding in fragrant resin, easily worked, susceptible of a beautiful polish, very durable in contact with the soil. One of the most valuable timber-trees of North America, largely cut for lumber and used for ship and boat building, fencing, flooring.

Cedar, Pencil, a name applied in Northern New South Wales and Queensland to *Dysóxylon Fraseriánum* and *D. Muélleri*, *Phyllánthus Ferdinándi* [See **Beech, White**], and *Podocárpus eláta*;

and in Cape Colony to the Coniferous *Widdringtónia juniperoídes*. In English commerce the name refers to *Juníperus virginiána* [See **Cedar, Red**]. *Dysóxylon Fraseriánum* Benth. (Order *Meliáceæ*), known also as "Rosewood" or "Bog-onion," 50—70 ft. in height and 3—4 ft. in diameter, yields a reddish, prettily-figured, fragrant, easily-worked wood, valued for furniture, turning, engraving and ship-building, being, in fact, a substitute for Mahogany. *D. Muélleri* Benth., known, from the smell of the wood when freshly cut, as "Turnip-wood," the "Kidgi-kidgi" of the aborigines, a taller tree, yields a rich red wood, equally valuable. *Podocárpus eláta* R. Br. (Order *Taxáceæ*), known also as "Pine, White" or "She Pine," or "Native Deal," is a fine tree, 50—130 ft. high and 2—5 ft. in diam. W 45·7. Though seldom cylindrical, it is free from knots, sometimes beautifully figured, soft, fine, close and silky in grain, easily worked, durable, termite- and teredo-proof, and valued for joinery and cabinet-work. *Widdringtónia juniperoídes*, growing above the winter range of snow on the Cederberg in the Clanwilliam district of Cape Colony, sometimes reaches 12 ft. in diam. It is known as "Ceder Boom" to the Dutch, and the demand for it exceeds the supply. The allied *W. Whítei* Rendle, of elevated kloofs in the Shiré Highlands, Nyassaland, is a fine tree, 150 ft. high, reaching 6 ft. diam., and yielding an ornamental, fragrant, light, yellow-brown wood, susceptible of a good polish, and suitable for building, pencils and other purposes. Though suitable for re-afforesting tropical highlands, this species is not at present abundant.

Cedar, Red, in Cape Colony (*Cunónia capénsis* L.: Order *Cunoniáceæ*). *Dutch* "Rood Els." Height 15—25 or even 60 ft.; diam. 1½—2 ft. Close-grained, tough, resembling Linden-wood, taking a good polish. Much used in cabinet-work and turnery and by wheelwrights.

Cedar, Pink or **Red**, of Sikkim tea-planters, used for tea-chests and furniture, is *Acrocárpus fraxinifólius* Wight (Order *Leguminósæ*), the "Mandania" of the natives.

Cedar, Red, in Australia (*Cedréla Toona*). See **Cedar, Moulmein.**

Cedar, Red, in North America (*Junîperus Virginiâna* L.: Order *Cupressîneæ*). "Florida Cedar, Savin, Pencil Cedar." *French* "Cèdre de Virginie," *Germ.* "Virginischer Sadebaum," "Virginische Wachholder," "Bleistift-holz." Throughout the coasts of the United States, but large only in the South. Height 40—50 or even 100 ft.; diam. 1—4 ft. S.G. 330. Sapwood broad, yellowish; heart rose-red to brown-red, fragrant; annual rings sinuous; pith-rays very fine; resin-ducts absent; light, soft, brittle, compact, fine-grained, strong, easily split, durable, especially in contact with the soil or water, and obnoxious to insects. One of the most valuable coniferous woods of America. Formerly much used by the Spaniards in Florida for ship-building, and in England, up to fifty years ago, for cabinets, work-boxes, etc., it is occasionally employed in the United States for railway-sleepers and fencing, in the Southern States for coffins, and in Philadelphia for cooperage. It is, however, generally too dear for any use but veneers and pencil-making, for which latter purpose several million cubic feet are cut annually. A useful paper for protection against moth is made from the refuse of the pencil factories.

Cedar, Rock (*Junîperus sabinoîdes* Sarg.), a native of Mexico and Texas. Height 20—40 ft.; diam. 1 ft. S.G. 690. W 43. Sapwood thin, nearly white; heart brown, often streaked with red, slightly fragrant, light, hard, close-grained, not strong, very durable in contact with soil. Used for sleepers, telegraph-poles, fencing, and fuel.

Cedar, Tasmanian (*Athrotâxis selaginoîdes* Don.: Order *Taxodîeæ*). Height about 45 ft. Light, yellow, straight, even-grained.

Cedar, West Indian (*Cedrêla odorâta* L.: Order *Meliâceæ*). "Cuba, Honduras" or "Mexican Cedar." The "Cedar" of English commerce. A tall tree capable of yielding timber 18—40 ft. long and 1—2 ft. square. S.G. 372—664. W 31—47. *p* 7600. *p'* ·7. *e'* 1·0. *f* 3·02. *fc* 1·98. *fs* ·362. *c* 2870. *c'* ·379. *v'* ·586. Sapwood narrow, reddish-white; heart cinnamon-brown; annual rings broad and distinct; pith-rays numerous and distinct; vessels very large, open, scattered, but more numerous, larger and partly filled with brown resin in spring wood. Fragrant, often beautifully marked and resembling the allied Mahogany; but very much softer, light, easily split, bitter in taste. Used mainly for cigar-boxes, but also for furniture.

Cedar, White, in the United States, a name applied to the similar coniferous woods of *Libocédrus decúrrens* [See **Cedar, Incense**], *Cupréssus Lawsoniána* [See **Cedar, Oregon**], *Thúya gigántea* [See **Cedar, Canoe**], *Cupréssus Thyoídes*, and *Thúya occidentális*. In Australia it is used of *Elæodéndron austrálé* and *Mélia compósita.*

Cupréssus Thyoídes L. *French* "Cèdre blanc," *Germ.* "Weisse Ceder, Ceder-Cypresse." Swamps on the east coast of the United States. Height 70—80 ft.; diam. 2—3 ft. Light greyish-brown, very light, soft, not strong, slightly fragrant, close-grained, easily worked, very durable in contact with soil, not warping. Used for sleepers, posts, shingles, cooperage, and boat-building.

Thúya occidentális L., known also as "Arbor-vitæ." *French* "Arbre de vie," *Germ.* "Lebensbaum." Eastern North American. Height 30—40 or 60 ft.; diam. 2—4 ft. S.G. 320. Light, soft, brittle, very fragrant, rather coarse-grained; but otherwise resembling the preceding in characters and uses.

Elæodéndron austrálé. See **Ash, Blue.**

Melia compósita. See **Bead-tree.**

Cedar, Yellow. See **Yellow-wood.**

Champak (*Michélia Chámpaca* L.: Order *Magnoliáceæ*). *Beng.* "Champa" or "Champaka." India, Ceylon, Moluccas. Height 30 ft.; diam. up to 4½ ft. S.G. 671. W 42. R 350 lbs. Light olive-brown, soft, but seasons well, taking a good polish, durable. Used for furniture, house- and boat-building. The allied *M. excélsa* Blume is the "Bara-Champ" of the Eastern Himalaya, the principal building and furniture wood of Darjîling, whilst *M. nilagírica* Zenk., the "Pila Champa" of Southern India and Ceylon, is also used.

Chaplash (*Artocárpus Chaplásha* Roxb.: Order *Moráceæ*).

India. "Lesser" or "Thorny Jack." *Hind.* "Chaplasha." Height 100 ft. S.G. 556. W 34·75. Yellow-brown, moderately hard, even-grained, durable, especially under water. Used for canoes, tea-chests, furniture, etc.

Chatwan (*Alstónia scholáris* R.Br.: Order *Apocynáceæ*). India, Ceylon, Moluccas. *Telugu* "Eda-kula." *Malay* "Kayu Gabas." A large tree, yielding white, close- but rather coarse-grained, soft, perishable, bitter wood. Used for writing-boards in schools, whence its name, for boxes, tea-chests, coffins, turnery, etc., and as a tonic.

Cheesewood (*Pittospórum undulátum* Vent.: Order *Pittospóreæ*). Eastern Australia. "Native Laurel," "Mock Orange." *Aborig.* "Wallundun-deyren." Height 30—50 or 90 ft.; diam. 1—2½ ft. W 61·25. White or whitey-brown, very close-grained, hard. Suitable for turning, rollers for mangles, and engraving, though inferior to Box. The name is also applied to *P. bícolor* Hook., the "Whitewood" of Tasmania, where it was used by the aborigines for clubs. This is a smaller tree with yellower wood. S.G. 874. Used for axe-handles, billiard-cues, etc.

Cherry (*Prunus Ávium* L. and *P. Cérasus* L., and probably *P. Pádus* L.: Order *Rosáceæ*). Europe, Northern and Western Asia. Sapwood reddish or yellowish white; heart light yellowish-brown, hard, heavy, firm, but not durable; annual rings distinct; pith-rays distinct, fine; vessels fine and equally distributed. The wood is valued by turners and for inlaying. After soaking for several days in lime-water it becomes a beautiful brownish-red, and can be used as a substitute for Mahogany. More important, however, is the "Perfumed Cherry," *P. Mahaleb* L., the perfumed brown or green-streaked wood of which is grown and manufactured in Austria into pipe-stems and walking-sticks.

Cherry, Brush (*Eugénia myrtifólia*). See **Myrtle, Native.**

Cherry, Brush (*Trochocárpa laurína* R.Br., See **Barranduna**

Cherry, Native (*Exocárpus cupressifórmis*) R.Br.: Order *Santaláceæ*). Australia. Height 10—20 or 40 ft.; diam. 6—20 inches. S.G. 756—845. Close-grained, handsome, hard, durable. Used for tool-handles, spokes, gun-stocks, cornice-poles, etc. The allied *E. latifólia* is sometimes called "Broad-leaved Cherry." See **Sandal-wood, Scrub.**

Cherry, Wild Black (*Prúnus serotína* Ehrh. : Order *Rosáceæ*). Eastern United States. Height 90—120 ft.; diam. 2—3 ft. Sapwood yellowish-white; heart pale reddish to brown, often with discoloured flaws, compact, fine- and close-grained, hard, heavy, strong, shrinking in drying, but taking a good polish, durable. Valued for cabinet-work and interior decoration; but scarce.

Chestnut (*Castánea vulgáris* Lamk.: Order *Cupulíferæ*). "Spanish Chestnut." *French* "Châtaignier," *Germ.* "Edelkastanie." A large tree, sometimes reaching an enormous girth, native to the continent of Europe and represented by a closely related variety, *americána*, in the Eastern United States. Sapwood yellowish-white or light brown; heart darker brown, resembling Oak, but distinguished by the absence of broad pith-rays; pores large, forming a broad circle in the spring-wood and bifurcating lines beyond; moderately hard, but much softer than Oak, light, coarse-grained, not strong, warping in drying, but durable when dry or wet. S.G. 450. E 85621. R 696 kilos. W 28—41. Used for fence-posts and rails, staves, vine-props, hop-poles, cabinet-work, and charcoal. The ancient roofs in England alleged to be of Chestnut are really of Oak.

Chestnut, Moreton Bay (*Castanospérmum austrálé* A. Cunn.: Order *Leguminósæ*). "Bean tree." North-east Australia, introduced into India. Height 80—90 or 130 ft.; diam. 2—3 or 6 ft. W 39·5. Prettily grained, streaked with dark brown, somewhat resembling walnut, soft, fine-grained, shrinking much in drying and so requiring thorough seasoning, taking a good polish, but not durable. Used for furniture and staves.

Chestnut, Wild, of South Africa (*Calodéndron capénsé* Thunb.: Order *Rutáceæ*). *Dutch* "Kastanie." Height 20—30 or 70 ft.; diam. 2—3 or 5 ft., the dimensions in Cape Colony exceeding those in Natal. White, soft, very light, but soft. Used for yokes, hoops of waggons, etc.

Chilauni (*Schíma Wallíchii* Chois.: Order *Camelliáceæ*). Bengal. Red, coarse grained, durable. Much used for bridges and sleepers.

China-berry, an American name for *Mélia compósita.* See **Bead-tree.**

Chinquapin, in the Eastern United States (*Castánea púmila* Michx.), in the Western States (*Castanópsis chrysophýlla* A. DC.), small trees allied to the Chestnut, with similar but slightly heavier wood.

Chittagong-wood, a name applied to several woods imported to Madras, from north-east Bengal, the best of which is *Cedréla Toóna* [See **Cedar, Moulmein**]. *Chickrássia tabuláris* A. Juss. (Order *Meliáceæ*), otherwise known as "Cedar, Bastard Cedar" or "Deodar." *Bengal.* "Chikrassi." *Sinh.* "Kulankik." *Tamil* "Kal-otthi," is also so-called. Height 80 ft.; diam. 2—3 ft. W 24. Yellowish-brown to reddish-brown, with a splendid satiny lustre, fragrant, hard, seasoning and working well, but warping and creaking in very hot dry weather. Used for furniture and carving.

Coach-wood (*Ceratopétalum apétalum* D. Don.: Order *Saxifragáceæ*). "Light-wood" or "Leather-jacket." New South Wales. Height 50—70 or 100 ft.; diam. 1½—2 or 3 ft. W 42. Soft, light, close-grained, exceedingly tough, with the fragrance of coumarin. Used for coach-building, tool-handles, cabinet-work, etc., and suggested for sounding-boards and stethoscopes. The name is also applied to *Schizoméria ováta* D. Don., an allied reddish wood of inferior character, known also as "Cork-wood, Beech" or "White Cherry."

Cocus. See **Ebony, Green.**

Coffee-tree (*Gymnocládus canadénsis* Lam.: Order *Leguminósæ*). *French* "Chicot, Gros fevier." *Germ.* "Amerikanischer Schusserbaum." *Span.* "Arbol de café falso." Eastern United States. Height 100 ft.; diam. 3 ft. S.G. 693. W 43·2. E 104822. R 771 kilos. Sapwood yellow or greenish-white; heart brown blotched with red, heavy, cross-grained, hard, strong, very stiff, taking a high polish, handsome, and durable. Used for fencing, building, and cabinet-work.

Cogwood (*Ceanóthus Chloróxylon* Nees: Order *Rhamnáceæ*). Jamaica. Hard, heavy, very elastic and durable under water. Used for cogs in sugar-mills.

Compass (*Kæmpássia malaccénsis* Maingay). Borneo. W 58. Red, heavy, tough, strong, coarse-grained, but liable to termite-attack and not durable.

Cooper's wood (*Alphitónia excélsa*). See **Ash, Mountain.**

Cork-wood tree of Missouri (*Leitnéria Floridána* Chapm.: Order *Leitneriáceæ*). S.G. 210. The lightest known wood.

Cork-wood tree of the Antilles (*Hibíscus tiliáceus* L.: Order *Malváceæ*). Grown throughout the Tropics. Nut-brown, very light. Used for floats for fishing-nets.

Cork-wood in Australia (i) (*Duboísia myoporoídes* R.Br.: Order *Solanáceæ*). Also known as "Elm." New South Wales and Queensland. Height 15—30 ft.; diam. 1—2 ft. W 30—30·75. White or yellowish, very soft, close-grained, and firm. Used for carving. Named from its bark resembling that of Cork Oak. The name is applied (ii) to *Schizoméria ováta* [See **Coachwood**], and (iii) to *Weinmánnia rubifólia* F. v. M. [See **Marrara**].

Cork-tree, Indian (*Millingtónia horténsis* L. fil.: Order *Bignoniáceæ*). Yellow-white, soft, taking an excellent polish. Used for furniture.

Cotton-tree (*Bombax Céiba* L.: Order *Bombáceæ*). Identical with *B. malabáricum* DC. Southern India, Burma, Northern Australia. "Malabar Silk-cotton," "Red-Cotton tree." *French* "Fromage de Hollande." *Hind.* "Shembal." Height 60 ft. or more; diam. 5 ft. W 20—32. Light, soft, coarse-grained, not durable. Used for planks, packing-cases, tea-chests, coffins, canoes, and fishing-floats.

Cotton-wood. See **Poplar** and **Dogwood**, in Tasmania.

Courbaril. See **Locust.**

Cowdie-pine. See **Kauri.**

Crab-wood (*Cárapa guianénsis* Aubl.: Order *Meliáceæ*). Guiana, Trinidad, etc. "Caraba, Carapo, Andiroba." Height 60 ft. and upward; diam. 1—2 ft. S.G. 894—349. W 46·25. *fc* 3·29. *fs* ·433. R 80 kilos. Reddish-brown, moderately heavy and hard, straight-grained, resembling inferior Mahogany, but affected by shakes and splitting in seasoning, taking a good polish, little attacked by insects. Used for furniture, internal fittings, masts, spars, staves, and shingles. *Cárapa prócera* DC., the "Toulou-couna" of Senegambia, is a very similar wood.

Crow's Ash. See **Flindosa.**

Cuamara. See **Tonka-bean.**

Cucumber-tree (*Magnólia acumináta* L. : Order *Magnoliáceæ*). "Mountain Magnolia." Eastern United States. Height up to 100 ft.; diam. 4 ft. S.G. 409. W 29·23. R 671 kilos. Sapwood white; heart yellowish-brown, soft, light, close-grained, moderately compact and durable, taking a satiny polish. Closely resembling and probably often confounded with Tulip-wood (*Liriodéndron tulipífera*), this wood is used for turnery, wainscot, packing-cases, and cheap furniture. [See also **Papaw.**]

Cudgerie. See **Flindosa.**

Cypre, Bois de (*Córdia Gerascánthus* Jacq. : Order *Borragíneæ*). Tropical America. "Spanish Elm," "Dominica Rosewood," "Bois de Rhodes." Dark, open-grained, soft. W 47·69. E 553. *f* 2·73. *fc* 2·16. *fs* ·428. Used in cabinet-work.

Cypress (*Cupréssus sempervírens* L. : Order *Cupressíneæ*). Mediterranean region, Asia Minor, and Persia. Height up to 100 ft.; diam. sometimes 7 ft. S.G. 620. Reddish, fragrant, moderately hard, very fine- and close-grained, and virtually indestructible. Used by the ancient Egyptians for mummy-cases; for the coffins of the Popes; in Assyria and in Crete for shipbuilding; for the gates of Constantinople destroyed by the Turks in 1453, eleven hundred years after their construction; and for the doors of St. Peter's, which were quite sound when replaced, about the same time and after a similar duration, by brass. Used, according to Evelyn, for harps and organ-pipes, and also for vine-props; but now seldom employed.

Cypress, Bald, Black, Deciduous, Red, Swamp or **White** (*Taxódium dístichum* Richard : Order *Taxodiéæ*). Swamps of the Southern United States. Height 80—100 or more ft.; diam. 6—8 or 13 ft., but tapering. Wood lighter and less resinous on low ground, and then termed "White Cypress," reddening on exposure, soft, straight- and fine-grained, not strong, but very durable in contact with the soil. Formerly used in Louisiana for canoes, water-pipes, and house-frames, and now for sleepers, fencing, and, on a large scale, for shingles.

Cypress, Himalayan or **Indian** (*Cuprḗssus torulṓsa* D. Don.), a light-brown, fragrant, moderately hard wood, used for building, etc.

Cypress, Japanese. See **Hi-no-ki.**

Cypress-Pine, the general name for the species of *Frenḗla* (Order *Cupressíneæ*), in Northern and Eastern Australia, especially the varieties of *F. robústa* A. Cunn. (= *Cállitris robústa* R. Br.), "Black, Common, Dark, Lachlan, Murrumbidgee" or "White Pine, Camphor-wood." *Aborig.* "Marung." Height 60—70 ft. diam. 1½—2 ft. Light to dark brown, often with pinkish longitudinal streaks and beautifully figured, with a camphor-like fragrance, straight-grained, but very full of knots, easily worked, shrinking and warping but little, and taking a good polish, largely teredo- and termite-proof. Much used for piles, building, and furniture. *Frenḗla robústa*, var. *microcárpa* A. Cunn., the "Coorung-coorung" of the aborigines, is a similar and valuable wood, but dark-coloured and somewhat brittle, used for telegraph-poles. *F. robústa*, var. *verrucósa* A. Cunn., sometimes known also as "Rock Pine," "Desert Cypress" or "Sandarac Pine," is also dark. S.G. 691. W 43—44·5. It is used for telegraph-poles and cabinet-making, its camphoraceous smell being said to be obnoxious to insects. *Frenḗla Endlícheri* Parlat., known as "Black, Red, Scrub" or "Murray Pine," a rich brown, beautifully mottled with darker brown, presenting a superb figuring, fragrant, fine-grained, susceptible of a high polish and durable, is a valuable wood, used for internal work and for piles, sleepers, etc. *Frenḗla rhomboídea* Endl., known also as "Light" or "Illawarra Mountain Pine," or, in Tasmania, as "Oyster Bay Pine," is close-grained, strong, easily worked, takes a good polish and is durable, but smaller than the varieties just mentioned. W 39·25. It is used for similar purposes.

Cypress-Pine, Mountain (*Frenéla Parlatórei* F. v. M.), also known as "Stringybark Pine." Height 40—60 ft.; diam. 1—2 ft. Light straw-colour, fragrant, close-grained, soft, easily worked. Much used for joinery.

Date, Kafir, or **Plum** (*Harpephýllum Cáffrum* Bernh.). Cape Colony. W 45·7. *f* 5·86. *fc* 2·94. Dull red mahogany-like, easily worked, and suitable for carpentry and cabinet-work.

Deal, a term properly describing soft (coniferous) wood sawn in thicknesses of 2—4 in., but often used with prefixes as to colour or country of origin. Thus Dantzic, Red or Yellow Deals are derived from the Northern Pine (*Pínus sylvéstris* L.) [See **Pine, Northern**], White Deals from the Spruce (*Pícea excélsa*), Canadian and New Brunswick Spruce Deals, mostly from *Pícea nígra*, narrow-ringed trees yielding "Black," wide-ringed ones "White Spruce."

Deodar. See **Cedar, Deodar.**

Dhaura (*Anogeíssus latifólia* Wall.: Order *Combretáceæ*). India. Height up to 200 ft.; diam. 3 ft. or more. Sapwood wide, grey or yellowish; heart purplish, hard, very strong and tough, but splitting in seasoning and only durable when kept dry. Used for axles, axe-handles, agricultural implements, furniture, etc.

Dilo. See **Poon.**

Dogo. See **Mangrove.**

Dogwood, in England (i) (*Córnus sanguínea* L.: Order *Cornáceæ*). Europe and Northern and Western Asia. Known also as "Cornel," "Prickwood." A mere shrub. Hard, horny, flesh-coloured, with minute evenly-distributed vessels, 1—4 together, without pith-flecks and with indistinguishable pith-rays. Used formerly for skewers and arrows, and to some extent for gunpowder charcoal. (ii) (*Rhámnus Frángula* L.: Order *Rhamnáceæ*). Europe, North Africa, and Siberia. Known also as "Berry-bearing Alder," and to gunpowder-makers as "Black Dogwood." A shrub 5—10 ft. high. Sapwood narrow light yellow; heart brilliant yellowish-red; vessels minute, not in flamboyant groups as in the allied Buckthorn (*R. cathárticus*); soft. Largely used in the manufacture of gunpowder charcoal.

Dogwood, in the West Indies (*Piscídia Erythrína* L.: Order *Leguminósæ*). W 56·89. *f* 3·74. *fc* 2·03. *fs* ·387. Used in building.

Dogwood, in Tasmania (*Bedfórdia salicína* DC.: Order *Compósitæ*), the "Cottonwood" of New South Wales. Height 12—30 ft. S.G. 896. Pale brown, often beautifully mottled, hard, close-grained, fetid when cut, brittle and difficult to season.

Dogwood, in Australia (i) (*Emmenospérmum alphitonioídes* F. v. M.: Order *Rhamnáceæ*). Height 130—170 ft.; diam. 2—2½ ft. Straight-grained and durable. Useful for tool-handles, oars, staves, boat and house-building. (ii) (*Jacksónia scopária* R. Br.: Order *Leguminósæ*). Height 10—15 ft.; diam. 3—12 in. W 55·25—56·5. Dark-yellow or brown, hard and polishing well, fetid when burning. Too small for much use. (iii) (*Myopórum montánum* R. Br.: Order *Myoporíneæ*), known also as "Waterbush" or "Native Daphne." Height 30—40 ft.; diam. 1 ft. W. 47. Soft, light, tough, straight-grained. Used in building. (iv) (*Myopórum platycárpum* R. Br.), known also as "Sandal" or "Bastard Sandal-wood," or "Sugar-tree." Small, light walnut colour, fragrant when cut, often with a bird's-eye mottling, especially in burrs, fine-grained, taking a fine polish. Suitable for veneers. (v) (*Synóum glandulósum*). See **Rosewood.**

Dongon (*Stercúlia cymbifórmis* Blanco). Philippine Islands.

Douglas Fir. See **Pine, Oregon.**

Dudhi (*Wríghtia tinctória* R. Br.: Order *Apocynáceæ*). India. *Hind.* "Kala Kudu." *Tamil* "Nila Pila." A small tree, 12—15 in. in diam. Wood white, close-grained and hard, resembling ivory. Used for turning.

Durmast. See **Oak, Durmast.**

Eagle-wood (*Aquilária Agállocha* Roxb. in Sylhet, and *A. malaccénsis* Lam. of Further India: Order *Aquilariáceæ*). "Lign-aloes, Aloes-wood, Agilawood, Black Agallocha." *French* "Bois d'Aigle." *Germ.* "Adlerholz." *Sanskrit* "Agaru." *Arab.* "Aqulugin." *Latin* "Agallochum." Large trees with whitish wood, containing an abundance of resin and an essential oil, much valued as a perfume and possibly the "aloes" of Psalm xlv. The wood retains its fragrance for years and is burnt in Indian temples and also used for inlaying and as a setting for jewels, selling at £30 per cwt. in Sumatra.

Ebony, a name for a very dense, hard, and generally black wood, mentioned by Herodotus and perhaps by Ezekiel, and originally applying to *Diospýros Ébenum* König (Order *Ebenáceæ*). *Latin* "Ebenus." *French* "Ebène." *Germ.* "Ebenholz." *Arabic* "Abnoos." A large tree, a native of Southern India and Ceylon. S.G. 1187. W 70 or more; the heart 75—80. *p* 756—1180. Sapwood dingy grey with black patches, flexible and very liable to insect attack; heart deep black, very heavy, hard, and fine-grained, the rings and pith-rays being scarcely recognizable, capable of a very high polish, but affected by weather, and, therefore, used largely as veneer, selling in England at from £5 to £10 per ton.

Ebony, Acapulco, Cuernavaca or **Mexican** (*Diospýros Ebenáster* Retz.). A native of India, cultivated in Mauritius, the Philippines and tropical America, the "Bastard Ebony" of Ceylon.

Ebony, American, Green, Jamaica or **West Indian** (*Brýa Ébenus* DC.: Order *Leguminósæ*). "Cocus" or "Kokra" of Jamaica, "Granadillo" of Cuba, "Billy Web" or "Chichipate" of Honduras. A small tree. S.G. 1206. W 61·45. E 1178 tons. *f* 9·10. *fc* 4·5. *fs* ·529. R 480 lbs. Heavy, black, durable. Used for inlaying flutes, flageolets, etc.

Ebony, Bombay, Ceylon and **Siam** (*Diospýros Ébenum* König, *Ebenáster* Retz., *melanóxylon* Roxb. and other species). *D. melanóxylon* Roxb., also known as "Coromandel" or "Godavery Ebony." *Hind.* "Tendu." A large tree. S.G. 978—1200. W 61·12. R 294 lbs. Sapwood pink; heart black, with beautiful purple streaks, irregular, heavy, very hard, strong. Used for building, shafts, carving, etc.

Ebony, Camagoon (*Diospýros pilosánthera* Bl.), the "Bolongnita" of the Philippine Islands.

Ebony, Cape or **Orange River** (*Eúclea Pseudébenus* E. Mey: Order *Ebenáceæ*). South-west Africa. Jet-black, hard.

Ebony, Cuba (*Diospýros tetraspérma* Jacq.). "Ebano Real." S.G. 1300. R 305 lbs. Black, hard, brittle. Used in cabinet-work.

Ebony, False, the "Corsican Ebony" of ancient Rome (*Cýtisus Labúrnum* L.: Order *Leguminósæ*). *French* "Faux ebénier." Sapwood broad, yellowish; heart dark-brown with a greenish tinge; rings and pith-rays distinct, hard, capable of high

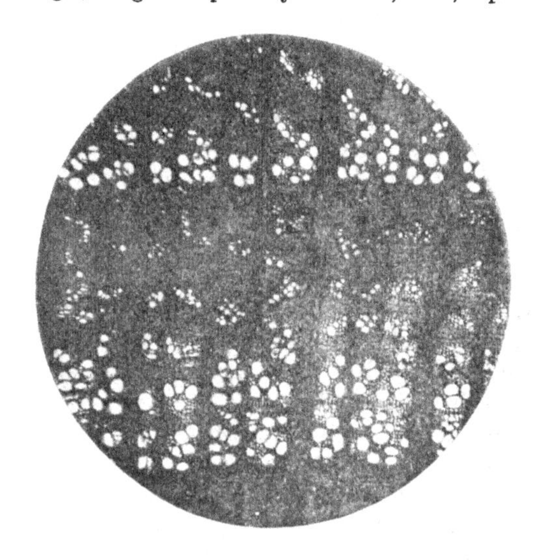

FIG. 51.—Transverse section of Laburnum (*Cytisus Laburnum*) "False Ebony."

polish, but not durable. Used in turnery and by the ancients as veneer.

Ebony, Gaboon, Lagos or **Old Calabar** (*Diospýros Dendo* Welw.). Tropical West Africa. Black.

Ebony, German, the wood of Pear or Yew, stained black.

Ebony, Green (*Diospýros chloróxylon* Roxb.). *Telugu* "Nella ulimera." Southern India. A large tree yielding a hard, useful wood. In Brazil the name is applied to *Tecóma leucóxylon* Mart. (Order *Bignoniáceæ*), known also as "Quirapaiba," a heavy, hard, dark-green, close-grained cabinet-wood yielding logs 14 ft. long and 14—16 in. square. S.G. 1220—1211. R 481 kilos. In the West Indies the name is applied to *Brýa Ebenus.* [See **Ebony, American.**]

Ebony, Madagascar, Macassar or **Zanzibar** (*Diospýros mespilifórmis* Hochst., *haplostýlis* Boiv. and *microrhómbus* Hiern, and also *Acácia glaucophýlla* Steud. and *Dalbérgia melanóxylon* Guill. and Perr.: Order *Leguminósæ*), all natives of tropical East Africa, with black heartwood.

Ebony, Manila (*Diospýros philippénsis* Gürke and *D. Ebenáster* Retz.).

Ebony, Mauritius (*Diospýros tesselária* Poir.).

Ebony, Mountain (*Bauhínia Hoókeri* F. v. M. and *B. Carrónii* F. v. M.: Order *Leguminósæ*). Also known as "Queensland Ebony." Eastern Australia. The former 30—40 ft. high; 1—1½ ft. in diam., dark-red, heavy: the latter light-brown to dark-brown. Suitable for veneers.

Ebony, Red (*Diospýros rúbra* Gärtn.). Mauritius.

Ebony, St. Helena (*Dómbeya melanóxylon* Roxb.: Order *Byttneriáceæ*). Height 10—15 ft. S.G. 1145. W 71·5. Almost exterminated by goats.

Ebony, White (*Diospýros Malacapái* A. DC.). Philippines.

Elder (*Sambúcus nígra* L.: Order *Caprifoliáceæ*). Europe, West Asia and North Africa. A small tree. Pith very large; pith-rays numerous and distinct; vessels more numerous in spring-wood; wood yellowish, hard, firm, difficult to dry, warping. Used in turnery.

Elder, Box. See **Maple, Ash-leaved.**

Elm, a name referring originally and mainly to species of the genus *Ulmus* (Order *Ulmáceæ*), broad-leaved trees with very large vessels in their spring-wood, and the vessels in the autumn-wood in wavy peripheral lines. *French* "Orme." *Germ.* "Ulm" or "Rüster." *Ital.* "Ulmo."

Elm, American, Water or **White** (*Úlmus Americána* L.). *French* "Orme parasol." Alluvial ground in Eastern North America. Height 100 ft. or more; diam. 6—7 ft. S.G. 650. W 40·5. R 852 kilos. Sapwood yellowish-white; heart light-brown, heavy, strong, tough, compact, but not durable; pores in spring-wood conspicuously large and almost entirely in a single row. Valuable for tool-handles, agricultural implements, wheel-hubs, cooperage, etc., and for fuel.

Elm, Canadian, Cliff, Cork, Hickory, Rock or **White** (*U. racemósa* Thomas). *French* "Orme à grappe." *Germ.* "Trauben Ulme, Felsen Ulme." Canada and Eastern United States. Height 80—150 ft.; diam. 2—3 ft. S.G. 726—765. W 45·26—47.

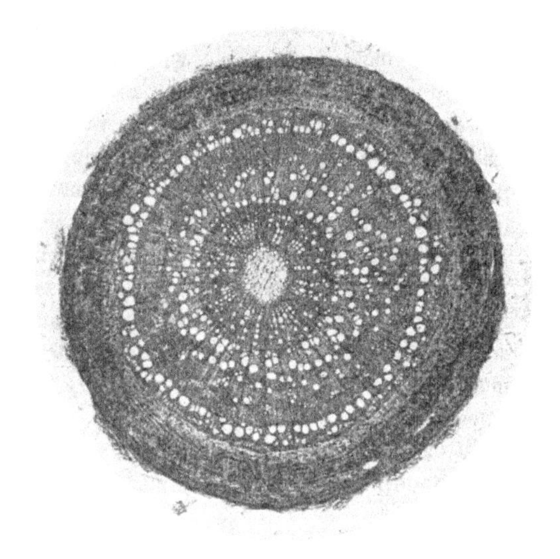

FIG. 52.—Transverse section of Common Elm (*Ulmus campestris*).

R 1066 kilos. *c* 9182. *c*′ 1·213. *v*′ 1·191. *e*′ 1·39. *p*′ 1·14. Sapwood greenish, not durable; pores in spring-wood small, those in summer-wood in fine rather distant lines; heavy, hard, compact, very strong, tough and elastic. Logs 20—40 ft. long and 11—16 in. square, liable to split in drying, and, therefore, preferably kept immersed; very durable under water. A valuable, but very slow-growing timber, making on an average only one inch of diameter in fourteen years. Largely used for the same purposes as the above-mentioned species, and for house and boat-building, and exported in large quantities to Liverpool and London for coach-building, wheels, piles, boat-building, etc., fetching £5 10s. per load.

Elm, Cork, Common or **English** (*U. campéstris* Sm.). *Germ.* "Korkulme, Rote Rüster." Height 80—90 ft.; diam. 2—3 ft. S.G. 542—909. W 34—56·7. E 445 tons per sq. in. *e'* ·56. *p* 14,400—13,489. *p'* ·49. *f* 3·5. *ft* 6·25. *c* 5,460. *c'* ·721. *fc* 4·6. *v'* ·757. *fs* ·62. Sapwood narrow, yellowish-white, as durable as the heart; heart dark-brown or brownish-red, heavy, hard, firm, elastic, very tough, very difficult to split, extremely durable if kept either dry or wet. The Rialto at Venice is said to be built on 12,000 elm piles. The wavy lines of pores in the summer-wood consist of single rows of pores and are interrupted: pith-rays hardly distinguishable: grain twisted. Though free from shakes, Elm timber is very liable to druxy knot, and, though not splitting, is liable to warp. On the Continent Elm is valued for gun-carriages and gun-stocks. In England it was used formerly for water-pipes and is now employed for coffins, butcher's-blocks, pulley-blocks, naves of wheels, pumps, ships' keels, coach-building, turnery, etc., though for most purposes inferior to Oak. (Fig. 51.)

Elm, Dutch or **Sand**, a large-leafed form, allied to the last-mentioned, grown only for ornament, its wood being subject to star-shake.

Elm, Indian (*U. integrifólia* Roxb.). *Hind.* "Papri." *Telugu* "Nalli." *Burm.* "Thalai." India, Ceylon, Burma. A large tree. Wood light yellowish-grey to red, moderately hard and strong. Used for door frames, cart-building, and carving.

Elm, Moose, Red or **Slippery** (*U. fúlva* Michx.). *French* "Orme gras." Southern Canada and North-eastern United States. Height 60—70 ft.; diam. 2 ft. S.G. 695. W 43·35. R 869 kilos. Brownish-red, heavy, hard, strong, compact, tough, more durable than other Elms; pores in spring wood forming a broad band of several rows, those in the summer wood in broken, slightly waved, narrow lines. It is more easily split than other

Elms, and is, therefore, much used for fence-rails, whilst its toughness and flexibility when steamed fit it for the ribs of boats.

Elm, Scotch, Wych, or **Mountain** (*U. montána* Sm.). Also known as "Chair-elm" and formerly as "Wych-hazel." *Germ.* "Bergrüster." Height 80—120 ft.; diam. up to 16 ft. In Britain most abundant north of the Trent. Wood lighter-coloured, softer, straighter-grained, and, therefore, more easily split than English Elm. Pores in the summer wood in complete bands. Used for the backs of Windsor chairs, shafts, and other purposes to which Ash is applied, and for boat-building.

Elm, Spanish. See **Cypre, Bois de.**

Elm, Spreading (*U. effúsa* Willd.). *Germ.* "Flatterrüster." Sapwood broad, yellowish; heart light-brown; vessels in spring wood in a single line, those in the summer wood in broad continuous wavy bands; less strong than the other species; but valued on the Continent on account of its markings for turnery, cabinet-work, gun-stocks, etc.

Elm, Winged, or **Cork-winged** (*U. aláta* Michx.). *Aborig.* "Wahoo." South-east United States. Small, heavier and more compact than *U. americána*, fine-grained; pores in spring wood small, in a single row, those in summer wood in broad, slightly wavy bands.

In Australia the name "Elm" is applied to *Duboísia myoporoídes* [See **Cork-wood**] and to *Aphanathé phillippinénsis* Planch. (Order *Urtíceæ*), known also as "Tulip-wood," and by the aborigines as "Mail," a tree introduced in the north-east, 80—90 ft. high and 1—1½ ft. in diam., yielding a light-coloured, close-grained wood, used for internal work in building.

Els, Klip or **Rock Ash** (*Rhus Thunbérgii* Hook.: Order *Anacardiáceæ*). Cape Colony. Hard, heavy, close-grained, and tough. Suitable for musical instruments or carving.

Els, Rood. See **Cedar, Red.**

Eng (*Dipterocárpus tuberculátus* Roxb.: Order *Dipterocarpáceæ*). Burma. *Burm.* "Eng." A large tree, 60 ft. in height and 3 ft. in diam. W 55. Reddish, hard. Used for house-posts, canoes, and planking. Other species, such as *D. grandiflóra* Wall. and *D. alátus* Roxb. [See **Gurjun**], are confused under the same name.

Engyin (*Shórea siaménsis* Miq. : Order *Dipterocarpáceæ*). Burma and Siam. Perhaps the same name as the preceding. A large tree, resembling the allied Sal. W 54—55. Heart very hard, very heavy and cross-grained : pith-rays finer than in Sal. Used in house-building, and for bows, etc.

Epel of Borneo may be "Ypil" of the Philippines, "Yepi, Apa" or "Epe" in Telugu, *Afzélia bíjuga* A. Gray [See **Shoondul**], *Bauhínia diphýlla* Buch. or *Hardwíckia bináta* Roxb. [See **Anjan**].

Essen-boom. See **Ash, Cape.**

Featherwood (*Polyósma Cunninghámii* J. J. Benn. : Order *Saxifragáceæ*). North-eastern Australia. "Hickory, Wineberry." *Aborig.* "Yaralla." Height 40—60 ft. ; diam. 1—3 ft. W 49·3. Yellow, close-grained, tough. Used for ladders, hand-spikes, etc.

Fiddle-wood (*Citharéxylum melanocárdium* Sw., *cinéreum* L., *surréctum* Griseb., and *quadrangularé* Jacq. : Order *Verbenáceæ*). West Indies. *French* "Bois fidèle, Bois de cotelet." Used for posts, shingles, etc.

Fig, the general name for the genus *Fícus*, few species of which yield timber of any value. That of *Fícus índica* L. (Order *Artocarpáceæ*) is used in Ceylon for common furniture ; but it, and that of other species, being soft and spongy, is readily charged with oil and emery for knifeboards or polishing purposes.

Fig, Leichhardt's Clustered (*Ficus glomeráta* Willd.). India, Burma, Northern Australia. *Aborig.* "Parpa." *Hind.* "Kith Gúlar." Height 40—60 ft.; diam. 1—3 ft. W 25—36. Greyish or straw-colour, coarse but straight-grained, light, soft, porous, moderately strong, not durable, except under water. Used for well-frames in India, and for furniture in Ceylon, and suggested for packing-cases.

Fig, Illawarra, Port Jackson or **Rusty** (*F. rubiginósa* Desf.). *Aborig.* "Dthaaman." Eastern Australia. Height 60—80 ft. ; diam. 4—5 ft. W 28·5. Light, soft, brittle, spongy. Sometimes used for packing-cases.

Fig, Large-leaved or **Moreton Bay** (*F. macrophýlla* Desf.). North-eastern Australia. Height 50—100 ft., diam. 3—6 ft. W 34. Pale-brown, with a beautiful wavy figure on a darker brown, but difficult to season, soft and not durable, so only occasionally used for packing-cases.

Fig, Prickly. See **Ash, Blueberry**.

Fir, a name very loosely used both in commerce and in botany, mostly for coniferous trees.[1] Thus Dantzic, Eliasberg, Memel, Norway, Red, Riga, Saldowitz, Scots, Stettin, Swedish and Yellow Fir are all *Pínus sylvéstris* [See **Pine, Northern**], named mainly from the port of shipment, Douglas or Oregon Fir is *Pseudotsúga Douglásii* [See **Pine, Oregon**], and White Fir is *Pícea excélsa* [See **Spruce, Norway**]. The name is preferably restricted to the genus *Ábies*, conifers distinguished by their flat leaves with two lateral resin-canals, and by their erect cones which fall to pieces when the seed is ripe. Their wood is generally without resin-ducts, coarse-grained, soft and perishable.

Fir, Balsam or **American Silver** (*Ábies balsámea* Miller). Wet ground in Eastern North America. Known also as "Balm-of-Gilead Fir." *French* "Sapin baumier." *Germ.* "Balsam-Tanne." Height 30—60 or 80 ft.; diam. 2 ft. S.G. 382. W 23·8. R 515 kilos. Yellowish, very light, soft, coarse-grained, not strong or durable. Sometimes used for staves for fish-barrels. The most valuable product of this species is Canada balsam, a resin collected in Quebec. The names "Balsam Fir" and "Black Balsam" are sometimes applied to *A. cóncolor* [See **Fir, White**].

Fir, Red (*A. nóbilis* Lindl.). Western United States. "Larch-fir." "Noble Fir." *Germ.* "Edel Weisstanne. Height 100—200 ft. or more; diam. 4—5 ft. or 9 ft. Light-brown, streaked with red, light, hard, strong, durable when seasoned. Used for internal work. The name is also applied to *A. magnífica* Murray, *Germ.* "Prächtige Weisstanne," a loftier species of the same region, with inferior timber, used for rough work or fuel.

[1] Among Anglo-Indians species of *Casuarína* are known as "Fir." [See **Oak, Swamp**].

Fir, Scots. See **Pine, Northern.**

Fir, Silver (*A. pectináta* DC.). Mountains of Central and Southern Europe. "Swiss Pine." *French* "Sapin des Vosges, Sapin de Lorraine." *Germ.* "Tanne, Edeltanne, Weisstanne, Silbertanne." Height 100—180 ft.; diam. 6—8 ft. Yellowish or pinkish-white without distinct heart, and with few or no resin-canals, with regular circular, well-defined rings, owing to the darker autumn wood, light, soft, porous, silky in lustre, strong, elastic, easily worked, not durable, taking glue well. Used by the ancient Romans for masts and ship-building (Virgil, *Georgics* ii., 68, Pliny. *Nat. Hist.* xvi.), and still so employed. One of the most sonorous of woods, and, therefore, imported into London as "Swiss Pine" for the sounding-boards of pianos and bellies of violins. Much used in toy-making, for carving, and for packing-cases, which are largely exported from Switzerland and the Tyrol. Used also, where it grows, for fence-posts, internal work, sluices, joists, planks, general carpentry, paper-pulp, and charcoal; but inferior to Spruce.

Fir, Colorado Silver (*A. cóncolor* Lindl. and Gordon). Western United States. "White Fir, Balsam Fir, Black Balsam." *Germ.* "Californische" or "Gleichfarbige Weisstanne." Height 100—250 ft.; diam. 4 ft. or more. Very light, soft, coarse-grained, neither strong nor durable. Used locally for butter-tubs, etc.

Fir, Great or **Tall Silver** (*A. grándis* Lindl.). North-western United States and British Columbia. "White Fir of Oregon." *Germ.* "Grosse Küstentanne." Height 250—300 ft.; diam. 3—5 ft. or more. Light, soft, easily worked, not strong. Used for indoor carpentry, packing-cases, cooperage, etc., and forming, with Oregon Pine, the chief lumber exported from the Pacific ports.

Fir, Indian Silver (*A. Webbiána* Lindl.). Himalayas. Height 120—150 ft.; diam. 3—5 ft. Whitish, scentless, non-resinous, open-grained, soft, easily worked, but not durable, if exposed. Used locally for shingles and building.

Fir, Western or **Lovely Silver** (*A. amábilis* Forbes). North-western United States and British Columbia. "White Fir."

Height 150—200 ft.; diam. 3—4 ft. Light, hard, but not strong.

Fire-tree (*Stenocárpus sinuátus* Endl.: Order *Proteáceæ*). North-eastern Australia. "Tulip-tree." *Aborig.* "Yiel-yiel." Height 60—70 ft.; diam. 2 ft. Nicely marked, close-grained, hard, susceptible of a good polish, durable. Used for staves and veneers.

Flindosa (*Flindérsia austrális* R. Br.: Order *Meliáceæ*). North-eastern Australia. "Crow's" or "Mountain Ash," or "Beech. *Aborig.* "Cudgerie." Height 80—100 ft.; diam. 2—4 ft. S.G. 936. W 44·8—77·8. E 960 tons. *f* 7·03. *fc* 4·54. *fs* ·62. Resembling Oak, with slight or no figure, very hard, close and strong, difficult to saw, but shrinking little in drying, very durable, not discoloured by iron. Used for staves and as a substitute for Beech, and suitable for railway construction.

Flintamentosa or **Wyagerie,** the product of a larger tree, reaching 150 ft. in height and 6 ft. in diam. in Northern New South Wales; may be the same or some other species of *Flindérsia.* It is used in house-building.

Fuchsia (*Fúchsia excorticáta* L. fil.: Order *Onagráceæ*). New Zealand. *Maori* "Kohutuhutu." Height 10—30 ft.; diam. sometimes 3 ft. Durable.

Fustic (*Chloróphora tinctória* Gaud., var. *Xanthóxylon*: Order *Moráceæ*). Tropical America. *French* "Bois d'orange." *Span.* "Fustete, Palo narango." *Port.* "Espinheiro branco, Amoreira de espinho." R 305 lbs. A large tree, yielding timber 20 ft. long and a foot or more wide. Yellow, light, durable. Used for spokes; but chiefly as a dye under the name of "Old Fustic," "Young" or "Zante Fustic" being the wood of *Rhus Cótinus.* [See **Sumach, Venetian.**]

Galaba or **Galba.** See **Santa Maria.**

Gangaw. See **Ironwood,** xviii.

Granadillo (*Ámyris balsamífera* L.: Order *Burseráceæ*). West Indies. Known also as "Rosewood," "Mountain Torchwood," "Lignum Rhodium," or "Funera." W 74—60. E 986—565 tons. *f* 6·7—4·7. *fc* 4—5·7. *fs* ·35—·43. Red, handsomely figured, aromatic, resinous, hard. Used for building and furniture, and exported. In Cuba the name is applied to *Brya Ébenus.* [See **Ebony, Green.**]

Grape, Sea-side (*Coccolóba uvífera* Jacq. : Order *Polygonáceæ*). Jamaica. W 65·34. E 637 tons. *f* 4·13. *fc* 2·52. *fs* ·428.

Greenheart (*Nectándra Rodiœ́i* Schomb. : Order *Lauráceæ*). North-eastern South America and the West Indies. *Aborig.* "Bibiru," "Sipiri." A large tree, yielding timber 24—70 ft. long and 1—2 ft. square. S.G. 1079—1210. W 58—76·5. E 1286 tons. *e′* ·97. *p′* 1·65. *f* 8·97. *c* 8820. *c′* 1·165. *fc* 5·17. *v′* 2·0. *fs* ·435. Dark-greenish or chestnut, often nearly black in the centre, fine, even, and straight-grained, the rings indistinguishable, very heavy, hard, tough, strong, elastic, and durable, the heart-wood being teredo-proof, though the similar sapwood is not so. This very valuable timber is liable to heart-shake. It is largely used for piles, bridge-building and the keelsons, beams and planks in ships, being classed in the second line in Lloyd's Register.

Grignon (*Búcida angustifólia* DC. : Order *Combretáceæ*). Guiana. Known in Surinam as "Wane." S.G. 714. Very large, straight-growing, pale red, rather less hard than Oak, even and straight in grain. Used for masts and for furniture.

Grignon fou (*Quálea cærúlea* Aubl. : Order *Vochysiáceæ*). Guiana. Known also as "Couaie." S.G. 800. Large, reddish, soft, straight-grained, very common, but of inferior quality to the preceding. Used for masts.

Gru-gru (*Astrocáryum* sp. and *Acrocómia sclerocárpa* Mart. : Order *Palmáceæ*). Trinidad. Height 20—30 ft.; diam. 1 ft. The outer part of the stem of these palms is hard, heavy, susceptible of a fine polish and durable. Used for walking-sticks.

Guarabu (*Terminália acumináta* Allem. : Order *Combretáceæ*, or *Peltogýné macrolóbium* Allem., or *P. confertiflóra* Benth. : Order *Leguminósæ*). Brazil. "Pao roxo." A large tree, yielding straight, dark-purple, fine-grained wood, with numerous pores filled with a hard white substance. Used in shipbuilding.

Guijo. See **Sal.**

Gum, a name referring mainly to the many and valuable

Australian and Tasmanian species of the Myrtaceous genus *Eucalýptus*, the identification and synonymy of many of which is much involved.

Gum, Apple-scented (*Eucalýptus Stuartiána* F. v. M.). Eastern Australasia. Frequently called "Turpentine" or "Peppermint tree"; in Tasmania "Red Gum"; in Victoria also "Mountain Ash" or "Apple-tree"; in New South Wales "Woolly Butt"; in Queensland "Box" or "Tea-tree." Introduced in the Punjaub. Height 40—90 ft.; diam. 2—4 ft. S.G. 834—1010. Light red-brown, wavy, hard, difficult to split, with gum-veins, weak, but said to be durable underground, polishing well. Used for ships' planks, sleepers, fence-posts, and rough furniture.

Gum, Bastard. See **Gum, Cider.**

Gum, Black, in North America (*Nýssa multiflóra* Wangenh. = *N. sylvática* Marsh: Order *Cornáceæ*). Known also as "Sour," "Yellow," or "Tupelo gum," or "Pepperidge." *Span.* "Tupelo." Chiefly in the Southern United States. Height up to 120 ft.; diam. 4 ft. S.G. 635. W 39·6. R 830 kilos. Sapwood light yellow; heart light brown, rather heavy, very strong and tough, cross-grained, hard to split, warping, not durable. Used for waggon-hubs, rollers, handles, sabots, turnery. *N. uniflóra* Wangenh. (= *N. tomentósa* Michx.), sometimes known as "Cotton Gum," is similar.

Gum, Black, in South-east Australia (*Eucalýptus stelluláta* Sieb.), known also as "White, Green," or "Lead Gum," "Sally," and "Box," from 12—50 ft. high and 1½—3 ft. in diam., but used only as fuel.

Gum, Blue (*E. salígna* Sm.). In New South Wales known also as "Flooded, Grey, White" or "Silky Gum," or "Grey Box." Height 40—120 ft.; diam. 2—7 ft. S.G. 1023. W 63·89. Warm red-brown, wavy, very heavy, close- and cross-grained, difficult to season, strong and durable, but liable to shakes. Excellent for sleepers, fencing, ships' planks, spars, and building.

Gum, Bastard Blue and **Scribbly Blue** are names of *E. leucóxylon*. See **Ironbark**.

Gum, Tasmanian or **Victorian Blue** (*E. glóbulus* Labill.). South-eastern Australasia. Introduced into India, South Africa, California and Southern Europe for sanitary purposes. Height 200—350 ft.; diam. 6—25 ft. S.G. 698—1108. W 52 5—71·87. *e'* 1·75. *p'* ·88. *c* 6048. *c'* ·798. *v'* ·915. Pale straw-colour, hard, heavy, moderately strong, tough, with curled and twisted grain, planing well, durable; pith-rays very numerous, fine; pores moderate-sized, round, grouped or in lines. Used for fence-rails, telegraph-poles, bridge-building, piles, and implements; extensively for carriage-buildings, and formerly for sleepers, for which *E. rostráta* is now preferred; and classed in the third line of Lloyd's Register for shipbuilding.

Gum, Brown. See **Mahogany, Swamp.**

Gum, Cabbage (*E. Sieberiána* F. v. M.). Known also as "Gum-top Stringybark, Ironbark" and "Blackbutt" in Tasmania, and as "Mountain Ash" in New South Wales. Height 40—150 ft.; diam. 1—5 ft. S.G. 896. W. 55·9. Buff, moderately heavy, rather coarse and cross-grained, very tough, easily worked, elastic, full of gum-veins, seasoning badly, generally soft, whence its name, and doubtfully durable. Used for posts, rails and fuel.

Gum, Cider (*E. Gúnnii* Hook. fil.). Tasmania and South-eastern Australia, being known in the latter district as "Sugar, White, Swamp, Yellow" or "Bastard Gum." Height 30 or exceptionally 250 ft. S.G. 802—1021. Light reddish-brown, hard, tough, with a few gum-veins, often crooked. Chiefly valuable for charcoal and for its sap.

Gum, Creek (*E. rostráta*). See **Gum, Red.**

Gum, Drooping (*E. vimindlis*) [See **Gum, Manna**] or (*E. pauciflóra*) [See **Gum, Mountain White**].

Gum, Flooded, a name applied to *E. Gúnnii* [See **Gum, Cider**], *E. pauciflóra* [See **Gum, Mountain White**], *E. rostráta* [See **Gum, Red**], *E. salígna* [See **Gum, Blue**], *E. tereticórnis* [See **Gum, Mountain**].

Gum, Giant (*E. amygdalína*). See **Ash, Mountain.**

Gum, Gimlet (*E. salúbris* F. v. M.). West Australia. Known also as "Fluted Gum." Height 120 or 150 ft. Tough, but easily worked. Used for shafts, implements, rough engraving and furniture.

Gum, Green (*E. stelluláta*). See **Gum, Black.**

Gum, Grey, a name applied to *E. crébra* [See **Ironbark, Grey**], *E. goniocályx* [See **Box, Bastard**], *E. largiflórens* [See **Gum, Slaty**], *E. punctáta* [See **Leatherjacket**], *E. resinífera* [See **Mahogany**], *E. salígna* [See **Gum, Blue**], *E. tereticórnis* [See **Gum, Mountain**] and *E. viminális* [See **Gum, Manna**].

Gum, Lead (*E. stelluláta*, Sieb.). See **Gum, Black.**

Gum, Manna (*E. viminális* Labill.). South-eastern Australasia. Known as "White" or "Swamp Gum" in Tasmania, as "Grey, Blue, Ribbony, Drooping" or "Weeping Gum," or as "Woolly Butt" in New South Wales, and as "Box" or "Peppermint Gum" in Victoria. Height up to 320 ft.; diam. up to 17 ft. S.G. 685—1003. W. 67·5. Buff to dull brick-colour or warm brown, moderately heavy, straight coarse-grained, full of gum-veins, weak, easily worked, requiring careful seasoning, only durable underground. Used for palings, shingles and rough building material. The name is also applied to *E. amygdalína.* [See **Ash, Mountain.**]

Gum, Mountain (*E. tereticórnis* Sm.), Eastern Australia. Known also as "Red, Flooded, Grey, Blue" or "Slaty Gum" and "Bastard Box." *Aborig.* "Mungurra." Height 40—150 ft.; diam. 1½—4 or 6 ft. S.G. 843. W. 52·5. Red-brown, resembling Cedar, with cross, curly grain, lustrous, heavy, very hard, tough, with some gum-veins, easy to dress, but difficult to season, very durable. Largely used for fencing, naves, felloes, sleepers, telegraph-poles, fuel, etc.

Gum, Mountain White (*E. pauciflóra* Sieb.). South-eastern Australasia. Known in Tasmania as "Weeping Gum," and in Australia as "White, Swamp, Drooping" or "Flooded Gum," "Peppermint" or "Mountain Ash." Height 100 ft.; diam. 2—4 ft. White or buff, soft, straight but short-grained, full of gum-veins. Used for fencing, and excellent for fuel.

Gum Nankeen (*E. populifólia* Hook.). North-east Australia. Also known as "White Gum, White, Red, Poplar," or "Bembil

Box." Height 50—60 ft.; diam. 2 ft. Grey or light brown, hard, heavy, close-grained, very tough, strong, hard to work, but susceptible of a fine polish, liable to gum-veins and often unsound, durable. Used for sleepers, posts, and building.

Gum, Peppermint. See **Gum, Manna.**

Gum, Red, a name applied in Australia to *Angóphora lanceoláta* [See **Apple-tree**], *Eucalýptus amygdalína* [See **Ash, Mountain**], *E. Gúnnii* [See **Gum, Cider**], *E. melliodóra* [See **Box, Yellow**], *E. punctáta* [See **Leatherjacket**], *E. resinífera* [See **Mahogany**], *E. tereticórnis* [See **Gum, Mountain**], and especially to *E. rostráta* and *E. calophýlla*. *E. rostráta* Schlecht. Eastern Australia. Known also as "Creek, River, Forest, Flooded, Blue, White Gum," or "Yellow-jacket." *Aborig.* "Yarrah," not to be confounded with Jarrah, though little inferior to it. Height 30—80 or 100 ft.; diam. 1—6 or 8 ft. S.G. 790—1045. W 53·5—62·5. Dark red, with a pretty curly figure, moderately heavy, exceedingly hard when dry, and therefore most difficult to work, liable to twists and shakes in seasoning, but can take a fine polish, very durable, termite- and teredo-proof. Highly valued for ships' beams, sleepers, piles, bridges, posts, building, fencing, and charcoal; but, owing to its hardness, only slightly for furniture. This is the chief wood used for paving in Melbourne, costing about £9 per 1000 blocks, or 14s. per 100 feet super. *E. calophýlla* R. Br. of South-west Australia, reaching a height of 150 ft., yields a tough but not durable wood, used for wheels, handles, and building.

Gum, Red, in Tasmania (*E. Stuartiána*). See **Gum, Apple-scented.**

Gum, Red, in the United States, a trade name for *Liquidámbar styracíflua*. See **Gum, Sweet.**

Gum, Rusty. See **Apple-tree.**

Gum, Scribbly (*Eucalýptus hæmástoma* Sm.). Queensland and New South Wales. Known also as "Spotted, White" or "Blue Gum," "Black-butt," "Mountain Ash," etc. Height, 60—120 ft.; diam. 2—3 ft. S.G. 1101. W. 68·75. Grey or reddish, wavy or stripy, often crooked, close, smooth, short-grained, brittle, easily-worked, not durable. Used for fuel and rough carpentry.

Gum, Slaty (*E. largiflórens* F. v. M.). Eastern Australia. Known also as "Cooburn, Black, Yellow, Bastard, or Grey Box," or as "Ironbark." Height 100—120 ft.; diam. 2—3 ft. Red, hard, tough, durable, especially underground. Used for fencing, sleepers, building, cogs, etc. The name is also applied to *E. tereticórnis.* [See **Gum, Mountain.**]

Gum, Sour. See **Gum, Black.**

Gum, Spotted, a name applied to *Eucalýptus capitelláta* [See **Stringybark, White**], *E. goniocályx* [See **Box, Bastard**], *E. hæmástoma* [See **Gum, Scribbly**], and *E. maculáta* Hook. This last mentioned species, native to Eastern Australia, reaches 100—150 ft. in height, and 3—8 ft. in diam. S.G. 1035—1170. Dark yellow to walnut-brown, sometimes with a wavy figure, heavy, close but very coarse in grain, with large gum-veins, strong, tough, durable. In great demand for paving, girders, bridge and ship building, shafts, naves, shingles, etc. It fetches 2s. a cubic foot in London.

Gum, Sugar (*E. corynocályx* F. v. M.). South Australia. Height 120 ft.; diam. 5—6 ft. Yellowish-white, very heavy, hard, strong and durable, termite-proof, not warping. Used for sleepers, piles, planks, fencing, wheels. The name is also applied to *E. Gúnnii.* [See **Gum, Cider.**]

Gum, Swamp. See **Gum, Cider, Manna,** and **Mountain White.**

Gum, Sweet (*Liquidámbar styracíflua* L.): Order *Hamameláceæ*). Eastern United States. "Bilsted" or "Red Gum," "Californian Red Gum" (though shipped from New Orleans). "Satin Walnut." *French* "Copalm," *Germ.* "Storaxbaum," *Span.* "Liquid-ambar." Height 100 ft. or more; diam. 4—5 ft. S.G. 591. W 36·8. R 651 kilos. Sapwood cream-white; heart irregular, reddish-brown, with dark false rings, rather heavy, close-grained, soft, tough, taking a satiny polish, warping and twisting badly in drying, unless first steamed. Used for furniture, veneers, turnery, shingles, and clapboards, and, though little suited for the purpose, for paving.

Gum, Water (*Tristánia neriifólia* R. Br.: Order *Myrtáceæ*). New South Wales. Height 80—100 ft.; diam. 1½—2 ft. W 66·5. Very close-grained and elastic, apt to split in drying. Used for handles, mallets, cogs.

Gum, Broad-leaved Water (*T. suavéolens* Sm.). Eastern Australia. Known also as "Swamp Mahogany" and "Bastard Peppermint." Height 50—60 ft.; diam. 1—1½ ft. Red, resembling Spanish Mahogany, close-grained, strong, elastic, tough, durable, but apt to split in drying, termite-proof. Extensively used for piles, sleepers, posts, handles, cogs, coach-frames, etc.

Gum, White, a name applied to *Eucalýptus amygdalína* [See **Ash, Mountain**], *E. gomphocéphala* [See **Tewart**], *E. goniocályx* [See **Box, Bastard**], *E. Gúnnii* [See **Gum, Cider**], *E. hæmástoma* [See **Gum, Scribbly**], *E. hemiphlóia* [See **Canarywood**], *E. leucóxylon* [See **Ironbark**], *E. pauciflóra* [See **Gum, Mountain White**], *E. populifólia* [See **Gum, Nankeen**], *E. redúnca* [See **Wandoo**], *E. resinífera* [See **Mahogany**], *E. rostráta* [See **Gum, Red**], *E. saligna* [See **Gum, Blue**], *E. stelluláta* [See **Gum, Black**], *E. Stuartiána* [See **Gum, Apple-scented**], and *E. viminális* [See **Gum, Manna**].

Gum, Yellow, a name applied in Australia to *E. Gúnnii* [See **Gum, Cider**], *E. punctáta* [See **Leatherjacket**], or *E. melliodóra* [See **Box, Yellow**], and in North America to *Nýssa multiflóra* [See **Gum, Black**].

Gum, York (*E. fœcúnda* Schauer = *E. loxophléba* Benth). Western Australia. Small, light pink, hard, heavy, close-grained, elastic. Used by the aborigines for spears; suitable for spokes.

Gurjun (*Dipterocárpus turbinátus* Gaertn. fil. or *D. alátus* Roxb.: Order *Dipterocarpáceæ*). Andaman Islands and Further India. Also known as "Wood-oil tree." *Burm.* "Ka-nyin" and sometimes "Éng." Height up to 250 ft.; diam. 8 ft. W 38. Light brown to red, dense, hard, but will not stand moisture. Excellent for house-building, posts, or planks.

Guru-kina (*Calophýllum tomentósum* Wight: Order *Guttíferæ*). Ceylon. W 62·6. E 657 tons. *f* 3·90. *fc* 2·41. *fs* ·423. Used for tea-chests and in building.

Hackberry (*Céltis occidentális* L.: Order *Ulmáceæ*). Eastern North America. "Sugarberry, Nettle-tree, False Elm." *French* "Micocoulier occidentale," *Germ.* "Abendlandischer Zürgelbaum," *Span.* "Almez Americano." Height 100 ft.; diam. 4—5 ft. S.G. 729. W 45·4. R 789 kilos. Sapwood yellowish or greenish; heart brown, often dark, rather heavy, not hard or strong, tough, fine-grained, working well, shrinking moderately, and taking a good satiny polish. Sometimes used for fencing, furniture, or wheelwrights' work, as a substitute for Elm; but scarce.

Hackmatack. See **Tamarack.**

Haldu (*Adína cordifólia* Hook. fil. and Thom. = *Naúclea cordifólia* Roxb.: Order *Rubiáceæ*). India, Ceylon, Burma. *Hind.* "Haldu." *Tamil* "Manja Kadamba." *Burm.* "H'nau." Height 75—80 ft.; diam. 2—3 ft. Pretty, yellow, light, close-grained, resembling Box, but soft, easily worked, not strong, much affected by weather, but durable if kept dry, cracking and warping, but polishing well. Used for masts, interior work in house-building, turnery, modelling, combs, etc.

Hannoki (*Álnus marítima* Nutt.: Order *Betuláceæ*). Japan. Height 40 ft.; diam. 1 ft. Used for gunpowder-charcoal.

Harra (*Terminália Chebúla* Retz: Order *Combretáceæ*). India, Ceylon, Burma. "Pilla murda wood." *Hind.* "Harra." *Sansk.* "Haritaka." *Tamil* "Pilla marda." Height 45—80 ft.; diam. 1½—4 ft. S.G. 682. W 42—53. Brownish-grey, with a greenish, yellowish, or reddish tinge, darkening externally on seasoning, hard, strong, smooth and close-grained, taking a good polish and seasoning well, but subject to dry rot and to the attacks of termites and carpenter-bees. Used for beams and other house-building purposes, agricultural implements, etc. Its fruits, Chebulic or Black Myrobalans, are largely exported for dyeing.

Hawthorn (*Cratægus Oxyacántha* L.: Order *Rosaceæ*). "Whitethorn," "May." Europe, North and West Asia, and North Africa. Height 10—20 ft.; diam. seldom large. Flesh-coloured, with numerous pith-flecks, hard, heavy, difficult to split, without lustre; vessels small, those in spring-wood not numerous; pith-rays indistinguishable. Used in turnery and for walking-sticks, and the best substitute for Box for engraving yet discovered; but slow-growing and seldom procurable of any size.

Hazel (*Córylus Avellána* L.: Order *Cupulíferæ*). Europe, North Africa, Temperate Asia. Height rarely 30 ft.; diam. generally less than 1 ft. Reddish-white, resembling Beech, without heart, soft, highly elastic, easily split, not durable; annual rings almost circular; pith-rays wide and narrow; vessels small, in radial lines. Used for barrel-hoops and walking-sticks.

Hemlock Spruce (*Tsúga canadénsis* Carr.: Order *Coníferæ*). Eastern North America. "Hemlock," "Hemlock Fir." *French* "Peruche," *Germ.* "Schierling Tanne." Height 80—100 ft.; diam. 3 ft. S.G. 244. W 26·4. R 736 kilos. Light reddish-grey or brown, with lighter sapwood; free from resin-ducts, light, soft, stiff, but brittle, usually coarse-grained, splintery, not very easily worked, shrinking and warping considerably in seasoning, retaining nails firmly, but wearing rough and not very durable.. Used for sleepers, laths, rafters, planks, fencing, etc. Its bark is valuable for tanning.

Hemlock, Western (*Tsúga Mertensiána* Carr.). Western North America. "Prince Albert's Fir." Height 180 ft. or more; diam. 9 ft. or more. Heavier and harder than the eastern form, but not strong. Used for rough lumber; but chiefly valued for its bark.

Hiba (*Thujópsis dolabráta* Sieb. and Zucc.: Order *Cupressíneæ*). Japan. *Japanese* "Hiba, Asu-Naro, Thuia." *French* "Thuia de Japon." *Germ.* "Hiba-Lebensbaum, Beilblätriger Lebensbaum." Height 7—40 ft.; diam. small. Yellowish-white, durable. Used in house, bridge, and boat-building.

Hickory, originally the name of the North American genus *Hicória* (Order *Juglandáceæ*), closely allied to the Walnuts. "So close an analogy exists in the wood of these trees that, when stripped of their bark, no difference is discernible in the grain, which is coarse and open in all, nor in the colour of the heartwood, which is uniformly reddish" (Michaux). In all, the sapwood is broad and white, the heart a reddish nut-brown, very heavy, hard, strong, proverbially tough, elastic, coarse, smooth and straight-grained; the pith large, pith-rays numerous but hardly discernible; with numerous fine peripheral lines of wood-parenchyma; the pore-circle narrow but with large pores, whilst the vessels in the autumn wood are small and scattered. The wood seasons slowly, shrinking and warping considerably; unlike Walnut, is very subject to the attacks of boring insects; and is not durable if exposed. It is, consequently, never used in house or ship building; but is specially valued for carriage-building, axles, the handles of implements, screws such as those of bookbinders' presses, bows, chair-making, coach-whips, gunstocks, hoops, fuel and charcoal. It is harder, heavier and tougher than Ash. In Australia the name has been applied to many species of *Acacia* and other genera, to which reference will be made after the description of the true Hickories.

Hickory Black, a name applied equally to *H. álba* [See **Hickory, Mocker-nut**] and to *H. glábra* [See **Hickory Pig-nut**].

Hickory Big-bud. See **Hickory, Mocker-nut.**

Hickory, Bitternut (*H. mínima* Britton = *Cárya amára* Nutt.). Eastern United States. "Swamp Hickory." *French* "Noyer amer." Height 70—80 or 100 ft.; diam. 2—3 ft. S.G. 755. W 47. R 1101 kilos. Less valuable than Shell-bark Hickory; but used for ox-yokes, hoops and fuel.

Hickory, Brown. See **Hickory, Pig-nut.**

Hickory, Mocker-nut (*H. álba* Britton = *Cárya tomentósa* Nutt.). Chiefly in the Southern United States. "Black, Big-bud" or "White-heart Hickory." Height 90 ft. or more; diam. 3 ft. or more. Heart-wood in young trees white, slow-growing, durable, except when attacked by borers.

Hickory, Pecan (*H. Pecán* Britton = *Cárya olivæfórmis* Nutt.). South-central United States. "Illinois Nut." Height 75 ft. or more; diam. 2 ft. or more. Brittle, not strong. Inferior to Shell-bark Hickory, and chiefly used for fuel.

Hickory, Pig-nut (*H. glábra* Britton = *Cárya glábra* and C. *porcína* Nutt.). Eastern North America. "Brown Hickory." *French* "Noyer de cochon." *Germ.* "Ferkelnusz." Height

80—100 ft.; diam. 3—4 ft. S.G. 822. W 51. R 1046 kilos. Perhaps the best of all the hickories for axletrees and axe-handles.

Hickory, Shell-bark (*H. ováta* Britton = *Cárya alba* Nutt.). Eastern United States. "Shag-bark Hickory," "White Hickory." *French* "Noyer tendre." Height 100—110 ft.; diam. 1½—3 ft. S.G. 837. W 52. R 1200 kilos. Deriving its French-Canadian name from its elasticity, this is the species most exported, especially for axe and hammer-handles, spokes, etc. It is also the best fuel.

Hickory, Big or **Thick Shell-bark** (*H. laciniósa* Sarg. = *Cárya sulcáta* Nutt.). Central United States. Height 70—100 ft.; diam. 3—4 ft. S.G. 810. W 50·5. R 1083 kilos. Similar in character and uses to the last mentioned.

Hickory, Water (*H. aquática* Britton = *Cárya aquática* Nutt.). South-eastern United States. The lightest, weakest and most useless species.

Hickory, in Australia, is applied to *Acácia binerváta* [See **Wattle, Black**], *A. doratóxylon* [See **Spearwood**], *A. falcáta* [See **Myall, Bastard**], *A. melanóxylon* [See **Blackwood**], *Eucalýptus punctáta* [See **Leatherjacket**], *E. resinífera* [See **Mahogany**], and *Polyósma Cunninghámii* [See **Feather-wood**].

Hinau (*Elæocárpus dentátus* Vahl.: Order *Tiliáceæ*). New Zealand. Small, light dull brown, very tough, strong and durable. Used for sleepers, fencing, etc.

Hinoki (*Cupréssus obtúsa* Koch: Order *Cupressíneæ*). Japan. "Japanese Cypress." *Germ.* "Feuercypresse, Sonnencypresse." Height 70—100 ft.; diam. 2½—3 ft. Sapwood yellowish-white; heart rose-red, fragrant, strong, fine-grained, taking a high polish. One of the best of Japanese timbers, held sacred by the followers of the Shinto faith, whose temples are built of it, as also are the palaces of the Mikado. It is also the best for lacquering.

Holly (*Ilex Aquifólium* L.: Order *Ilicíneæ*). Central Europe and West Asia. *French* "Houx." *Germ.* "Stechbaum, Hulse, Christdorn." Height 10—40 or 80 ft.; diam. 1—4 or 5 ft. W 47·5. White to greenish white, fine-grained, with fine but distinct rings and pith-rays, vessels scarcely visible, approaching ivory in colour and texture more than any other wood, hard, heavy, susceptible of a high polish, but shrinking and warping very much. Used, in the round, for engraving, especially in calico-printing; for staining as imitation Ebony, as in the wooden handles of metal tea-pots; in veneers, especially for white or stained strings in inlaying, as in Tunbridge ware; and, when small, for walking-sticks.

Holly, American (*Ilex opáca* Ait.). Eastern United States. Height 50 ft.; diam. 3—4 ft. S.G. 582. W 36. R 686 kilos., Similar to the European species and similarly used.

Holly, Smooth (*Hedycárya angustifólia* A. Cunn.: Order *Monimiáceæ*). Eastern Australia. "Native Mulberry." Small, very light, close-grained, and tough. Used by the aborigines for fire-sticks and spears, and fit for cabinet-work.

Honey, Locust. See **Locust, Honey.**

Honeysuckle, a general name in Australasia for species of *Bánksia* (Order *Proteáceæ*), especially *B. marginúta* and *B. serráta.*

B. margináta Cav. (=*B. austrális* R. Br.). South-eastern Australasia. Height 10—20 or 40 ft. S.G. 598—610. W 38. When fresh cut resembling raw beef, with reddish-white sapwood, light, soft, porous, twisting and warping; but, when thoroughly seasoned, hard, susceptible of a fine polish, and beautifully figured. Used for cabinet-work.

B. serráta Linn. fil. Eastern Australasia. S.G. 803. W 39—50. Dark red, mahogany-like, handsome, finely figured, coarse and open-grained, strong, requiring careful seasoning, much bored by beetles. Used for window-frames and boats'-knees and might be used for furniture.

Honeysuckle, Coast (*Bánksia integrifólia* L.). Eastern Australia. Known also as "Beefwood." Height 20—30 ft.; diam. 1 ft. S.G. 799. W 50—39. Pink, beautifully grained, moderately dense, tough, durable when not exposed. Used for boats'-knees, etc.

Honeysuck, Silvery. See **Beefwood.**

Honeysuckle in New Zealand (*Knightia excélsa*). See **Rewa-rewa**.

Honeysuckle-wood in the United States (*Plátanus occidentális*). See **Plane**.

Honoki (*Magnólia hypoléuca* S. and Z.: Order *Magnoliáceæ*) Japan. Height 50 ft.; diam. $2\frac{1}{2}$ ft. A dense, hard, ornamental wood, used for tables, wooden shoes, pencils, and for charcoal.

Hornbeam (*Carpínus Bétulus* L.: Order *Coryláceæ*). Central Europe and West Asia. *French* "Charme." *Germ.* "Weissbuche," "Hainbuche." Height 40—50 sometimes 70 ft.; diam. 1, rarely reaching $3\frac{1}{2}$ ft. S.G. 1038—759. W 47·5. *c* 6405. *c'* ·846. *v'* 1·087. Yellowish-white, close-grained, heavy, hard, very tough, strong, difficult to split, somewhat lustrous, and very durable if kept dry: pores minute, in radial lines: broad pith-rays lighter than the rest of the wood: annual rings very sinuous, bending outward between the broad pith-rays. Used for handles, mallets, lasts, etc., unequalled for cogs and bearers for printers' rollers, and excellent for fuel, and imported from France at 9d. or 10d. per cubic foot.

Hornbeam, American (*Carpínus Caroliniána* Walt.). Eastern North America. Known also as "Blue" or "Water Beech" and "Ironwood." Height sometimes 50 ft.; diam. 1—2 ft. S.G. 728. W 45·4. R 1149 kilos. Used to a small extent for handles, mallets, levers, and hoops; but apparently slightly inferior to the European species.

Hornbeam, Hop (*Ostrýa virgínica* Willd.: Order *Coryláceæ*). Eastern North America. "Iron-wood, Lever-wood." *French* "Bois dur." *Germ.* "Amerikanische Hopfenbuche." Height 50 ft.; diam. 2 ft. S.G. 828. W 51·6. R 1134 kilos. Sapwood whitish; heart dull brownish, heavy, hard, very strong and tough, durable. Used for fence-posts, handles, levers, etc., but scarce: excellent fuel.

Horse-chestnut (*Æsculus Hippocástanum* L.: Order *Sapindáceæ*). Supposed to be a native of Asia; but largely grown for shade throughout Europe and the United States. *French* "Marronier d'Inde." *Germ.* "Roszkastanie." Height sometimes 80 ft.; diam. 3—4 ft. W 60—35·5. White, or slightly yellowish or reddish, soft, close-grained, warping little, not durable, being deficient in tannin and resin; annual rings wide, circular; pith-rays narrow, numerous, indistinct; vessels small, numerous, uniformly distributed, 1—7 together; pith large, round. Wood similar in character to Willow and Poplar. Used for flooring, cart-linings, barrows, packing-cases, blind-wood in cabinet-making, moulds for castings; and, in France, for sabots.

Huon Pine (*Dacrýdium Franklínii* Hook. fil.: Order *Coníferæ*). Tasmania. Also known as "Macquarie Pine." Height 60—80 or 100 ft.; diam. 3—5 ft. Light yellow, very beautifully marked with dark wavy lines and small knots, light, close-grained, tough, easily worked, susceptible of a good polish, durable, noxious to insects. Used for boat-building, carving, and bedroom furniture, and burns briskly with an aromatic fragrance; but is now quite scarce.

Illupi. See **Mahwa.**

Ipil. See **Epel.**

Iroko (*Chloróphora excélsa* Benth. and Hook. fil.: Order *Moráceæ*). Tropical West Africa. "Iroko" in Yorubu land, "Odum" in Guinea, "Mbundu" in Uluguru, "Muamba-Camba" in Angola. Yellowish or brownish, with dark zones, resembling Satin-wood or wavy Maple, handsome, very strong, and durable, termite-proof. Perhaps constituting part of the "African Mahogany" of commerce, valuable for building or cabinet-work.

Iron-bark, a name applied to various species of *Eucalýptus* (Order *Myrtáceæ*). In Tasmania, E. *Sieberiána* [See **Gum, Cabbage**]. In Australia, *E. largiflórens* [See **Gum, Slaty**], *E. macrorrhýncha* [See **Stringybark**], and especially *E. leucóxylon* and *E. siderophlóia*. *E. leucóxylon* F. v. M. South-eastern Australia. Known also as "Black" or "Red Ironbark, Black Mountain Ash, White, Bastard," or "Scribbly Blue Gum." Height up to 200 ft.; diam. 2—5 ft. S.G. 1173—908. W 73·26—63·5. Light brown, yellowish or pale pinkish-white, close- and straight-grained, hard, very strong, tough, and durable, both in water and in the ground, slightly greasy, which renders it suitable for cogs. It is also used for naves and felloes, sleepers, piles, planks, telegraph-poles, fence-posts, axe-handles, beams, rafters, treenails, and screws. An allied form, *E. sideróxylon* A. Cunn. in New South Wales, has darker and heavier wood, similarly employed, and furnishing one of the best fuels in the country. *E. siderophlóia* Benth. New South Wales and South Queensland. Known also as "Red" or "Broad-leaved Ironbark." *Aborig.* "Tanderoo." Height 70—100 ft.; diam. 1½—4 ft. S.G. 1171—936. W 71·5—64. *e′* 2·16. *p′* 1·74. *c* 8377. *c′* 1·106. *v′* 1·348. Deep red, very hard, heavy, strong, rigid, and difficult to work, plain and straight-grained, liable to heart and star shake; pores very minute, filled with a hard, white, brittle secretion. Used for beams, keelsons in shipbuilding, and in engineering works, being one of the strongest and most durable of Australian timbers.

Ironbark, Grey (*E. crébra* F. v. M.). Eastern Australia. Known also as "White, Red" or "Narrow-leaved Ironbark," or "Grey Gum." Height 70—90 ft.; diam. 1½—3 ft. S.G. 1119—1211. Dark purplish or brown, hard, very heavy, tough, cross-grained, hard to work, durable. Used for sleepers, piles, fence-posts, spokes, etc.

Ironbark, White (*E. paniculáta* Smith.). Eastern Australia. Known also as "Red, Pale" or "She Ironbark," and as "Blood-wood." Height 70—150 ft.; diam. 3—4 feet. Brown, heavy, very hard, tough, strong, seasoning and working well, durable. Perhaps the most valuable Ironbark: much used for sleepers and other railway work, fence-posts, beams, etc.

Ironwood, a name applied to many widely different timbers in various countries; but to a greater variety in Australia than elsewhere.

(i) *Acácia excelsa* Benth. (Order *Leguminósæ*). Queensland. Height 70—80 ft.; diam. 2—3 ft. Violet-scented, ornamental, hard, close-grained, tough, elastic. A cabinet wood.

(ii) *A. stenophýlla* A. Cunn. Eastern Australia. Known also as "Dalby Myall." Height 40—60 ft.; diam. 1—2 ft. Dark, beautifully marked, very hard, heavy, close-grained, taking a fine polish.

(iii) *Casuarína equisetifólia* Forst. (Order *Casuaríneæ*). See **Oak, Swamp**.

(iv) *Melaleúca genistifólia* Sm. (Order *Myrtáceæ*). North-eastern Australia. Known also as "Ridge Myrtle." Height 30—40 ft.; diam. 1½—2 ft. Greyish, close-grained, hard, durable.

(v) *Mýrtus gonocláda* F. v. M. (Order *Myrtáceæ*). North Queensland. Very hard, and suggested for engraving.

(vi) *Notelǽa ligustrína* Vent. (Order *Oleáceæ*). South-eastern Australasia. Known also as "Heartwood" in Tasmania, and as "Spurious Olive" or "White Plum" in Victoria. Height 30 ft.; diam. 1 ft. S.G. 925. Irregularly figured, like Olive-wood, exceedingly hard and close-grained. Used for mallets, blocks, and turnery.

(vii) *Ólea paniculáta*. See **Marblewood**.

(viii) *Tarriétia argyrodéndron*. See **Silver-tree**.

In Borneo and the Straits Settlements (ix) *Eusideróxylon Zwagérii* is so called. See **Billian**.

In Burma (x) *Xýlia dolabrifórmis* goes by this name. See **Acle**.

In Cape Colony (xi-xv) *Ólea laurifólia* Lam. (Order *Oleáceæ*), 40—70 ft. high and 2—3 ft. in diam., *O. unduláta* Jacq., *O. capénsis* L., *O. exasperáta* Jacq., and *O. verrucósa* Link., "Olyvenhout" or "Umguma," all very similar and nearly equal to Lignum-vitæ, are known as "Black Ironwood"; whilst (xvi) *Toddália lanceoláta* Lam. (Order *Xanthoxyláceæ*) is known as "White Ironwood" [See **Umzimbit**], and (xvii) *Sideróxylon inérmé* L. (Order *Sapotáceæ*), which occurs along the east coast of Africa from the Cape to Zanzibar, a very heavy, hard, close-grained, durable, greyish-yellow wood, with brownish-red markings, used in ship and bridge-building, and for telegraph-poles, is known as "White Ironwood of Mauritius," and also as "Soft" or "White Milk-wood" in South Africa.

In Ceylon and India the name Ironwood is applied to (xviii) *Mésua férrea* L. (Order *Guttíferæ*), also known as "Indian Rose-chestnut." *Hind.* "Nagesar." *Andaman* "Gangaw." Height 20 ft. or more; diam. 1—2 ft. W 69—72. Dark red, extremely hard and difficult to work, taking a high polish. Used for gun-stocks, handles, wood-paving, and building. "Black Ironwood" is here (xix) *Condália férrea* Griseb. (Order *Rhamnáceæ*). S.G. 1300. W 85.

The Ironwood of China and Japan, used for rudders and anchors, is believed to be (xx) *Metrosidéros véra* Rumph. (Order *Myrtáceæ*), occurring in the Malay archipelago, and known in Amboyna as "Nani": that of Guiana and Honduras is (xxi) *Laplácea Hæmatóxylon* Camb. (Order *Camelliáceæ*), also known as "Blood-wood" and used for cogs; whilst in Jamaica (xxii) *Sloánea jamaicénsis* Hook. (Order *Tiliáceæ*), known also "Break-axe," and (xxiii) *Erythróxylon areolátum* L., also known as "Red-wood," are also so named.

Ironwood, Morocco (xxiv). See **Argan**.

In Natal, besides the White Ironwood [See **Umzimbit**], there is (xxv) *Olea laurifólia* Lam., known as "Black Ironwood," *Zulu* "Tamboti." W 64·68. E 896 tons. *f* 7·84. *fc* 4·79.

In New Zealand the name is applied to (xxvi—xxviii) *Metrosidéros robústa* and *M. lúcida* Menz. [See **Rata**] and to *M. tomentósa* A. Cunn., *Maori* "Pohutakawa." This last yields timber 10—20 ft. long, and 9—16 in. square, with S.G. 1200—858, dark red or walnut-brown, very heavy, hard, close-grained, strong and durable, suitable for ship-building.

In the United States (xxix) *Carpínus caroliniána* is sometimes called Ironwood [See **Hornbeam, American**]; but in New Mexico the name is applied to (xxx) *Olneya Tesota*, A. Gray (Order *Leguminósæ*). In Persia (xxxi) *Parrótia pérsica* (Order *Hamamelidáceæ*); and in the island of Réunion (xxxii) *Stadmánnia sideróxylon* DC. (Order *Sapindáceæ*) is so called.

Jacaranda, the Brazilian name for various species of *Dalbérgia* and *Machærium* (Order *Leguminósæ*), known in English commerce as Rosewood, including **Jacaranda cabiuna** (*Dalbérgia nígra* Allem.), **Jacaranda roxa** (*Machærium fírmum* Benth.), and **Jacaranda preto** (*M. legálé* Benth.). See **Rosewood**.

Jack (*Artocárpus integrifólia* L.: Order *Moráceæ*). India. *Beng.* "Kanthal." *Sinh.* "Kos." *Brazil.* "Jaqueira." Sometimes known as "Orangewood." Height 80—100 ft.; diam. 2—5 ft.

S.G. 554—676. W 35·6—42. Yellow or orange, darkening on exposure to a dull red or mahogany colour, somewhat coarse and crooked in grain, moderately hard, requiring thorough seasoning to check warping, taking a good polish; but brittle when dry and not tolerant of alternations of dryness and damp. Used as a yellow dye, for boat-building, furniture, musical instruments, grain-measures, and in England for cabinet-work, marquetry, turning, and the backs of brushes.

Jack, Jungle. See **Angelly**.

Jack, Long (*Flindérsia Oxleyána* F. v. M.: Order *Meliáceæ*). North-eastern Australia. Known also as "Light Yellow Wood." Height 80—100 ft.; diam. 2—3½ ft. Yellow, often pretty, fine-grained, strong, durable, almost termite-proof. Used in boat-building and cabinet-work as a substitute for Cedar and often coming to market as "Beech," *i.e.* *Gmelína Leichhárdtii* [See **Beech**]; but not so valuable.

Jambolana (*Syzygium Jambolana* DC.: Order *Myrtáceæ*). India, Ceylon, Mauritius, Australia. *Port.* "Jambu." *Hind.* "Jamoon." *Beng.* "Jam." *Mahr.* "Jambool." *Tam.* "Nagal." *Austral. aborig.* "Durobbi." Height 80—100 ft.; diam. 2—3 ft. W 49. Flesh-colour or red, hard, firm, close-grained, durable, shrinking little in drying, resisting the action of water or termites. Used for sleepers, building, carts, well-work, and agricultural implements.

Jarrah (*Eucalýptus marginàta* Sm.: Order *Myrtáceæ*). South-western Australia. Sometimes known as "Mahogany" or "Bastard Mahogany." Height 100—150 ft.; diam. sometimes 10 ft. S.G. 837—1120. W 54—76. E 620 tons. *e'* ·66. *p'* ·85, *f* 413. *ft* 7·20. *fc* 3·04. *c* 2940. *c'* ·388. *v'* ·937. Straight-grown, and, even when unsound in the centre, yielding timber 20—40 ft. long and 1—2 ft. square, red, mahogany-like in colour, sometimes exhibiting a ray of light across the grain and a beautiful mottling, and sometimes curled in grain, very heavy, hard, close-grained, working smoothly, taking a good polish, and, when sound, extremely durable, resisting the action of damp, water, earth, rust, termites or ship-worms, and very uninflammable. This most valuable of Australian timbers is stated to cover 14,000 square miles; but the best timber grows only on the ironstone ridges. It should be cut when the sap is at its lowest ebb and banded if in the round, or seasoned one month for every inch in thickness if in scantlings. Its durability is due to from 16—20 per cent. of a powerfully astringent gum, mainly consisting of an acid allied to tannic, which is present in the heartwood when sound. Burrs are sometimes formed on the trees, from 6—10 ft. across and equal to those of Oak or Walnut in their figure. For shipbuilding Jarrah is classed in line 3 of Lloyd's Register: it can be used without copper-sheathing: while cheaper in India than Teak when in the log and only half its price in scantlings, roof-shingles made of it are water-tight and almost uninflammable; and it is largely used for sleepers, telegraph-poles, piles, dock gates, and keelsons, but especially for wood paving-blocks, for which purpose it is unequalled. Its price in England is about £7 per ton, or from £9 10s. to £13 10s. per 1000 blocks, the freight alone being 50—60 shillings a ton. The ornamental varieties are valued for furniture, in spite of their great weight; and the wood also yields an excellent charcoal.

Jarul (*Lagerstrœmia flos-regínæ* Retz.: Order *Lythráceæ*). India, Burma, and Ceylon. "Queen Lagerstroemia." *Sansk.* "Stotulari." *Hind.* "Jarul." *Burm.* "Pym-mah." *Sinh.* "Muruta." Height 30 ft. to first branch; diam. 4 ft. S.G. 744. W 41—46·5. E 544 tons. *f* 5·22. *fc* 2·76. *fs* ·337. R 822 lbs. Light red, hard, lustrous, durable under water. The most valuable timber of North-east India and second only to Teak in Burma. Chiefly used in boat-building, often yielding compass-timber suitable for knees; but also for naves, felloes, waggon-frames, gun-carriages, and building.

Jhand (*Prosópis spicígera* L.: Order *Leguminósæ*). Persia, Afghanistan, Western India. A moderate-sized tree, yielding timber 9 in. square, purplish-brown, straight-grained, very hard, tough and strong, easily worked, but not durable. Used for wheels, carts, agricultural implements, weavers' shuttles, furniture, and building.

Judas-tree (*Cércis Siliquástrum* L.: Order *Leguminósæ*). Southern Europe and Warmer Temperate Asia. Known also as "Love-tree." *French* "Arbre de Judée, Gainier." *Germ.* "Judas-baum." Sapwood white; heart brownish-yellow, veined with black, handsome, hard, taking an excellent polish; rings distinct; pith-rays moderately broad; vessels in spring-wood very large, those in the summer-wood much smaller, 1—8 together. (Fig. 37.)

Juniper (*Juníperus commúnis* L.: Order *Cupressíneæ*). Europe, Northern Asia, and North America. *Amer.* "Ground Cedar." *French* "Genévrier." *Germ.* "Wachholder." Height 15—20 ft. or more; diam. seldom considerable. S.G. 660. Sapwood narrow, yellowish; heart light yellowish-brown, fragrant, close-grained, with no resin-ducts, no distinguishable pith-rays, wavy annual rings marked by narrow reddish-brown zone of autumn wood, tolerably heavy, soft, difficult to split, very durable. Used, on the continent of Europe, for whip-handles, vine-stakes, and turnery. Very similar in character to the wood of *Cupréssus* and *Thúja*.

Juniper, Greek or **Tall** (*Juníperus excélsa* Bieb.). Greek Archipelago, Asia Minor, Syria, Persia, Afghanistan, Biluchistan, Himalaya to Nepal. "Himalayan Pencil-Cedar." Height 30—45 or 70 ft.; diam. 1 ft. or more. Very fragrant, deep red, easy to work, durable. Often the only valuable timber, as near Quetta. Used for building and carpentry.

In Australia the name "Native Juniper" is applied to *Myopórum serrátum* R. Br. (Order *Myoporíneæ*), known also as "Blueberry, Native Currant, Native Myrtle," and "Cockatoo Bush." S.G. 809—819. White, hard, durable when protected, but small. Used for inlaying.

Kaddam (*Stephegýne parvifólia* Korth.: Order *Rubiáceæ*). India, Burma, Ceylon. Height 70—80 ft.; diam. 2—5 ft. W 37. Light pinkish-brown or deep yellow, easily worked, taking a good polish, durable if kept dry. Used for building, furniture, carving and turnery.

Kafir Date or **Plum**. See **Date, Kafir.**

Kahikatea. See **Pine, White,** of New Zealand.

Kamassi (*Gonióma Kamássi* E. Mey.: Order *Apocynáceæ*). Cape Colony. "Knysna Boxwood." Height 16—20 ft.; diam. 1—1½ ft. Close-grained, hard, tough, heavy. One of the finest woods in South Africa, but small. Used for cabinet-work, planes and other carpenters' tools, and suitable for engraving.

Kapor. See **Camphor, Borneo.**

Karamatsu. See **Larch, Japanese.**

Karri (*Eucalýptus diversícolor* F.v.M.: Order *Myrtáceæ*). South-west Australia. Sometimes known as "Blue Gum." Height 300—400 ft.; diam. 3—12 ft. S.G. 1023—885. W 60—63. E 760 tons. *e'* 2·10. *p'* 1·05. *ft* 6·20. *fc* 2·92. *c* 7070. *c'* ·934. Reddish, heavy, slightly wavy or curled in grain, but without ornamental figure, hard, tough, strong, not so easily wrought as Jarrah, subject to star-shake and gum-veins, durable under water or when exposed to alternate drought and wet, but not between wind and earth. Much used locally for wheels, ship-building, and planks, being classed in the third line of Lloyd's Register, suited for piles and bridges, and coming into use for paving-blocks and for furniture. Stated to cover 2300 miles of country.

Katsura (*Cercidiphýllum japónicum* S. and Z.: Order *Magnoliaceæ*) Japan. Height 80 ft.; diam. 3 ft. Used in building, carpentry, and turnery.

Kauri (*Agathis austrális* Salisb.: Order *Araucaríneæ*). North Island, New Zealand. "Kauri" or "Cowdie Pine." Height 120—200 ft.; diam. 4—10 or 20 ft. at base. S.G. 498—595. W 37·4. E 470 tons. *e'* 1·78—1·39. *p'* 1·01—·79. *f* 2·16. *fc* 2·03. *c* 4543. *c'* ·6. *v'* ·769. Sapwood 3—5 in. wide, very resinous: heart yellowish-white to brown, clean, fine, close and straight in grain, moderately hard for Pine, very firm, strong and elastic, generally sound or with slight heart-shake, shrinking very little in seasoning, planing up well, with a beautiful silky lustre like the plainest Satinwood, taking a good polish, wearing even without splintering, and more durable than any other Pine, except where exposed to the Teredo. It is sometimes richly mottled.

Unrivalled for masts and spars, valuable for the decks of yachts owing to its freedom from knots and regularity of grain, used also for sleepers, telegraph-posts, house-building, and joinery. It is the most valuable forest-tree of New Zealand and the soft wood of the country; but the supply is limited, and, though there is a considerable export trade to Australia, the cost of freight limits its employment elsewhere. Its price in London auctions is from 2s. 6d. to 3s. 4d. per cubic foot.

Kaya (*Tórreya nucífera* S. and Z.: Order *Taxáceæ*). Japan. "Japanese Torreya." *French* "Porte-noix Torréya." Height 20—30 or 80 ft.; diam. up to 4—5 ft. Yellowish-white, hard, straight-grained, strong. Much valued for building and cabinet-making.

Keurboom (*Virgília capénsis* Lam.: Order *Leguminósæ*). Cape Colony. Height 15—20 ft.; diam. $1\frac{1}{2}$—2 ft. Light, soft. Occasionally used for rafters, spears, etc.

Keyaki (*Zelkówa acumináta* Planch.: Order *Ulmáceæ*). Japan. Used for common lacquered ware.

Kharpat (*Garúga pinnáta* Roxb.: Order *Burseráceæ*). India and Burma. *Mahr.* "Kooruk." *Telug.* "Garuga." Height 40 ft. to first branch; diam. 3 ft. W 52. Reddish, moderately hard, seasoning well, but not durable. Occasionally used for building or for fuel.

Khat (*Cátha édulis* Forsk.: Order *Celastríneæ*). East Africa. Seldom more than a shrub, but yielding a beautiful reddish-white wood, with zones of darker red, very hard and heavy.

Kiamil (*Odína Wódier* Roxb.: Order *Anacardiáceæ* = *Calésium grándé* O.K.). India, Burma, Ceylon. *Burm.* "Na-bhay." *Tam.* "Ooday." *Telug.* "Goompana." Height 50 ft. to first branch; diam. 4 ft. S.G. 656. W 41—65. Light-red when first cut, darkening to red-brown on exposure, close-grained, moderately hard, seasoning well but slowly, requiring two or three years, not warping, not very durable. Used for spear-shafts, scabbards, spokes, oil-presses and rice-pounders, suitable for cabinet-work.

Kirni (*Mímusops Kaúki* L.: Order *Sapotáceæ*). India, Burma, and represented by a variety, *Browniána* A.DC. in tropical

Australia. *Port.* "Poma d'Adæo." *Hind.* "Kirni." *Malay.* "Manil kara." "Ironwood" in commerce. Red, fine-grained, very hard, easily worked.

Kizi (*Paulównia imperiális* S. and Z.: Order *Scrophulariáceæ*). Japan. A moderate-sized tree, yielding a soft, white, very light wood, employed for the finest lacquer ware.

Kola (*Cóla acumináta* R. Br.: Order *Sterculiáceæ*). West Africa; introduced into tropical America. Height 40 ft. Whitish, light, porous, Poplar-like wood, obnoxious to insects. Used in boat and waggon-building and for tables.

Kolavu (*Hardwíckia pinnáta* Roxb.: Order *Leguminósæ*). South-west India. Brown, moderately hard. Used in building.

Kosum (*Schléichera trijúga* Willd.: Order *Sapindáceæ*). India, Burma, Ceylon. "Ceylon Oak." *Beng.* "Koon." *Mahr.* "Kusoombh." *Sinh.* "Kon." *Tam.* "Kulu" or "Puvu." Height 50 ft.; diam. 1—4 ft. W 70. Red, heavy, hard, strong, durable, but not large. Used for axles, spokes, pestles, sugar-crushers, and screw rollers for mills.

Kranji (*Diálium índum* L.: Order *Leguminósæ*). Java, Borneo. "Tamarind Plum." S.G. 956—1067. *e′* 3·4. *p′* 1·83. *c* 10,920. *c′* 1·442. Red, very heavy, hard, very tough and strong, close-grained, resembling Spanish Mahogany, but without figure. Used in Borneo for ship and house-building.

Kurumi (*Júglans mandshúrica* Maxim.: Order *Juglandáceæ*). Japan. Resembling the European Walnut in characters and uses.

Laburnum. See **Ebony, False.**

Lacewood. See **Plane.**

Ladle-wood (*Hartógia capénsis* L.: Order *Celastríneæ*). Cape Colony. Resembling Mahogany, hard. Suitable for turnery or cabinet-work.

Lancewood in Honduras, etc. (*Guattéria virgáta* Dun: Order *Anonáceæ*). "Yaya." Yellow, light, elastic. Used for shafts, bows and arrows, and imported in spars fetching about 7s. each. In Guiana the allied "Yariyari" (*Duguétia quitarénsis* Benth.) is exported under the same name. [See also **Myrtle, Scrub** and **Shad-bush**.]

Larch (*Lárix europǽa* DC.: Order *Abietíneæ*). Alps of Central Europe, and represented by a variety in Siberia. *French* "Mélèze." *Germ.* "Lärche." *Ital.* "Larice." Height 80—100 or 120 ft.; diam. 2—4 ft. at base. S.G. 809—519. W over 68, when green 32—38. E 400—600 tons. *e'* 1·45. *p'* ·78. *f* ·43. *ft* 4—5·5. *c* 4203. *c'* ·555. *fc* 2·5. *v'* ·783. *fs* ·75. Yellowish-white, generally straight and even, but sometimes rather coarse in grain, soft, tough, strong, very easily split and very durable, being rich in tannic and phenolic antiseptic subtances, shrinking excessively and warping in seasoning, but lustrous and working up tolerably

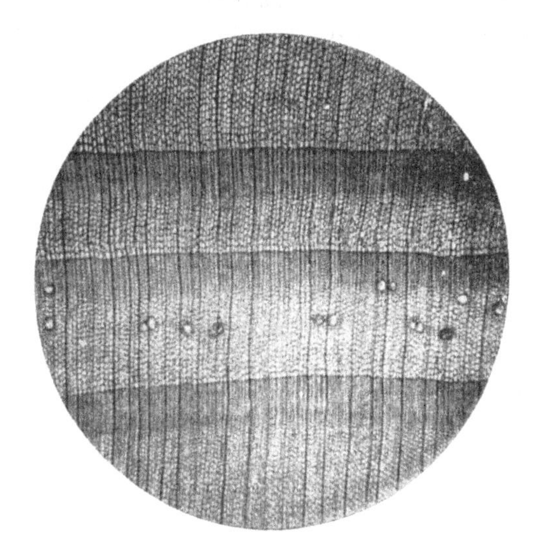

FIG. 53.—Transverse section of Larch (*Lárix europæa*).

well. In its native cold uplands, though there may be an inch of yellowish-white sapwood, the heart is reddish-brown and harder. The pith is small; the pith-rays with tracheids with bordered pits above and below, and parenchyma with simple pits in the middle; resin-ducts smaller and fewer than in *Pinus*; knots irregularly distributed; annual rings wide, defined by a broad dark zone of autumn-wood, finely sinuous. (Fig. 53.) Its durability rendered

Larch a favourite wood in ancient Rome. Cæsar styles it "lignum igni impenetrabile." Augustus built his forum with it; Tiberius brought this timber for the repair of bridges from the forests of Rhætia and preserved one tree, which was 120 feet long and 2 feet in diameter throughout, as a curiosity; and Vitruvius attributes the decay of the buildings erected in Rome at the time to the disuse of Larch on the exhaustion of the forests near the city. Much of Venice is built on Larch piles, which, after ages of exposure to alternate wet and drought, are still sound. Being of rapid growth, Larch is much used for scaffold-poles, ladders, pit-props, sleepers, and fencing; and, being more free from knots than Spruce, is much prized by carpenters and wheelwrights. In ship-building, though its durability is in its favour, its shrinking is against it; but it is classed with Douglas, Huon, Kauri, and Pitch Pines, in the eighth line of Lloyd's Register. Larch is not largely imported; that from Italy being small, crooked, and coarse-grained, that from Poland rather larger and straighter, and that from Northern Russia the largest. When growing in the plains the Larch has proved so susceptible to the fatal attacks of the fungus *Peziza Willkómmii* that it seems likely to be replaced as an object of cultivation by the Douglas Spruce.

Larch, American, Black, or **Red.** See **Tamarack**.

Larch, Chinese or **Golden** (*Pseudolárix Kæmpferi* Gord.: Order *Abietíneæ*). China. Height 120—130 ft. Very heavy and hard.

Larch, Himalayan (*Lárix Griffithii* Hook. fil.). Eastern Himalayas. Height 40—60 ft. Of small dimensions, white, soft, easily split, but durable. Of little value.

Larch, Japanese (*L. leptolépis* Gord.). Central mountains of Japan. "Toga, Kara-matsu, Fuji-matsu." Height 60—80 ft. or more; diam. 1½—2½ ft. Heart red-brown, heavy, hard, and strong; but little used, as it grows at altitudes of 5000 or 6000 ft.

Lasrin (*Albízzia odoratíssima* Benth.: Order *Leguminósæ*). India, Burma. Dark brown with darker streaks, very hard, seasoning and polishing well, and fairly durable. Used for wheels, oil-mills, and furniture.

Lauan (*Anisóptera thurífera* Blume = *Dipterocárpus thúrifer* Blanco: Order *Dipterocarpáceæ*). Philippines. Does not split with shot. Used formerly for shipbuilding.

Laugoussi (*Terminália Taniboúca* Rich.: Order *Combretáceæ*). French Guiana. S.G. 1226—922. R 250 kilos. Very commonly used for canoes and curved timbers.

Laurel is a name not applied to any timber tree in Europe. In Australia it is applied (i) to *Pánax élegans* and (ii) to *Cryptocárya austrális*. *Pánax élegans* F. v. M. (Order *Araliáceæ*). North-eastern Australia. Also known as "Light" or "White Sycamore" or "Mowbulan Whitewood." Height 30—40 ft.; diam. 1 ft. W 31. White, with a pretty grain, light, soft, easily split, not durable, warping and cracking unless very carefully seasoned. Might do for cricket-bats or blind wood.

Cryptocárya australis Benth. (Order *Lauríneæ*). North-eastern Australia. Also known as "Moreton Bay Laurel" and "Grey Sassafras." Height 80—100 ft.; diam. 1—1½ ft. White, light, easily wrought, obnoxious to insects, not durable if exposed.

Laurel, Alexandrian. See **Poon.**

Laurel, Big. See **Magnolia, Large-flowered.**

Laurel, California (*Umbellulária califórnica* Nutt.: Order *Lauráceæ*). California. Also known as "Myrtle." Light brown, heavy, hard, susceptible of a high polish. A local substitute for Oak.

Laurel, Native. See **Cheesewood.**

Laurel, White. See **Beech, She.**

Laurier Cypre (*Ocotéa cérnua* Mez. = *Oreodáphné cérnua* Nees: Order *Lauríneæ*). West Indies. Durable, useful timber of moderate size.

Laurier Madame (*Nectándra sanguínea* Rottb.: Order *Lauríneæ*). West Indies. Light, used for staves and planks.

Laurier marbré (*N. concínna* Nees). A cabinet wood.

Leatherjacket, a name applied in Australia to (i) *Alphitónia excélsa* [See **Ash, Mountain**], (ii) *Ceratopétalum apétalum* [See **Coachwood**], (iii) *Cryptocárya Meissnérii* F. v. M. (Order

Laurineæ). North-eastern Australia. Height 80—100 ft.; diam. 2—3 ft. White, close-grained, tough. Used for staves.

(iv) *Eucalýptus punctáta* DC. (Order *Myrtáceæ*). New South Wales. Also known as "Hickory, Turpentine, Bastard Box," "Grey, Red" or. "Yellow Gum." Height 40—100 ft.; diam. 1—2 ft. Sapwood yellow; heart pale reddish-brown, heavy, hard, close-grained, tough, with gum-veins, difficult to split, but seasoning well and very durable. Used for sleepers, fence-posts, ship and house-building, wheelwrights' work, and fuel.

(v) *Weinmánnia Bentkámii* F. v. M.: Order *Saxifragáceæ*). North-eastern Australia. Height 50—60 ft.; diam. 1½—2 ft. Close-grained, firm, easily wrought. Used for staves and inside work.

Lein or **Lienben** (*Terminália bialáta* Wall.: Order *Combretáceæ*). India, Burma, and the Andaman Islands. Height 80 ft. to the first branch; diam. 4 ft. W 39. Brown, beautifully mottled, moderately hard.

Lemon Wood (*Psychótria eckloniána* F. v. M.: Order *Rubiáceæ*). Cape Colony. "Lanumi." Height 20—30 ft.; diam. 2—3 ft. Hard, tough, useful.

Leopard or **Letter-wood** (*Brósimum Aublétii* Poepp. = *Piratinéra guianénsis* Aubl.: Order *Moráceæ*). Guiana. "Snake-wood." *French* "Lettre moucheté." *Germ.* "Lettern-holz." *Port.* "Pao de letras." *Aborig.* "Buro-koro." S.G. 1333—1049. R 340 kilos. Sapwood yellow, not used; heart squaring 20 inches, but only exhibiting its characteristic dark mottling for about 6 in., very hard, heavy, compact, taking an excellent polish, but difficult to work and full of defects. Imported for inlaying and walking-sticks. Microscopically this wood is remarkable in having its large vessels filled with tyloses of very thick-walled cells.

Letter-wood, Red or **Striped** (*Amanóa guianénsis* Aubl.: Order *Euphorbiáceæ*). Guiana. *French* "Lettre rouge" or "rubanné." S.G. 1175—1038. R. 317 kilos. Sapwood whitish; heart brown-red with blackish veins.

Lightwood. See **Coachwood.**

Lign-Aloes. See **Eagle-wood.**

Lignum-vitæ (*Guaiácum officinálé* L.: Order *Zygophylláceæ*). Colombia, Venezuela, Jamaica, Cuba, Hayti, but chiefly St. Domingo and the Bahamas. *Cuba* "Guayacan." Height 20—40 ft.; diam. 1—2 ft. S.G. 1393—1248. W 60—83. E 508—498 tons. *f* 4·88—7·18. *ft* 7·14. *fc* 3·4—4·4. *fs* ·447—1·246. R 246 lbs. Sapwood dingy yellow, ¼—1 in. broad, as durable as the heart, some of it being, therefore, left on to preserve the rest from splitting; heart blackish with a greenish tint; pith-rays not recognizable and annual rings scarcely so; very heavy, hard, strong, and close-grained, with fibres running obliquely both radially and tangentially, so that it can hardly be split, and containing 25 per cent. of gum-resin, which renders it almost imperishable. It is liable to cup-shake when more than 10 in. in diameter. Imported in lengths of 6—12 ft. up to 10 in. diam. and 3—6 ft. when of greater diam., realizing £5—18 per ton. Used for ships' blocks, pestles, mortars, skittle-balls, rulers, string-boxes, etc. Among the ancients and in France the name has been applied to *Tetraclínis articuláta*. [See **'Arar**.]

Lignum-vitæ in Australia (i) *Acácia falcáta* [See **Myall, Bastard**] (ii) *Mýrtus semenioídes* [See **Myrtle, White**], (iii) *Eucalýptus polyánthema* [See **Box, Red**], (iv) *Phyllánthus Ferdinándi* [See **Beech, White**], (v) *Vítex lígnum-vitæ* A. Cunn. (Order *Verbenáceæ*). North-eastern Australia. Height 50—70 ft.; diam. 1½—2 ft. Blackish, hard, close-grained. Useful, but not yet known to cabinet-makers. (vi) *Dodonǽa viscósa* L. (Order *Sapindáceæ*). Found throughout the tropics. "Switch-sorrel" of Jamaica. Sapwood white, heart dark-brown, in some varieties greenish-black, streaked with rose, very hard, close-grained, and durable. Used in India for engraving, turning, tool-handles and walking-sticks, and suited for all the uses of the true Lignum-vitæ.

Lignum-vitæ in British Guiana (*Ixóra férrea* Benth. = *Sideredéndron triflórum* Vahl: Order *Rubiáceæ*). Known also as "Hackia," or "West Indian" or "Martinique Ironwood." Height 30—60 ft.; diam. 1—2 ft. W 65·8. E 1027 tons. *f* 6·72. *fc* 4·85. *fs* ·457. Dark-brown and hard. Valuable for cogs, shafts, or furniture.

Lignum-vitæ in New Zealand (*Metrosidéros scándens* Banks & Sol.: Order *Myrtáceæ*). Known also as "Akibaum." A creeper "growing up the stem and over the tops of the tallest trees in the New Zealand forests . . . so exceedingly like the Rata (*M. robústa*) in wood, bark, leaf, and flower, that I could never distinguish any difference between them" (Laslett).

Lilac, Persian. See **Bead-tree.**

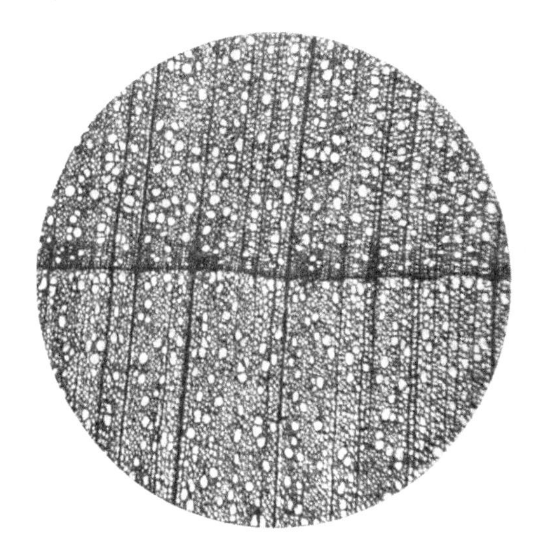

FIG. 54.—Transverse section of Linden (*Tília parvifólia*).

Lilly-pilly (*Eugénia Smíthii* Poir.: Order *Myrtáceæ*). Eastern Australia. Height 80—120 ft.; diam. 1—3 ft. S.G. 935—898. Close-grained, but liable to split in seasoning and to dry rot. Used for axe-handles.

Lima-wood. See **Peachwood.**

Lime, a corruption of **Line,** also known as **Linden** (*Tilia parvifólia* Ehrh., *platyphýllos* Scop. and *argéntea* Desf.: Order *Tiliáceæ*). Europe, the last-named only in the south-east. *French* "Tilleul." *Germ.* "Linde," *T. parvifólia* being known as "Winterlinde," *T. platyphýllos* as "Sommerlinde." *Russ.* "Lipa." Height 20—90 ft.; diam. 1—4 ft. S.G. 794—522. Pale yellow, white or reddish-white, light, soft, close-grained, easily split, with a feebly silky lustre, not very durable, being liable to become "worm-eaten"; annual rings indistinct; pith-rays fine but distinct; vessels scarcely visible, equally distributed (Fig. 54). Though not durable where exposed to the weather, Linden wood stands fairly well when thoroughly dried and kept in a uniform atmosphere or protected by paint or varnish. Used for sabots, as blind-wood in pianofortes and other furniture, in turnery, especially for druggists' boxes, for carving, as in the beautiful work of Grinling Gibbons, for leather-cutters' planks, and for gunpowder-charcoal. Imported from the Baltic for the piano trade.

Lime, American. See **Bass-wood.**

Locust, in Guiana and the West Indies (*Hymenæa Courbaril* L. : Order *Leguminósæ*). "Simiri," "Courbaril." *Span.* "Algarrobo." Height 60—80 ft.; diam. 8—10 ft. S.G. 1191—904. W 49—59. E 549—1018 tons. *f* 4·54—8·13. *fc* 2·23—5·41. *fs* ·49—·585. R 333 kilos. Fine reddish-brown, streaked with veins, hard, compact, close and even grained, easily worked, elastic, taking a beautiful polish, not splitting or warping in seasoning. Used for ships' planks, engineering, tree-nails, furniture, and cabinet-work.

Locust or **Black Locust** in the United States are names used indiscriminately for the allied *Robínia pseudacácia* and *Gledítschia triacánthos,* the former known distinctively as the Yellow, the latter as the Honey or Sweet Locust.

Robínia pseudacácia L. (Order *Leguminósæ*). Southern United States. Height 75 ft. or more; diam. 3 ft. or more. Sapwood very narrow, comprising generally only five rings, yellowish-white; heart yellowish-brown, with shades of red and green, very heavy, hard, strong, tough, firm, offering the greatest resistance to compression in the direction of the fibres, elastic, shrinking considerably in seasoning, but very durable, especially in contact with soil. Vessels all plugged with thin-walled tyloses and appearing as clear yellow spots: those in the spring-wood very large, forming a broad pore-circle, those in the autumn-wood often in peripheral lines. Used for fence-posts, waggon-hubs, and tree-nails in America; and in Europe, where it is considerably grown, under the name "Acacia," especially on railway-banks to protect forests from sparks, for vine-props, wheel-spokes, turnery, and cabinet-work.

Gledítschia triacánthos L. (Order *Leguminósæ*). Central United States. Known also as "Three-thorned Acacia." *French* "Fevier à trois épines." *Germ.* "Dreidorniger Honigdorn." *Span.* "Algarrobo de miel." Height 90—100 ft.; diam. 3—4 ft. S.G. 674. W 42. R 923 kilos. Sapwood broad, yellowish to greenish-white; heart rose or brownish-red, heavy, hard, strong and very durable, especially in contact with the soil, resembling *Robínia* in character and uses; but with open vessels, *i.e.* without tyloses.

Logwood (*Hæmatóxylon campechiánum* L.: Order *Leguminósæ*). Central America, naturalized in Jamaica and introduced into India. "Campeche." *Germ.* "Blauholz." Height 40 ft.; diam. 1½ ft. S.G. 995. Deep dull brownish-red, very heavy and hard. The heartwood is used exclusively as a red or black dye. We import 40,000 to 60,000 tons annually, and it fetches from £5 to £10 per ton. It comes to market in logs about 3 feet long.

Loquat (*Eriobótrya japónica* Lindl.: Order *Rosáceæ*). Japan. "Biwa." *Germ.* "Wellenmispel." Hard. Used for musical instruments.

Magnolia, Large-flowered (*Magnólia grandiflóra* L.: Order *Magnoliáceæ*). Southern United States. "Big Laurel," "Bull Bay." Height 70 ft. or more; diam. 3 ft. "White, heavy, soft, not strong. Suitable for cabinet-work and interior finish.

Mahogany originally (*Swieténia Mahágoni* L.: Order *Meliáceæ*). Central America, Mexico, Cuba, and other West-Indian islands. "Spanish" or "Cuba Mahogany," "Mexican Mahogany," "Honduras Mahogany" or "Bay-wood," "Nassau," and "St. Domingo" are local names. *French* "Acajou." *Germ.* "Mahagoniholz." *Span.* "Caoba." Height 5—50 ft. to the branches; diam. 1—4 ft. S.G. 612—880. W 35—53. E 560—650 tons. *e*′ 1·11—1·9.

p' ·97—1·06. f 4·46. ft 4—7. c 2998—3791. c' ·396—·5. fc 3·3—3·5. v' ·772—·953. When freshly felled light reddish-brown, soon darkening on exposure to light; vessels equally distributed; annual rings distinct; pith-rays fine, but distinct; sapwood only ¾—1 in. thick; heart generally heavy, hard, close and straight in grain, difficult to split, susceptible of a very high polish, with a beautiful satiny lustre, and sometimes with a wavy figure that much enhances its value as a furniture wood. It does not, as a rule, shrink or warp, and is superior to all other woods in taking a firm hold of glue: it is also durable. Of the varieties above enumerated Cuba or Spanish Mahogany yields accurately squared timber 18—35 ft. long and 11—24 in. square, very solid at centre, rarely affected by cup or star-shake and with insignificant heart-shake. Its specific gravity ranges from 720—817. W 53. e' 1·71. p' 1·06. c 3791. c' ·5. v' ·953. It has been used as an Oak-substitute in ship building for beams, planks, and stanchions, whilst figured logs demand high prices for furniture. St. Domingo Mahogany is very similar in quality, but much smaller, generally 8—10 ft. long and 12 or 13 in. square, though occasionally 25 ft. long and 15 in. square. It is very hard, almost horny, a stress of 4300 lbs. per square in. being required to indent it $\frac{1}{20}$ in. transversely to the fibres. It is mostly figured, presenting a rich curl or feather at the bases of its branches. It is entirely used for cabinet-work, especially for veneers. Nassau Mahogany is similar, but even smaller, measuring 3 or rarely 5 ft. in length and 6 or rarely 12 in. square. It is used in turnery. Honduras Mahogany, reaching 50 ft. at its first branch and 3 ft. in diam., yields logs 25—40 ft. long and 12—24 in. square, or even larger. It is seldom figured, becomes somewhat brittle on drying, and is apt to develop deep star-shakes. S.G. 644—684. W 35. e' 1·11. p' ·99. c 2998. c' ·396. v' ·806. Stress required to indent it $\frac{1}{20}$ in. transversely to its fibres 1300 lbs. It is known commercially as "Baywood," and, besides being used as an Oak-substitute in ship-building, is largely used in cabin fittings and by cabinet-makers, turners, and carpenters. Some Mahogany sold as Honduras is really Guatemalan. Mexican Mahogany reaches the largest dimensions, sometimes squaring 48 in., but generally coming to market in logs 18—30 ft. long and 15—36 in. square. It is generally somewhat soft and spongy at the centre, often affected by star-shake, and plain in figure; but that shipped from Tabasco is largely free from those defects. S.G. 612—790. *e*′ 1·9. *p*′ ·97. *c* 3427. *c*′ ·451. *v*′ ·772. We import some 75,000 loads of Mahogany annually, the value of which is about £730,000.

Mahogany, African (*Kháya senegalensis* A. Juss.: Order *Meliáceæ*), also known as "Gambia Mahogany" and "Caïlcedra-wood," and probably also *Chloróphora excélsa* Benth. and Hook. fil. (Order *Moráceæ*), known as "Odum" in Guinea and "Muamba-Camba" in Angola. Yellowish or brownish, with darker zones, very strong and termite-proof. Among the most valuable of African woods, fetching 2d.—6d. per foot; but sometimes so finely figured as to realize 7 to 10 shillings per superficial foot for veneers. Perhaps identical with Niger, Lagos, and Benin Mahogany. [See **Iroko.**]

Mahogany, Australian. See **Jarrah.**

Mahogany, Bastard (*Eucalýptus botryoídes* Sm.: Order *Myrtáceæ*). Known also as "Swamp" or "Gippsland Mahogany," "Blue Gum," "Bastard Jarrah," "Woolly Butt," and "Bangalay." South-eastern Australia. Height 40—100 or 160 ft.; diam. 2—4 or 8 ft. S.G. 891. W 55·59. Light dull red to warm rich brown, heavy, hard, tough, close, even and straight in grain, easy to work, but somewhat subject to gum-veins and shakes, durable. Valuable for ship and waggon-building, yielding compass-timber suitable for ships' knees. The name is also applied to Jarrah.

Mahogany, Bay (*Cercocárpus ledifólius* Nutt.: Order *Rosáceæ*). California. Dark-coloured, hard and heavy.

Mahogany, Borneo or **Penagah** is probably *Calophýllum inophýllum.* See **Poon.**

Mahogany, East India (*Sóymida febrifúga* A. Juss.: Order *Meliáceæ*). "Bastard Cedar, Indian" or "Coromandel Redwood." *Hind.* "Rohuna." *Telug.* "Somida." Central and Southern India. A large tree yielding logs 17—20 ft. long and 1—1½ ft. diam. S.G. 378. W 54·8. Dark blood-red, heavier than water when fresh, very hard, close and straight-grained, easily worked, durable underground, termite-proof, but splitting on exposure and becoming very brittle when seasoned. Used in boat-building, well -work, plough-shares, tables, and carved work in temples.

Mahogany, Forest or **Red** (*Eucalýptus resinífera* Sm. : Order *Myrtáceæ*). North-east Australia. Also known as "Red, Grey" or "Botany-bay Gum, Hickory" and "Jimmy Low." Height 80—130 ft.; diam. 1½—5 ft. Light brown or dark or very dark red, very heavy, close and smooth in grain, very strong, not shrinking, affected with gum-veins, but very durable in air, water, or soil, teredo-proof. Used for ships' knees, piles, fence-posts, rafters, and shingles. The name "Forest Mahogany" is locally applied also to *E. microcórys* [See **Tallow Wood**].

Mahogany, Horseflesh (*Cæsalpínia* sp. : Order *Leguminósæ*). Bahamas. Very strong and durable. Used for ships' knees. This name, or that of Horseflesh-wood, is also applied to the allied species *Swártzia tomentósa* DC., which has S.G. 1020 and is known in Venezuela as "Naranjillo." Horseflesh Mahogany is exported from the Bahamas to England as "Sabicu."

Mahogany, Indian. See **Cedar, Moulmein.**

Mahogany, Madeira (*Pérsea índica* Spreng. : Order *Lauríneæ*). Teneriffe. Known also as "Veñatico" or "Viñacito."

Mahogany, Mountain (*Bétula lénta* [See **Birch, Cherry**] and *Cercocárpus parvifólius* Nutt. (Order *Rosáceæ*).

Mahogany, Swamp, a name applied in Australia to (i) *Eucalýptus botryoídes* [See **Mahogany, Bastard**], (ii) *Tristánia laurína* [See **Box, Bastard**], (iii) *T. suavéolens* [See **Gum, Broad-leaved Water**], and (iv) *Eucalýptus robústa* Sm. (Order *Myrtáceæ*). This last species, a native of New South Wales, is known also as "White Mahogany" and "Brown Gum." Height 100—150 ft.; diam. 2—4 ft. S.G. 1098—889. W 58·5. Light-brown to dark-red, generally containing some gum-veins, often cross-grained, difficult to split, seasoning well, but with some warping, and becoming rather brittle, durable in damp situations and obnoxious to insects, probably owing to its containing no less than 19 per cent. of kino-red, the astringent gum-resin so characteristic of this genus, a higher percentage than in any other species. It is valued for ship-building, shingles, inside work, wheelwrights' work, mallets, rough furniture, and fuel.

Mahwa (*Illipé latifólia* Eng. and *I. Malabrórum* König = *Bássia latifólia* Roxb. and *B. longifólia* L. : Order *Sapotáceæ*). The former in Northern India, the latter in the south and Ceylon. *Hind.* "Mahwa." *Tam.* "Illupa." *Sinh.* "Mee." These trees are rarely felled, being valued for their edible flowers. In Central India the wood is pinkish, weak, invariably rotten at the heart, so as only to square 4—6 in., though approaching 2 ft. in diam.; but in the Upper Provinces it is harder, strong and tough, and is used for the naves of wheels, furniture, and sleepers. The southern species is light-reddish, hard, close grained, flexible, and durable. W 61. It is apt to split on exposure to wind and sun; but is used for spars, keels, treenails, bridges, house-building, etc.

Maire, Black (*Olea Cunninghámii* Hook. fil. : Order *Oleáceæ*). New Zealand. Height 40 ft. Light-brown, very heavy and hard, dense and durable, averaging 11 fairly even rings to the inch. Used for wheels and mill machinery. The strongest wood in the Colony.

Makita (*Parinárium laurínum* A. Gray : Order *Rosáceæ*). Fiji Islands. Height 50 ft. Very hard, durable, and tough. Used for spars for canoes.

Mammee-apple (*Mámmea americána* L. : Order *Guttíferæ*). West Indies. Also known as "Wild" or "St. Domingo Apricot." Height 50 ft. W 59—61. E 763—857 tons. *f* 6·95—7·4. *fc* 2·2—3·5. *fs* ·36—·55. White or reddish, light, durable under ground or water. Used in building and carpentry.

Mammoth-tree. See **Big Tree.**

Mangachapui (*Shórea Mangáchapoi* Blum. : Order *Diptero-carpáceæ*). Philippines. S.G. 671. W 42. Used in ship-building at Manila, and classed in the third line in Lloyd's Register.

Mangeao (*Tetránthera calicáris* Hook. fil. : Order *Lauríneæ*). New Zealand. Height not exceeding 40 ft. Tough, close-grained. Used for ships' blocks.

Mango (*Mangífera índica* L. : Order *Anacardiáceæ*). Tropical and sub-tropical Asia. Introduced in the West Indies. *Hind.* "Am." *Telug.* "Mamidi." S.G. 597. Dull grey, porous, becoming a light chocolate colour, harder, closer-grained, and more durable in the centre of very large old trees, holding a nail faster than any other wood, and standing exposure to salt water, but not to fresh. It could readily be creosoted. Used for solid cart-wheels, canoes, rough furniture, planking in the interior of houses, for packing-cases, as blind wood and as ground for veneers, being the cheapest light wood obtainable in Madras.

Mangosteen, False or **Wild** (*Sandóricum índicum* Cav.: Order *Meliáceæ*). Southern India, Burma, Philippines, and Moluccas. "Indian sandalwood." *Burm.* "Theit-to." A large tree yielding a timber with white or grey sapwood; heart reddish, dense, hard, susceptible of a high polish. Used in cart and boat-building.

Mangrove (*Rhizóphora mucronáta* Lam.: Order *Rhizophoráceæ*). Maritime tropical swamps from Zanzibar to the Fiji Islands. *East Africa* "Mkonko." *Fiji* "Dogo." *Telug.* "Ponna." *Malay* "Mangi-mangi," "Api-api." Height 15—25 ft. W 70·5. Light-coloured, red or brown-red at the centre with darker zones often nearly black, very heavy, hard, close-grained, tough and durable.

Maple, originally *Acer campéstré* L. (Order *Aceríneæ*). England, Central Europe, Northern Asia. "Common" or "Field Maple." *French* "Erable champêtre." *Germ.* "Gemeiner, Feld" or "Kleiner Ahorn." *Welsh* "Masaran." Height 10—20 or 40 ft.; diam. 9—12 in. W 61·5 when green, 52 when dry. Light-brown or reddish-white, hard, fine-grained, compact, tough, with a beautiful satin-like lustre, sometimes containing dark pith-flecks, and not uncommonly curled or speckled ("Bird's-eye Maple"); annual rings slightly wavy; pith-rays fine but distinct; vessels minute. (Fig. 55.) Curled or mottled specimens were prized in former days for "mazer-bowls," which were mounted in silver. These when cut into veneers, as by the rotary saw, are equal to American Bird's-eye Maple. In France the wood is sought after by turners and cabinet-makers. It makes excellent fuel and the very best charcoal.

In Australia the same "Maple" is applied to *Villarésia Móorei* F. v. M. (Order *Olacíneæ*) of New South Wales, also known as "Scrub Silky Oak." *Aborig.* "Belbil." Height 80—120 ft.; diam. 3—6 ft. W 41·36. White, close-grained, prettily figured, durable. Suitable for bedroom furniture.

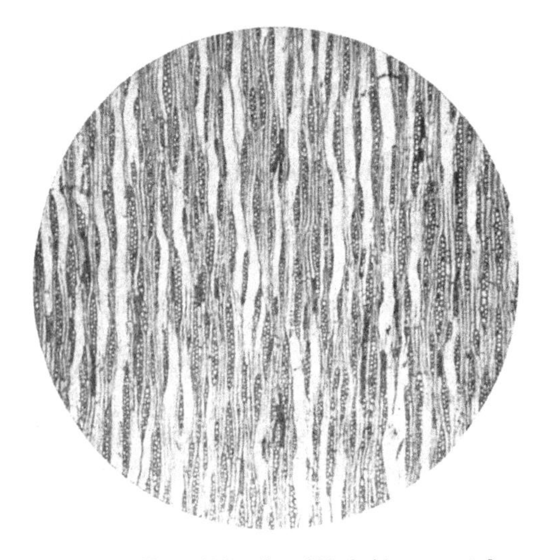

FIG. 55.—Tangential section of Maple (*Acer campestré*).

Maple, Ash-leaved (*Negúndo aceroídes* Mœnch.: Order *Aceríneæ*). South Central United States, California, and Manitoba. Also known as "Box Elder" or "Black Ash," "Negundo." *French* "Erable à giguieres." Height 50 ft. or more; diam. 2 ft. or more. S.G. 438. W 27. R 529 kilos. White or yellowish, light, soft, not strong or durable. Of inferior quality, used to some extent for interior finish and cooperage, and for fuel, but chiefly for paper-pulp.

Maple, Bird's-eye or **Pin.** See **Maple, Rock.**

Maple, Black. See **Maple, Rock.**

Maple, Blister. See **Maple, Rock.**

Maple, Broad-leaved. See **Maple, Oregon.**

Maple, California. See **Maple, Oregon.**

Maple, Great or **Sycamore.** See **Sycamore.**

Maple, Hard. See **Maple, Rock.**

Maple, Himalayan. Several species of *Acer* are of some importance in this region; viz. (i) *A. Campbélli* Hook. fil. and Thbg., in the Eastern Himalayas, a greyish-white, hard and dense wood, used for tea-boxes and planking; (ii) *A. oblóngum* Wall., in Nepal and Kumaon, a reddish-brown, hard and dense wood, used for agricultural implements; (iii) *A. Lóbeli* Tenore, growing from the Eastern Mediterranean to the Himalayas, a brownish-white, hard, dense, elastic wood; (iv) *A. lævigátum* Wall., in Nepal, reaching 30—40 ft. in height and 3—4 ft. in diam., and used for beams, rafters, etc.

Maple, Japanese (*A. pictum* Thunb., *A. polymórphum*, etc.). "Itaya-Kayede." Several ornamental species in Japan yield timber similar to that of the European species.

Maple, Norway (*A. platanóides* L.). Norway to Switzerland. "Plane Maple." *French* "Erable plane." *Germ.* "Spitz Ahorn." Height 60—70 ft.; diam. 2—3 ft. W 43. White, but inclining to grey, when mature, easily worked and taking a fine polish, with the satin-like lustre of the pith-rays characteristic of the group, and slightly wavy annual rings, hard, heavy, tough, cracking and warping, but durable if kept dry. Used in turnery, for musical instruments, gun-stocks, etc., being practically identical with the wood of the Sycamore.

Maple, Oregon (*A. macrophýllum* Pursh.). Pacific slope of North America. Also known as "California" or "Broad-leaved Maple." Height 90 ft. or more; diam. 4 ft. or more. Rather light, hard, and strong. Used in Oregon for axe and broom-handles, snow-shoe frames and furniture, and on radial sections exhibiting occasional curly figure, which is highly prized for cabinet-work. This figure is produced by an unexplained waviness or spiral twist in the elements of the wood, which is rarely recognizable in the growing tree, but produces transverse corrugations on the surface of the log when the bark is removed, these curls or corrugations varying in number in different trees from one or less than one to several to the inch. Though scarcely visible in a transverse section, this beautiful figuring is seen on all longitudinal or oblique sections, appearing on the planed surface so like the light and shadow on an undulating surface that it is difficult to believe it smooth. It is sometimes called "Fiddle-backed Maple" from its resemblance to a variety of Sycamore used for the backs of violins.

Maple, Red (*A. rúbrum* L.). Canada and Eastern United States. Known also as "Soft, Swamp" or "Water Maple." Height 70—90 ft.; diam. 3—4 ft. S.G. 618. W 38·5. R 811 kilos. Sapwood brownish-white with a small irregular brown or reddish-brown heart, moderately heavy, hard and elastic, close-grained, compact, taking a very smooth polish, not strong or durable. Used in turnery, chair-making, for wooden dishes, shovels, and other small ware. Occasionally curled, when it is valued for gunstocks and for veneers. The bird's-eye figure is rare in this species. This wood does not enter into American export trade.

Maple, Rock (*A. barbátum*, Michx.). Eastern North America. Also known as "Hard" or "Sugar Maple," and a variety as "Black Maple." Height 50—100 ft. or more; diam. 1—4 ft. S.G. 691. W 43. R 1149 kilos. White when first cut, becoming rosy on exposure, and, when seasoned; sapwood light yellowish; heart brown, heavy, very hard, fine and close in grain, compact, strong, tough, taking a fine polish, with a satiny lustre. The most valuable species, but not durable if exposed, requiring two or three years' seasoning. Used in New England as an Oak-substitute, in preference to Beech, Birch or Elm, for house-frames, ships' keels, axles, spokes, chairs and other furniture, flooring and interior finish, wooden bowls, considered the best in the market, shoe-lasts, pegs, saddle-trees, etc., and also for fuel. It is but little imported, except when figured. "Blister" or "Landscape," "Bird's-eye" or "Pin" and "Curly" or "Fiddleback" figures all occur in this species, the first two being almost confined to it. The Blister figure is produced by wart-like prominences on the wood beneath the bark and is cut tangentially by a rotary lathe in a veneer the length of the log and running spirally inwards to the centre of the tree. The name "Landscape Maple" is appropriate in that this figure much resembles a contoured map. The Bird's-eye or Pin figure, which is more common, is produced by pittings, which are visible on the bark. These appear in transverse section as "pins," or structures resembling branches, radiating nearly from the centre of the tree; and in tangential section as "eyes." Veneers of these varieties are largely imported at Liverpool.

Maple, Silver (*A. saccharínum*, Wang., often known as *A. dasycárpum*, Ehrh.). Eastern North America. Otherwise known as "Soft" or "White Maple." Height 90—100 ft.; diam. 4—5 ft. S.G. 527. W 32·8. R 1019 kilos. Sapwood delicate creamy white; heart reddish-brown, lighter and softer than Rock Maple, brittle, not very strong or durable if exposed, shrinking moderately, seasoning and working well, taking a fine polish, wearing smoothly. Used for flooring, cheap furniture, interior finish, turnery, wooden ware, and fuel. Sometimes curled.

Maple, Striped (*A. Pennsylvánicum*, L.). Eastern North America. "Moose-wood, Whistle-wood, Striped Dog-wood." Height 30—40 ft.; diam. 8—10 ins. S.G. 530. W 33. Sapwood wide, light brown; heart rather darker, light, soft, close-grained. Not used as timber.

Maple, Sycamore. See **Sycamore.**

Marblewood (*Olea paniculáta* R. Br.: Order *Oleáceæ*). New South Wales and Queensland. Known also as "Ironwood" and "Native Olive." Height 50—70 ft.; diam. 1½—2 ft. Whitish, darkening towards the centre, prettily mottled, hard, tough, close-grained, durable. Used for staves and suitable for turning or engraving.

Marblewood, Andamanese (*Diospýros Kurzii* Hiern: Order *Ebenáceæ*). Andaman Islands. Handsomely streaked with black and grey, very heavy, hard, close-grained and durable. Used for handles and furniture, and valuable as a substitute for the scarce Calamander wood of Ceylon.

Margosa (*Azadiráchta índica* A. Juss.: Order *Meliáceæ*). India, Burma, Ceylon, Java. Planted in East Africa. *Hind.* "Neem." S.G. 739. W 46. Light reddish-brown, beautifully mottled when old, heavy, hard, compact, and durable, resembling Mahogany. Used for furniture, images, flooring, etc.

Marrara (*Weinmánnia rubifólia* F. v. M.: Order *Saxifragáceæ*). North-east Australia. "Corkwood." Height 80—100 ft.; diam. 2—3 ft. Close-grained, tough, easily wrought. Not much used.

Massaranduba (*Mímusops eláta* Fr. Allem.: Order *Sapotáceæ*). Brazil. Height 80 ft.; diam. 10 ft. Very hard. Used in construction and in joinery.

Matai. See **Pine, Black.**

Mee. See **Mahwa.**

Me-matsu. See **Pine, Japanese Red.**

Mesquite (*Prosópis juliflóra* DC.: Order *Leguminósæ*). Southern United States and Northern Mexico. Height 30 ft. Dark brown to red, very heavy and hard, durable. Used in house-building and for furniture.

Messmate, a name applied in Australia (i) to *Eucalýptus amygdalína* [See **Ash, Mountain**], (ii) to *E. oblíqua* [See **Stringybark**], (iii) to *E. piperíta* [See **Peppermint**], from their association with the Stringybarks.

Milkwood. See **Ironwood** (xvii).

Milkwood, Red (*Mímusops obováta* Sond.: Order *Sapotáceæ*). South Africa. Height 15—20 ft.; diam. 1½—2 ft. W 52·4. E 876 tons. *f* 9·52. Moderately heavy, close-grained and tough. Used for felloes of wheels and other parts of waggons.

Milla or **Mililla** (*Vítex altíssima* L. fil.: Order *Verbenáceæ*). Ceylon and Southern India. *Sinh.* "Milla," "Mililla." W 60·9. E 721 tons. *f* 6·59. *fc* 3·12. *fs* ·448. A large tree yielding a grey or olive-brown, heavy, very hard, fine and close-grained wood, which in Ceylon is one of the most valuable for building purposes.

Mirabow (*Afzélia palembánica* Baker: Order *Leguminósæ*). Andaman Islands, Malay Islands, Borneo. "Merban," "Pyn-kadoo." A large tree yielding timber 30—40 ft. long, and 1½—

2½ ft. in diam. W 52. Dark yellow, darkening and reddening with age, prettily figured, moderately heavy, of fine even grain, very tough, durable, termite-proof, working freely and taking a fine polish, thus resembling Mahogany. An esteemed furniture-wood.

Miro. See **Pine, Black.**

Molavé (*Vítex altíssima* [See **Milla**] and *V. geniculáta* Blanco: Order *Verbenáceæ*). Philippines. S.G. 819. W 51·2. *e'* 1·87. *p'* 1·54. *c* 7812. *c'* 1·032. Straw-coloured, heavy, hard, close-grained, strong, with a figure resembling Satin-wood, not shrinking or splitting in seasoning, very durable even when exposed. Used extensively for all kinds of work and considered almost equal to Teak in building, while it might prove useful in cabinet-work.

Monkey-pot (*Lécythis grandiflóra* Aubl.: Order *Myrtáceæ*). Guiana. "Wadadura." *French* "Canari Macaque." Height 100 ft.; diam. 2—3 ft., yielding logs 20—50 ft. long and squaring 16—28 in. S.G. 1032. Light brown, very heavy, hard, close-grained, tough, working easily, taking a fine polish and very durable. Used for staves and furniture.

Moose-wood. See **Maple, Striped.**

Mora (*Dimorphándra excelsa* Baill.: Order *Leguminósæ*). Guiana and Trinidad. Height 100—150 ft., frequently 60—70 ft. to the first branch; diam. 2—2½ ft., yielding logs 18—35 ft. long, squaring 12—20 inches. S.G. 1075—1094. W 57—72·3. E 1068—1465 tons. *e'* 1·05. *p'* 1·64. *f* 6·87—9·42. *c* 9240. *c'* 1·220. *fc* 3·14—5·34. *v'* 1·117. *fs* ·456—·655. Chestnut-brown or red, very heavy, hard, straight-grained, tough, strong, sometimes with a beautiful curled figure, taking a good polish, durable, but very liable to star-shake. Suitable for keelsons, beams and planking in ship-building, classed in line 2 of Lloyd's Register, and also fitted for house-building and as a substitute for Rosewood or Mahogany for furniture and cabinet-work, especially when figured.

Moricypre (*Byrsónima spicáta* Rich.: Order *Malpighiáceæ*). Brazil and West Indies. Height 30—40 ft.; diam. 2 ft. Used in building and cabinet-work.

Morrel (*Eucalýptus macrocárpa* Hook.: Order *Myrtáceæ*). West Australia. Very hard. Used for spokes, shafts, and furniture.

Morung Sal. See **Sal.**

Moutouchi (*Pterocárpus Dráco* L. = *Moutóuchi suberósa* Aubl.: Order *Leguminósæ*). Guiana. Introduced into India in 1812. S.G. 1018—875. R 255 kilos. With long streaks of pale violet, light brown and white, easily sawn or worked.

Mountain Ash. See **Rowan** and **Ash, Mountain.**

Mulato-wood. See **Bois Mulatre.**

Mulberry (*Mórus álba* L.: Order *Moráceæ*). Said to be a native of China introduced into Europe in the 15th century. *French* "Mûrier." *Germ.* "Maulbeerbaum." *Span.* "Moral." Height 20—30 ft.; diam. 1—2 ft. Sapwood narrow, yellowish-white; heart yellowish-brown, becoming reddish, like old Mahogany, on exposure, heavy, hard, lustrous, difficult to split; vessels in the spring wood very large forming a very broad and distinct pore-circle, most of them open, but a few filled with a white secretion; those in the autumn wood regularly distributed, very minute; pith-rays fine but very distinct. A durable wood, similar to that of the Black Locust (*Robínia*). Used chiefly in veneers and inlaying.

Mulberry, Indian (i) *M. índica* L., a native of the Himalayas, India, China and Japan, has a wood very similar to that of *M. álba*, used for tea-boxes and furniture, (ii) more generally used of *Morínda citrifólia* L. [See **Canary-wood.**]

Mulberry, Native. See **Holly, Smooth.**

Mulberry, Red (*Mórus rúbra* L.). Eastern United States. Height 60—70 ft.; diam. 3—4 ft. S.G. 589. W 36·76. R 775 kilos. Sapwood very narrow, whitish; heart, orange-yellow, of moderate weight and hardness, tough, coarse-grained, strong, taking a satiny polish, and very durable in contact with the soil. Largely used for fence-posts and cooperage, and to some extent for agricultural implements and in boat-building.

Mulga (*Acácia aneúra* F. v. M.: Order *Leguminósæ*). Australia. "Myall." Height 20—30 ft.; diam. 9—12 in. Dark brown, very hard. Used for fence-posts, bullock-yokes, boomerangs, spears and "mulgas," narrow wooden shields.

Musk-tree (*Márlea vitiénsis* Benth.: Order *Cornáceæ*). Fiji islands, introduced in Australia. Height 20—30 ft.; diam. 6—12 in. Bright yellow, with black centre, pretty curl, musk-like odour, and close grain. An excellent cabinet-wood.

Muskwood (*Oleária argophýlla* F. v. M.: Order *Compósitæ*). South-east Australasia. Height 20—30 ft.; diam. 1½—3 ft. S.G. 642. W 40. Brownish-yellow, beautifully mottled at the butt end, working well, taking a good polish and fragrant. Highly prized for cabinet-work and turnery.

Mutton-wood (*Myrsíné variábilis* R. Br.: Order *Myrsíneæ*). East Australia. Also known as "Jemmy Donnelly," a name also applied to the totally dissimilar *Euroschínus falcátus*. Height 45—50 ft.; diam. 12—15 in. S.G. 714. Yellowish or pinkish, hard, tough, somewhat resembling Oak in figure, easily worked, durable.

Myall, a name applied in Australia to various species of *Acácia* (Order *Leguminósæ*), especially (i) *A. péndula* A. Cunn. "True" or "Weeping Myall," "Violet-wood." *Aborig.* "Boree." North-east Australia. Height 20—30 ft.; diam. 6—12 in. Rich dark brown, beautifully marked, heavy, hard, close-grained, violet-scented as long as it is unpolished. Used for boomerangs, veneers, fancy boxes, and especially for tobacco-pipes, and often imitated by artificially scenting the wood of other species, a perfume which does not last.

(ii) *A. homalophýlla* A. Cunn. Also known as "Spear-wood." *Aborig.* "Gidya." South-east Australia. A similar tree, similarly employed. Used also in turnery and formerly for spears. S.G. 1124.

(iii) *A. acumináta* Benth. West Australia. Known also from its fragrance as "Raspberry Jam." Height 38—40 ft. Similar, reddish-brown, but with a perfume resembling raspberries. Used for weapons and fence-posts; but suited for ornamental work and now coming into use for furniture.

(iv) *A. aneúra.* See **Mulga.**

(v) *A. glaucéscens* Willd. Also known as "Mountain Brigalow" and "Rosewood." South-east Australia. Height 30—45 ft.; diam. 1—1½ ft. W 54. Resembling Walnut or Rosewood, prettily marked, slightly fragrant, hard, close-grained, tough. Used for spears and handles and suitable for turnery or veneers.

Myall, Bastard (i) (*A. falcáta* Willd.). Also known as "Hickory, Lignum-vitæ" and "Salee." East Australia. Height 20—30 ft.; diam. 6—12 in. Sapwood yellow; heart light brown, heavy, hard, tough. Used for whip-handles and coach-building.

(ii) *A. Cunninghámi* Hook. East Australia. Height 20—30 ft.; diam. 9—12 in. W 46·75. Dark, resembling the wood of Red Cedar (*Cedréla Toóna*), but heavier, close-grained, and taking a good polish. A useful cabinet wood.

Myall, Brigalow (*A. harpophýlla* F. v. M.). Also known as "Brigalow" [See **Myall**]. South Queensland. Brown, strongly violet-scented, heavy, hard, elastic, durable, splitting freely. Used for turnery, vine-stakes, spears and boomerangs.

Myall, Dalby. See **Ironwood** (ii).

My Lady, a West Indian wood, perhaps *Nectándra sanguínea* Rottb. [See **Laurier Madame.**]

Myrobalan-wood (*Terminália belérica* Roxb.: Order *Combretáceæ*). India, Ceylon, Burma. "Babela." *Sansk.* "Bahira." *Pers.* "Beleyleh." *Tam.* "Tandi." Height 50—80 ft.; diam. 2—4 ft. W 40. Yellowish-grey, hard, improved by steeping, but is liable to become worm-eaten and does not appear durable. Used for packing-cases, coffee-boxes, grain-measures, canoes and planking. [For allied species see **Arjun**, **Harra**, **Lein**, and **Saj**.]

Myrtle, a name not applied to any useful wood in the Northern Hemisphere. In Tasmania and Victoria it refers generally to *Fágus Cunninghámii* [See **Beech, Evergreen**]; but in New South Wales and Queensland to *Syncárpia leptopétala* F. v. M. (Order *Myrtáceæ*). and to *Backhoúsia scadióphora* F. v. M.; belonging to the same Order. *Syncárpia leptopétala*

(= *Metrosidéros leptopétala* F. v. M.) which is known also as "Ironwood" or "Brush Turpentine," reaches a height of 50—60 ft., with a diameter of 2 ft., is heavy, hard and durable, and is used in turnery. *Backhoúsia scadióphora,* 80—90 ft. high, with a diameter of 2 ft., is hard, close-grained and prettily marked; but, though possibly useful for engraving, is not yet in use.

Myrtle, Black (*Cargíllia pentámera* F. v. M.: Order *Ebenáceæ*). North-east Australia. Known also as "Grey Plum." Height 80—100 ft.; diam. 2—3 ft. Reddish, close-grained, tough, durable. Used occasionally for tool-handles and for flooring.

Myrtle, Brush. See **Barranduna.**

Myrtle, Drooping (*Eugénia Ventenátii* Benth.: Order *Myrtáceæ*). North-east Australia. Height 40—60 ft.; diam. 2—3 ft. Grey or pinkish, beautifully marked, heavy, hard, close-grained, tough. Used for handles, ribs of boats, and floors of verandahs.

Myrtle, Grey. See **Myrtle, Scrub.**

Myrtle, Native or **Red** (*Eugénia myrtifólia* Sims: Order *Myrtáceæ*). North-east Australia. Known also as "Brush Cherry." Height 50—80 or 100 ft.; diam. 1—2 ft. W 47·75. Light reddish or yellowish, strong, elastic, seasoning and working well. Used for boomerangs, shields, staves, oars, boat-building and tool-handles. [See also **Myrtle, Scrub,** and **Juniper.**]

Myrtle, Ridge. See **Ironwood** (iv).

Myrtle, Scrub (*Backhoúsia myrtifólia* Hook and Harv.: Order *Myrtáceæ*). North-east Australia. Known also as "Grey" or "Native Myrtle" and as "Lancewood." Height 20—40 ft.; diam. 9—12 in. Light yellow, often prettily marked with dark brown, Walnut-like stains, hard, close-grained, tough, durable. Used for mallets, handles, bows, and suitable for turnery and perhaps engraving.

Myrtle, Three-veined. See **Turpentine, Brush.**

Myrtle, Water. See **Gum, Water.**

Myrtle, White (i) (*Mýrtus acmenioídes* F. v. M.: Order *Myrtáceæ*). North-east Australia. Height 60—70 ft.; diam-

1—1½ ft. W 61. Light-coloured, close-grained, very hard, tough and durable. Used by coach-builders.

(ii) (*Rhodámnia argéntea* Benth., in the same Order.) North-east Australia. *Aborig.* "Muggle-muggle." Height 80—100 ft.; diam. 2—3 ft. Close-grained, hard and durable; but seldom used.

Nagesar. See **Ironwood** (xviii).

Nani. See **Ironwood** (xx).

Narango, Palo. See **Fustic.**

Narra. See **Sanders, Red.**

Neem. See **Margosa.**

Needle-bush. See **Pin-bush.**

Nettle-tree (*Céltis austrális* L.: Order *Ulmáceæ*), Mediterranean. *French* "Micocoulier." *Germ.* "Zürgelbaum." Height 30—50 or 70 ft.; diam. 6—12 in. Yellowish, heavy, hard, compact, elastic, taking a high polish; vessels in spring wood few but large, the smaller, later ones arranged dendritically; pith-rays fine but distinct (Fig. 32). When cut obliquely it resembles Satin-wood. Used for furniture, carving, turnery, whip-handles, walking-sticks, flutes, etc. [See **Hackberry.**]

Nettle-tree, Giant (*Lapórtea gigas* Wedd.: Order *Urticáceæ*). North-east Australia. Height 80—100 or 160 ft.; diam. 2—8 ft. W 16—17. Brownish, soft, spongy. Useless.

Nettle-tree, Small-leaved (*L. photiniphýlla* Wedd.). A smaller tree, from the same region, yields an even lighter wood. W 13·8. It might be used for floats for fishing-nets.

Nicaragua-wood. See **Peachwood.**

Nispero. See **Sapodilla.**

Nut. See **Hazel.**

Nut, Queensland (*Macadámia ternifólia* F. v. M.; Order *Proteáceæ*). North-east Australia. *Aborig.* "Kindal-kindal." Height 30—50 ft.; diam. small. Reddish, firm, fine-grained, prettily figured. Used for staves, bullock-yokes, shingles, cabinet-work and veeners.

Oak, originally *Quércus Róbur* L. (Order *Cupulíferæ*), the principal hardwood of Europe, afterwards extended to other species of the genus in Southern Europe, North America, the

Himalayas and Japan, and to various other entirely unrelated timber-trees, chiefly species of *Casuarina*, in Australia. It will be convenient to depart from the strictly alphabetical arrangement of the many kinds of "Oak" in use, in favour of a geographical enumeration. Beginning, therefore, with the Common Oak of Europe, we will then describe the other European and North African forms, taking those of North America next, and then those of the Himalayas and of Japan, and relegating the so-called Oaks of Australia to the last.

Oak, Common, British or European (*Q. Róbur* L.) Syria, Mt. Taurus and Mt. Atlas to 60° N. lat. *French* "Chêne." *Germ.* "Eiche." *Span.* "Roble." *Japan.* "Gashi." Height 60—100 ft.; diam. 1—22 ft., often with a straight stem 30—40 ft. high and 2—4 ft. in diam. S.G. 1280 when fresh cut, to 780 or 597 when seasoned. "It must be borne in mind, however, that these weights refer to the wood as a structure, and do not give the specific gravity of the wood substance itself. This latter may be obtained by driving off all the air and water from the wood, and is found to be 1560" (Marshall Ward). W 62—43. E 535—800 tons. *e'* Mr. Laslett takes the mean elasticity of British Oak as unity for the comparison of other woods. Other specimens of the species range from ·64—1·41. *p* 6500—11,300. *p'* Here too Oak is unity, its range being ·6—1·06. *f* 5·27. *ft* 3·4—8·8. *c* 7571—8102. *c'* English Oak being taken as unity, French-grown Oak is 1·071. *fc* 2·7—4·5. *v'* English Oak unity, Dantzic Oak, probably the same species, ·99, French 1·04. *fs* 1—1·03. R 10,000—13,600 lbs. Sapwood narrow, yellowish; heart of various shades, from greyish or yellow-brown (fawn-colour) to reddish or very dark brown, darkening on exposure. "Oak is neither the hardest and heaviest, nor the most supple and toughest of woods, but it combines in a useful manner the average of these qualities. Good oak is hard, firm and compact, and with a glossy surface, and varies much; young oak is often tougher, more cross-grained, and harder to work than older wood" (Marshall Ward). A stress of 1900 lbs. per square inch is stated as the average requisite to indent Oak $\frac{1}{20}$ in. transversely to its fibres. Oak timber is apt to be affected by star and cup shakes, especially in certain districts; and, though it can be readily seasoned, it is very liable to warp and shrink during the process.

When Oak was largely in use in our Royal dockyards the rules as to specifications were that only logs would be accepted 10 ft. or more in length that would side 9 inches and upwards in proportion to their length; and that each piece was measured for contents by calliper measurement as far as its spire (or top-end) "will hold 12 inches in diameter." It was then found by experience that 30 inches calliper will yield sided timber of about 21 inches, 24 inches calliper 18½ in the side, or on an average a "siding" of about two-thirds of the calliper measurement. For fencing or staves, Oak splits easily, with a moderately smooth surface; and, for ornamental purposes, it is susceptible of a high polish.

The sapwood is very liable to insect-attack and cannot be termed durable; but the heart, whether under ground, under water, or exposed to alternations of drought and damp, is remarkably so, few woods changing so little when once seasoned. The "life" of a railway-sleeper of young Oak is stated to be from 7 to 10 years if not treated with any preservative, or 16 years if treated with zinc-chloride. The piles of Old London Bridge taken up in 1827, sound after six and a half centuries' use, are a striking instance of these lasting qualities; whilst the "Bog-oak" blackened by the action of the iron-salts in peat-mosses on the tannin it contains—a natural ink—remains sound after far longer periods. The durability of Oak timber is undoubtedly affected by the time of year at which it is felled, the best season being winter, when there is least water and sap or fermentable matter in the wood. The greater amount of tannin in the bark and the greater ease in stripping it in the spring have, however, often led to the trees being felled at that time. Incipient decay often shows itself in the heartwood of ancient oak-trees as "foxiness," a warm deepening of the colour that actually enhances the value of the wood for some ornamental purposes.

The minute structure of Oak has already been to some extent described and fully illustrated in Part I. The pith, at first white, then brown, is pentangular and from 1—4 mm. across: the pith-rays are of two kinds, very broad, lustrous, light-coloured ones—the "silver-grain"—sometimes $\frac{2}{5}$ in. apart, and others, far more numerous—about 300 to the inch—very fine and less straight. The annual rings undulate slightly, bending outwards between the broad pith-rays: they vary in width from 1—8 or more millimetres, and they are conspicuous owing to the pore-circle of very large vessels in the spring-wood, which is only a single row when the rings are narrow or four rows when they are wide. Into the autumn-wood there radiate outwards straight or bifurcating bands of finer vessels, tracheids and cells (Figs. 19 and 27). Numerous, very narrow, wavy, peripheral lines ("false-rings") of wood-parenchyma, recognizable by their contents, but seldom more than a single row of elements each, are generally visible, especially when the annual rings are broad.

Owing to the large proportion that the pores bear to the fibre when the annual rings are narrow, such slow-growing unthrifty Oak, growing on poor soil or in severe climatic conditions, is, though often beautifully marked, softer than the broad-ringed, thrifty, quick-grown wood of good soils and a favourable climate. They may differ to the extent of their specific gravities—a fair criterion of their hardness and strength—varying from 691 to 827 respectively.

Quércus Róbur is a somewhat variable species, three somewhat inconstant types being recognized as British, viz. *pedunculáta*, *sessiliflóra* and *intermédia*. *Quércus Róbur pedunculáta* derives its scientific name from the long stalks to its acorns, for which reason also the Germans call it "Stieleiche," whilst from the situations in which it grows they call it "valley oak" (Thaleiche), and from its early production and shedding of its leaves it is called "early oak" (Früheiche) and "Sommereiche." It is generally quick-growing, but does not, perhaps, produce so great a length of clear stem as *sessiliflóra*. Its wood may be lighter in colour, whence apparently it gets its French name, "Chêne blanc"; but it is generally more compact, denser and tougher, and therefore better for purposes where strength is a primary consideration.

Q. Róbur sessiliflóra, known, from a supposedly greater resemblance in its wood, as "chestnut oak," by the French as "Chêne rouge," and by the Germans as "Traubeneiche" "red, (Rotheiche) hill, (Bergeiche) late" (Spateiche) or "winter oak," has long stalks to its leaves, but not to its acorns, and is apparently generally less dense in its timber. It is also, perhaps, more liable to shakes; but it must be admitted that, in the absence of any record as to the source of the logs or of any exact measurement of specific gravity, timber-dealers cannot discriminate the wood of these two varieties. Stunted specimens, grown on rocky hill-sides, produce crooked, hard, knotty wood, difficult to split, formerly of considerable value in shipbuilding; and coppice oak is of a similar character.

Q. Róbur intermédia, the Durmast Oak, is not common. It has short stalks to both leaves and acorns and its leaves are downy on their under surfaces. It has a broad sapwood and a dark-brown heart, and is considered of inferior quality.

There is, perhaps, greater difference between the woods of *Q. Róbur* imported from various parts of the continent than there is between these home-grown varieties. French Oak, largely *Q. Róbur pedunculáta* grown in Brittany and Normandy, is generally smaller, shorter and more tapering than English; but, with S.G. 992—720, e' 1·39—1·41, p' 1·01—1·06, c 8102, c' 1·071, v' 1·04, and shrinking and splitting less in seasoning than English, it would appear, in spite of some former prejudice, to be better all round, always presuming that a good sample be selected. Dantzic Oak, shipped partly from Memel and Stettin, mostly brought down the Vistula from Poland, which comes to market as staves, in logs 18—30 ft. long and 10—16 in. square, or in planks about 32 ft. long, 9—15 in. wide and 2—8 in. thick, is brown, straight and clean-grained and free from knots. It would seem to be largely *Q. Róbur sessiliflóra*, and is sometimes so figured as to be classed as "wainscot-oak." It has S.G. 897—768.

e' ·43. p' ·59. c 4214. c' ·556. v' ·99, and is, therefore, decidedly inferior in strength to good English-grown Oak. It is carefully sorted or "bracked" for market, the planks of best or "crown" quality being marked W, those of second best or "crown brack" quality WW. Riga Oak, a very similar wood, also probably *sessiliflóra*, only comes to market in "wainscot logs" of moderate dimensions, for furniture or veneers, for which purposes it is the finest quality in the trade.

From Italy and Spain a variety of Oak timbers were formerly imported to our dockyards, partly the produce of varieties of *Q. Róbur*, but partly apparently from the evergreen Cork and Holm Oaks (*Q. Suber* L. and *Q. Ilex* L.). Most of this wood was comparatively small, curved, brown, hard, horny, tough, difficult to saw or work, and very liable to shakes and, therefore, unsuitable for boards.

Oak, Turkey (*Quércus Cérris* L.). Middle and Southern Europe and South-west Asia. Known also as "Adriatic, Iron, Wainscot" or "Mossy-cupped Oak." *Germ.* "Zerreiche." A tall species with straight, clean stems, hard-wooded in the south and in plains, softer in the north or on hills, very liable to ring and star shakes. Sapwood broader than in *Q. Róbur*; heart a redder brown; broad pith-rays more numerous. On the whole inferior.

Throughout Europe, and more especially in Britain, Oak was employed for every purpose both of naval and civil architecture until about the beginning of the eighteenth century, when Pine was first largely imported from the Baltic and North America. In our dockyards Oak continued to be in large demand until about 1865, all other hard and heavy woods used in shipbuilding being compared with it as a standard and described as "Oak-substitutes." Oak has, however, one serious drawback in this connection, viz., the presence of a powerful wood acid which exerts a rapidly corrosive action upon any iron in contact with it, this rusting being apt in turn to react upon the timber, producing rot. With the introduction of armour-plating and steel ships, wood of any kind has become far less important in shipbuilding and Teak has largely superseded Oak. In Lloyd's Register, however, English, French, Italian, Spanish, Portuguese and Adriatic Oak, and Live Oak, *Q. vírens* of the United States, are classed together on Line 2. Though the greater cheapness and lightness of coniferous wood have led to its being now generally preferred in building, Oak is still in request where strength and durability are objects. Large quantities are used for palings, shingles, staves, parquet-floors, wheelwright's work, wainscot, furniture and carving. For these last three purposes the softer, more figured wood is preferred, whilst for gate-posts, doors, stair-treads, door-sills, etc., the harder sorts are employed. The ancient Romans are said to have used the evergreen Holm Oak (*Q. Ilex*) for axles, and hard Oak is still used for this purpose on the Continent. Walking-sticks are also made of Oak and it furnishes an excellent charcoal. Excellent Oak is imported from Roumania.

Oak, Zeen (*Quércus Mibéckii* Durien). North-west Africa. Height 100—110 ft.; diam. up to 6 ft. S.G. when green 924. Breaking-weight per sq. millimetre 7·4 kilos., as against 4·7—7·2 kilos. for European Oak. Yellowish or rose-coloured; pith-rays numerous, broad, close; heavy, horny, straight-grained, very durable, but liable to shakes and warping. Used for sleepers, bridge-girders, piles, and wine-barrels; and, when winter-felled and seasoned for six or twelve months, is one of the most valuable timbers of Tunis, where it covers about 26,500 acres. In 1883 it fetched on the spot 3s. 9d. per cubic metre; in 1888 8s. 9d.

Oaks in America are somewhat numerous, three well-marked kinds, White, Red or Black, and Live Oak being distinguished in commerce. The evergreen or Live Oak (*Q. vírens*) of the Southern United States, formerly much employed in shipbuilding, though smaller than White Oak, is one of the heaviest, hardest and most durable timbers of the country. White Oak is more compact, tougher, stronger and more durable than Red Oak. We will, however, briefly describe the various species alphabetically.

Oak, Baltimore. See **Oak, White.**

Oak, Basket (*Q. Michaúxii* Nutt.). South-eastern States. Known also as "Cow" or "Swamp Chestnut Oak." *French* "Chêne de panier." *Germ.* "Korb-Eiche." *Span.* "Roble de canasto." Height 100 ft. or more; diam. 3 ft. or more. Sapwood white; heart fawn-colour: rings fairly broad: pores in about two rows in spring wood: very heavy, hard, tough, very strong, very durable in contact with soil. Largely used for agricultural implements, cooperage, fencing, baskets, and fuel.

Oak, Black. See **Oak, Red,** and **Oak, Yellow.**

Oak, Burr (*Q. macrocárpa* Michaux). Canada and the North-eastern and Central States, westward to the Rocky Mountains. Known also as "Mossy-cup" or "Over-cup Oak." *French* "Chêne à gros gland." *Germ.* "Grossfrüchtige Eiche." *Span.* "Roble con bellotas musgosas." Height 100 ft. or more; diam. 4—7 ft. S.G. 745. W 46·45. R 982 kilos. Sapwood pale buff, heart rich brown; rings fairly broad; pores in about three rows in spring wood; heavy, hard, strong, tough, rather more porous than White Oak, more durable, in contact with soil, than any other American Oak. Classed with and used as White Oak.

Oak, Chestnut (*Q. Prínus* L.). Southern Ontario and North-eastern United States. Known also as "Rock Oak." *French* "Chêne de roche." *Germ.* "Gerbereiche," "Felsen Eiche." Height 80 ft. or more; diam. 3—4 ft. or more. S.G. 750. W 46·7. R 1031 kilos. Sapwood brownish white; heart rich brown; rings narrow; pores hardly more than a single row; heavy, hard, rather tough, strong, durable in contact with the soil. Chiefly valued for its bark but used for fencing, railroad-ties and fuel. The name is also applied to *Q. Muhlenbérgii* [See **Oak, Chinquapin**]. The Californian Chestnut Oak is *Q. densiflóra* [See **Oak, Tan-bark**], and the Swamp Chestnut Oak, *Q. Michaúxii* [See **Basket-Oak**].

Oak, Chinquapin (*Q. Muhlenbérgii* Engelm.). Eastern United States. Known also as "Chestnut Oak" and "Yellow Oak." *French* "Chêne jaune." *Germ.* "Kastanien Eiche." Height

80 ft.; diam. 3—4 ft. S.G. 860. W 53·6. R 1238 kilos. Sapwood brownish white; heart rich brown; rings of moderate width; pores in 1—2 rows; heavy, hard, strong durable in contact with soil. Valued for railway-ties, cooperage, furniture, fencing and fuel.

Oak, Cow. See **Oak, Basket.**

Oak, Duck. See **Oak, Water.**

Oak, Iron. See **Oak, Post.**

Oak, Live (*Q. virens* Ait.). Southern States. *French* "Chêne vert." *Germ.* "Lebenseiche, Immergrüne Eiche." Height 60 ft. or more; diam. 5 ft. or more. Sapwood light-brown; heart dark brown; rings of moderate width; pores very few and small; pith-rays distinct and bright; very heavy, compact, hard, tough, strong, fine but somewhat twisted in grain and consequently very difficult to work, durable. Seldom yielding large straight timber, but with many crooked pieces, it was formerly much used for knees in shipbuilding. It is, perhaps, stronger than any known Oak, and is now used by wheelwrights, millwrights and toolmakers.

Oak, California Live (*Q. chrysolépis* Liehm.). Pacific States at altitudes of 3000—8000 ft. Known also as "Thick-cup Live Oak, Maul Oak," and "Valparaiso Oak." Height 80 ft. or more; diam. 5 ft. or more. Very heavy, hard, tough, very strong. Considerably used in waggon-building and for agricultural implements.

Oak, Mossy-cup. See **Oak, Burr.**

Oak, Peach. See **Oak, Tan-bark** and **Oak, Willow.**

Oak, Pin (*Q. palústris* Du Roi.). South-central States. Known also as "Swamp Spanish" and "Water Oak." *French* "Chêne marécageaux." *Germ.* "Sumpf Eiche." No distinct heart; rings wide, very wavy; pores very numerous, forming a wide zone; light brown.

Oak, Possum. See **Oak, Water.**

Oak, Post (*Q. obtusilóba* Michaux). Eastern and Southern States. Known also as "Iron Oak." *French* "Chêne poteau." *Germ.* "Pfahl Eiche, Posteiche, Eiseneiche." Height 60 ft. or more; diam. 3 ft. or more. Sapwood light brownish; heart sharply defined, dark brown; rings rather narrow; pores small, in about three rows; very heavy, hard, very durable in contact with soil. Used chiefly for railroad-ties, fencing and fuel; but occasionally for cooperage and carriage-building.

Oak, Punk. See **Oak, Water.**

Oak, Quebec. See **Oak, White.**

Oak, Quercitron. See **Oak, Yellow.**

Oak, Red (*Q. rúbra* L.). Canada and North-eastern States. Known in commerce as "Canadian Red" and as "Black Oak." *French* "Chêne rouge." *Germ.* "Rotheiche." Height 80—100 ft. or more; diam. 4—6 or 7 ft. S.G. 654. W 40·76. R 990 kilos. Sapwood almost white; heart light brown or reddish; rings wide; pores numerous, in a wide zone; pith-rays indistinct; heavy, hard, strong, but inferior to White Oak, coarse-grained and so porous as to be unfit for staves for liquor casks, shrinking moderately without splitting, easy to work. Used for flour and sugar barrels, clapboards, chairs and interior finish, and imported from Canada to London and still more to Liverpool for furniture-making. It is valued for its bark. [See also **Oak, Spanish.**]

Oak, Rock. See **Oak, Chestnut.**

Oak, Scarlet (*Q. coccínea* Wang.). Eastern United States. Height 100 ft. or more; diam. 3—4 ft. S.G. 740. W 46. R 1054 kilos. Sapwood whitish; heart ill-defined, pinkish-brown, heavy, hard, strong; rings narrow, wavy; pores in 3—4 rows making a rather broad zone; pith-rays prominent. Used in cooperage, chair-making and interior finish, being treated in trade as Red Oak and of small value.

Oak, Spanish (*Q. falcáta* Michx.). Eastern and Southern States. Known also as "Red Oak." Height 70 ft. or more; diam. 4 ft. or more. Heavy, very hard and strong, but not durable. Valued for its bark; but used in building and cooperage, and as fuel.

Oak, Swamp Spanish. See **Oak Pin.**

Oak, Tan-bark (*Q. densiflóra* Hook and Arn.). Pacific

coast. Known also as "Peach" or "California Chestnut Oak." Height 60—70 ft.; diam. 2—3 ft. Heavy, hard, strong. Classed as an inferior White Oak; but valued chiefly for its bark.

Oak, Water (*Q. aquática* Walt.). Central, southern and south-eastern States. Known also as "Duck, Possum" or "Punk Oak." Height 70—80 ft.; diam. 3—4 ft.; heavy, hard, strong. Sapwood whitish; heart ill-defined, light brown; rings of moderate width, wavy; pores in 1—2 rows, graduating into those of the autumn wood; pith-rays numerous and prominent, but not very wide. Used in cooperage, but chiefly as fuel. The name is also applied to *Q. palústris*. [See **Oak, Pin**.]

Oak, White (*Q. álba* L.). South-eastern Canada, Eastern United States. Height 70—130 ft.; diam. 6—8 ft. S.G. ·1054—695. W 46·35. *e*′ 1·19—1·58. *p*′ 1—·9. *c* 7021—3832. *c*′ ·927—·506. *v*′ ·912—·771. R 905 kilos. Sapwood whitish; heart defined, reddish brown; heavy, hard, tough, straight-grained, strong, durable in contact with soil; rings narrow, slightly wavy; pores in spring wood in 1—2 rows, those in summer wood very fine; pith-rays numerous and prominent; wide radial groups of dense woody-fibre extending across the summer wood crossed by several concentric lines of fine pores. One of the most generally useful of American hard-woods, being so elastic that "planks cut from it may, when steamed, be bent into almost any form," shrinking and splitting very little in seasoning, but liable to some twisting, free from knots, and shipped in logs from 25—50 ft. long and 11—28 inches square, or in thick-stuff or planks. Largely used in shipbuilding, house-frames, interior finish, door-sills, staves for wine-casks, railway and other carriage-building, agricultural implements, fence-posts, sleepers, piles, furniture and fuel. Though beautifully marked when quarter-sawn, it is inferior to the best European Oak. "Quebec Oak" is the trade name of an excellent quality and "Baltimore Oak" that of a somewhat inferior one, both named from their port of shipment, and realizing from 1s. 4d. to 2s. per cubic foot in London. The name "White Oak" is applied in the southern states to *Q. Durándii* Buckley, and in the west to *Q. garryána* Dougl.

Oak, Swamp White (*Q. bícolor* Willd.). Eastern Canada and United States. *French* "Chêne de marais." *Germ.* "Sumpf Weisseiche, Zweifarbige Eiche." Height 75—100 ft.; diam. 5 ft. S.G. 766. W 47·75. R. 909 kilos. Sapwood whitish; heart defined, pinkish brown, heavy, hard, tough and strong, resembling *Q. aquática* but with more defined heart and wide rings and pith-rays. Classed in trade as "White Oak"; but appearing inferior.

Oak, Weeping or **Western** (*Q. lobáta* Née). California. *Germ.* "Westliche Weisseiche." The largest-growing species on the Pacific coast. Classed as "White Oak."

Oak, Willow (*Q. Phéllos* L.). Eastern States. Known also as "Peach Oak." Heavy, hard, very elastic, but small.

Oak, Yellow (*Q. tinctória* Bartram). Eastern United States. Known also as "Black" or "Quercitron Oak." *French* "Chêne jaune." *Germ.* "Färber Eiche." Height 80 ft. or more; diam. 3 ft. or more. Sapwood white; heart reddish-brown, heavy, hard, coarse-grained, porous, strong, but not tough; rings narrow, wavy; pith-rays numerous; pores in spring wood in 3—5 rows. Valued for its bark and used as a substitute for White Oak in building, cooperage, etc., and for fuel. [See also **Oak, Chinquapin.**]

In the Himalayas there is a considerable variety of species of Oak, most of which are evergreen. The wood of these species is often hard, durable and valuable, resembling English Oak, but not having distinct annual rings, these being replaced by partial zones of wood-parenchyma or "false rings." Among them are:

Oak, Brown (*Q. semecarpifólia* Sm.). Afghanistan to Bhotan, at altitudes of 8000—10,000 ft. Wood large, reddish-grey, very hard. Used for all kinds of building and for charcoal.

Oak, Green (*Q. dilatáta* Lindl.). Afghanistan and the North-west. Wood large, hard, seasoning well without warping, durable. Used for building. The name is also applied to *Q. glaúca* Thunb., which grows from Kashmir to Bhotan and in Japan, and yields a brownish-grey, very hard wood, used in house and bridge-building.

Oak, Grey (*Q. incána* Roxb.). From the Indus to Nepal, at altitudes of 8000—3000 ft. Known also as "Himalayan Ilex" or "Ban," and in Kumaon as "Munroo." Heartwood reddish-brown, very hard, but warping and splitting considerably in building. Used in building.

Oak, Holm (*Q. Ílex* L.), the same species that occurs in Southern Europe, occurs also in the North-west.

Oak, Ring-cupped (*Q. annuláta* Sm.). Sikkim, up to altitudes of 10,000 ft. A well-marked, handsome, but not durable wood.

Q. fenestráta Roxb., of the Eastern Himalaya, from Sylhet to Burmah, and of the Khasia Hills, growing down to 50 ft. above the sea, yields a red, very hard, good and durable heartwood, somewhat inferior to English Oak.

Q. Griffíthii Hook. fil. and Thom., of Bhotan, Sikkim and the Khasia Hills, yields a brown, very hard, strong wood, much resembling English Oak, used in building.

Q. lamellósa Sm., occurring from Nepal to Bhotan, has a grey-brown wood with a beautiful silver grain, used in building, but not very durable if exposed.

Q. lanceæfólia Roxb., of the Garrow Hills and Assam, yields a light-coloured wood, resembling English Oak, but harder and very durable.

Q. lappácea Roxb., of the Khasia Hills, has a strong wood, resembling English Oak, but hard and more close-grained.

Q. pachyphýlla Kurz, of the Eastern part of the range, at altitudes of 8000—10,000 ft., yields a greyish, very durable, damp-resisting timber, used for fencing, shingles and planks.

Q. serráta Thunb., which ranges from the Himalaya into China and Japan, yields a brown, very hard, building wood, resembling that of *Q. Griffíthii*.

Q. spicáta Sm., the range of which extends from the Himalayas to Malacca and the Sunda islands, yields a reddish, very hard and durable wood, used in India for building.

In Southern Japan several species of evergreen Oak occur, including *Q. acúta* Thunb., "Aka-gashi," with a dark red-brown, very hard and heavy heartwood, used in waggon-building; the lighter-coloured *Q. gílva* Bl. "Ichii-gashi"; and the greyish-white *Q. vibrayeána* Tr. and Tav., "Shira-gashi," and *Q. myrsinæfólia* Bl., "Urajiro-gashi," used in ship-building and waggon-building. In Northern Japan occurs *Q. grosserráta* Bl., "O-nara," the wood of which is employed for building and furniture.

In Australia, where there are no true Oaks, many very diverse species are so named; but the name is chiefly applied to species of *Casuarína* (Order *Casuarínæ*), from a fancied resemblance in the colour and broad pith-rays of their wood to that of true Oak. These woods have been known in English trade as "Botany Oak" and used in veneer and inlaying.

Oak, Bull (*Casuarína glaúca* Sieb.). Also known as "Swamp-oak, Desert" or "River She-oak." *Aborig.* "Billa." Height 40—50 ft.; diam. 1—2 ft. Red with small darker veins, somewhat resembling *Quércus Ílex*, the Holm Oak, close-grained, strong. Used for staves, shingles and fence-rails, but not suited for posts. The name is also applied to *C. equisetifólia*. [See **Oak, Swamp**.]

Oak, Forest (*C. torulósa* Ait.). North-east Australia. Known also as "Beef-wood," "River" and "Mountain Oak." Height 60—80 ft.; diam. 1½—2 ft. W 64. Heart well defined, prettily marked, close-grained. Much used for shingles and fuel, and also used for furniture, either solid or in veneer. The name is also applied to *C. equisetifólia* [See **Oak, Swamp**] and to *C. suberósa* [See **Oak, Erect She**.]

Oak, River, a name applied to *Callistémon salígnus* [See **Bottle-brush, White**], *Casuarína Cunninghamiána* [See **Oak, Scrub She**], *C. distýla* [See **Oak, Stunted She**], *C. strícta* [See **Oak, Shingle**], and *C. torulósa* [See **Oak, Forest**].

Oak, She, a usefully distinctive name for the five following species of *Casuarína*, viz.: *C. strícta* [See **Oak, Shingle**], *C. glaúca* [See **Oak, Bull**], *C. suberósa*, *Cunninghamiána* and *distýla*.

Oak, Erect She (*C. suberósa* Ott. and Dietr.). Central and Eastern Australasia. Known also as "Beef-wood, Forest, Swamp" or "Shingle Oak." Height 30—50 ft.; diam. 1—2 ft. W. 59·6. Reddish, beautifully marked, very apt to split in drying. Used for shingles, handles, mallets, etc., and formerly for boomerangs; but would be valuable for veneers.

Oak, Scrub She (*C. Cunninghamiána* Miq.). North-east Australia. Known also as "River Oak." Height 60—70 ft.; diam. 2 ft. Prettily marked, hard, close-grained. Used for shingles, staves and fuel.

Oak, Stunted She (*C. distýla* Vent.). Southern and Western Australasia. Known also as "River Oak." Height 40—60 ft.; diam. 1½—2 ft. Brown to deep-red, light, tough, strong. Used for bullock-yokes.

Oak, Shingle (*C. stricta* Ait.). South-east Australasia. Known also as "Coast She-Oak," "Salt-water Swamp Oak" and "River Oak." Height 20—30 ft.; diam. 9—15 in. S.G. 1037—935. W 57—63. Reddish with dark longitudinal bands giving a beautifully mottled appearance to the outer part of the heart, the darker centre being less handsome, heavy, close-grained, very hard, tough, working up splendidly, but not durable. Used for shingles, staves, spokes, axe-handles, turnery and furniture.

Oak, Silky (i) (*Stenocárpus salígnus* R. Br.; Order *Proteáceæ*). North-east Australia. Known also as "Silvery Oak" and "Beef-wood." *Aborig.* "Melyn." Height 30—50 or 80 ft.; diam. 1—3 ft. W 44·25. Red-brown, sometimes dark, with a beautiful wavy figure, hard, close-grained, splitting and working readily, durable. A most beautiful wood, used for furniture, veneers and walking-sticks; but becoming scarce. (ii) (*Grevíllea robústa* A. Cunn.; Order *Proteáceæ.*). North-east Australia and successfully introduced into Ceylon. Height 70—80 or 100 ft.; diam. 2—3 ft. S.G. 564. W 35·25—38·8. Light-coloured but prettily marked, especially where knots are present, moderately hard, elastic, working well, durable. Largely used for staves for tallow-casks and now becoming scarce. (iii) (*Orítes excélsa* R. Br.; Order *Proteáceæ*). North-east Australia. Known also as "Red Ash."

Height 40—70 or 80 ft.; diam. 2—3 ft. Grey, prettily marked, hard, susceptible of a good polish, durable. Used for cask-staves, shingles and farm implements.

Oak, Swamp (*Casuarína equisetifólia* Forst.). North-east Australia; introduced near Madras. Known also as "Beef-wood," "Forest" and "Bull Oak," and in the South Seas as "Ironwood." *Anglo-Indian* "Fir," from an external resemblance to Larch. *Madagascar* and *Mauritius* "Filaof." *Dekhan* "Sarv" (cypress). *Indian Archipelago* "Aroo." *Tam.* "Chouk." Height 50—70 ft.; diam. 1—1½ ft. W 55—63. Reddish, coarse-grained, beautifully-marked, hard, tough, strong, straight in growth and very durable. Used for fencing and shingles and largely for fuel, for which it is excellent. The name is also applied to *C. glaúca* [See **Oak, Bull**] and *C. suberósa* [See **Oak, Erect She**].

Oak, White (*Lagundria Patersóni*). See **Tulip-tree**.

In addition to these the name Oak is applied in Ceylon to *Schléichera trijúga* [See **Kosum**], and in New Zealand to *Aléctryon excélsum* Gaertn. (Order *Sapindáceæ*), the "Titoki" of the Maoris, which is used in building.

Oak, African. See **African.**

Oak, Indian, a name sometimes applied to *Barringtónia acutángula* Gaertn. (Order *Myrtáceæ*), a species ranging from the Seychelles to Queensland and Northern India. *Hind.* "Samandar-phal, Hijjul." *Tam.* "Radami." *Telug.* "Kanapa" or "Kanigi." *Burm.* "Kyai-tha." A large tree, yielding red, fine-grained, hard timber, said to be equal to Mahogany, and used in boat and cart-building, well-work and cabinet making. W 46.

Olive (*Ólea europǽa* L.: Order *Oleáceæ.*). Mediterranean region; introduced into California, India, and other countries. *French* "Olivier." *Germ.* Oelbaum," "Olivenholz." Height seldom more than 20 feet. Very close and fine-grained, light yellowish-brown, with irregularly wavy dark lines and mottlings, especially near the root, resembling Box in texture, but not so hard and rather brittle, taking an excellent polish, with no distinguishable rings or pith-rays and minute, evenly distributed vessels. Used chiefly in turnery and carving for small articles, fancy boxes, paper-knives, etc.

Olive or **Wild Olive** in Cape Colony (*Ólea verrucósa* Link.) *Boer* "Olivenhout." Also known as "Olina wood." *Zulu* "Umguna." Height 14—16 ft.; diam. 8—15 in. W 68·95. E 669 tons. *f* 6·65. *fc* 3·90. *fs* ·8. Dark, very hard, heavy, dense, taking a good polish. Used in waggon-building and for furniture.

Olive, Indian (*Ólea dióica* Roxb.). Silhet and Assam southward. White, compact, strong. Used in building and might be creosoted. Other Indian species are *O. glandulífera* Wall., light-brown, dense, hard, susceptible of a good polish, and durable, used in building; and *O. cuspidáta* Wall., resembling the Common Olive.

Olive, Mock. See **Axe-breaker.**

Olive, Native, in Australia (*Noteléa ováta* R. Br.; Order *Oleáceæ*). Eastern Australia. *Aborig.* "Dunga-runga." Small and crooked in growth, light-coloured, with irregular dark-brown blotches, fine, close and even in grain, hard, firm, working easily and taking a good polish, but requiring careful seasoning. W 60·3. Used for tool-handles. [See also **Marblewood.**]

Olive, Native, in North America (*Osmánthus americánus* Gray: Order *Oleáceæ*). Also known as "Devil-wood." *French* "Olivier d'Amerique." *Germ.* "Amerikanischer Oelbaum." *Span.* "Madera del diablo." Resembling Box, very hard and durable, with evenly circular rings and fine vessels in dendritic lines across the entire ring.

Olive, Spurious. See **Ironwood** vi., and for other allied species, **Ironwood** vii, xi—xv and xxiv.

Omatsu. See **Pine, Japanese Black.**

Orange (*Cítrus Aurántium* L.: Order *Aurantiáceæ*). Probably a native of India, cultivated for its fruit in most tropical and subtropical countries. *Sansk.* "Nagranga." *Arab.* and *Pers.* "Naranj." *Hind.* "Naringi." *Span.* "Naranja." *French* "Oranges." *Germ.* "Pomeranzen." Small, light yellow, close-grained, hard. Imported from Algeria for walking-sticks, and used in the West Indies in cabinet-work.

Orange, Black. See **Broom.**

Orange, Mock. See **Cheesewood.**

Orange, Native, in Australia (i) (*Cítrus austrális* Planch.). Height 30—40 ft.; diam. 9—12 in. Resembling the Common Orange; (ii) from the shape of the fruit, (*Cápparis Mitchelli* Lindl.: Order *Cappárideæ*), known also as "Small Native Pomegranate." Height 14—20 ft.; diam. 1 ft. Whitish, hard, close-grained, closely resembling Lancewood. Suitable for engraving; (iii) *Endiándra vírens* F. v. M.: Order *Lauráceæ*). North-east Australia. Known also as "Bat and ball," "Native pomegranate," and "Ullagal Mabbie." A tall shrub, with grey, close-grained, firm, apparently useful wood.

Orange, Osage (*Maclúra aurantíaca* Nutt.: Order *Moráceæ*). Arkansas and Texas. Known also as "Bow-wood." *French* "Bois d'arc." Height 50 ft. or more; diam. 2 ft. Sapwood yellow, heart brown transversely, yellow longitudinally, soon turning greyish on exposure, very heavy, hard and strong, not tough, flexible, of moderately coarse texture, shrinking considerably in drying, very durable in contact with soil. Formerly used for bows and wheelwrights' work, now for fence posts, railway-ties, waggon-building and paving-blocks; but suitable for turnery and carving.

Pader or **Padri** (*Stereospérmum chelonoídes* DC: Order *Bignoniáceæ*). India and Burma, Ceylon and Sunda islands. Height to first branch 30 ft.; diam. 1—2 ft. Grey, reddish-brown or orange wood, hard, elastic, easy to work, moderately durable. Used in house-building, for canoes, furniture and tea-chests.

Padouk (*Pterocárpus índicus* Willd.: Order *Leguminósæ*). Burma, Andaman, Sunda and Philippine Islands and Southern China. Known also as "Andaman Redwood," "Burmese Rosewood," or "Tenasserim Mahogany." *Fiji* "Cibicibi." Height to first branch 35 ft.; diam. 2—5 ft., yielding timber 15—30 ft. long. W 60. R 1000 lbs. Dark red, beautifully variegated and darker near the root, resembling Mahogany, but heavier, slightly aromatic, very heavy, moderately hard, coarse but close-grained, working fairly well, taking two years to season, termite-proof, and susceptible of a high polish. It is used by the Burmese for musical instruments and cart-wheels; in India for gun-carriages and furniture; and is recommended as a Teak-substitute for railway carriages.

Pahautea. See **Cedar, New Zealand**.

Pai'cha (*Euónymus europǽus*, var. *Hamiltoniánus* Wall.: Order *Celastríneæ*). Ning-po. Perhaps also known as "Tu chung mu." Yellowish-white, very hard, close and fine-grained. Inferior to Box; but one of the best substitutes yet found for it as an engraver's wood.

Palisander-wood (i) *Jacaránda brasiliána* Pers.: Order *Bignoniáceæ*). Brazil. Sapwood very narrow, grey; heart dark chocolate-brown, marked by deep black veins and bands, very heavy, hard, difficult to split, almost brittle; rings scarcely visible; pith-rays invisible; vessels large, appearing like strings of pearls on longitudinal, and as light red spots on transverse sections. A valuable wood, chiefly used in pianofortes. (ii) possibly *Dalbérgia nígra* Allem., or some species of the allied genus *Machǽrium* (Order *Leguminósæ*) may be the source, in whole or in part, of this wood. *Dalbérgia nigra*, sometimes apparently known as "Jacaranda cabiuna," is dark-coloured, porous and open-grained. S.G. 768—841. It is a valuable furniture wood.

Palo Maria. See **Poon**.

Palo Mulato. See **Bois Mulatre**.

Palo Narango. See **Fustic**.

Panacoco (*Robínia Panacóco* Aubl. = *Swártzia tomentósa* DC.: Order *Leguminósæ*). French Guiana. Height up to 50 ft.; diam. 8 ft., imported in logs squaring 17 in. and upwards of 32 ft. long. S.G. 1231—1181. R 400 kilos. Sapwood white; heart black, more lustrous than ordinary Ebony, very compact and durable. Used in fencing; but most valuable for cabinet-work.

Pao d'arco (*Tecóma speciósa* D.C.; Order *Bignoniáceæ*). Brazil. Height 100 ft.; diam. 10 ft. or more. Very hard, compact and elastic.

Pao precioso (*Mespilodáphné pretiósa* Nees; Order *Lauríneæ*).

Brazil. Very hard, compact, with beautiful grain, fragrant. Used in building and in perfumery.

Papaw (*Asimína tríloba* Dunal: Order *Anonáceæ*). Middle, southern and western United States. Also known as "Custard-apple" and "Cucumber-tree." *French* "Asiminier." *Germ.* "Dreilappiger Flachenbaum." *Span.* "Anona." Height 15—30 ft. S.G. 397. W 24·74. Sapwood narrow, heart dark green, light, soft, weak, coarse-grained.

Papri. See **Elm, Indian**.

Parcouri (*Clúsia insígnis* Mart.; Order *Clusiáceæ*). Guiana. Known in Demerara as "Wild Mammee, Coopa" and "Cowassa." S.G. 816 for the black, 784 for the yellow variety. Fine, compact and even in grain.

Partridge-wood in South America may be *Ándira inérmis* [See **Angelin**]. In Northern Australia it is the outer part of the Palm *Livistóna inérmis* R. Br., also known as "Cabbage-palm." Height 14—40 ft.; diam. 1 ft. Light grey, streaked with a darker colour, the fibro-vascular bundles, producing a beautiful effect, very hard and taking a good polish.

Patawa (*Œnocárpus Bataúa* Mart.; Order *Palmáceæ*). French Guiana. With parallel veins of black and white. Suitable for walking-sticks, umbrella-handles and inlay.

Peach-wood (*Cæsalpínia echináta* Lam.: Order *Leguminósæ*). Central and South America. Known also as "Lima, Nicaragua or Pernambuco-wood," or "Bresil de St. Martha." Small and little known, hard, heavy and susceptible of a good polish. Valuable as a red, orange, or peach-coloured dye; but inferior to Brazil-wood.

Pear. (*Pýrus commúnis* L.: Order *Rosáceæ*). Europe and Western Asia; cultivated as a fruit tree in other temperate climates. *French* "Poirier." *Germ.* "Birnbaum." *Span.* "Peral." Height 20—50 ft.; diam. 1—2 ft. No true heart, but sometimes a darker more chocolate-brown in the centre, light pinkish-brown, moderately heavy, hard, close-grained, tough, firm, difficult to split, but easily cut in any direction, taking a satiny polish, and very durable, if kept dry. Rings recognizable by the dark zone of the autumn-wood; pith-rays and vessels not visible to the naked eye. Highly esteemed for turnery, cabinet-work, T-squares and other drawing-instruments, calico-printing blocks, coarse wood-engraving and, when "ebonized" or stained black, for picture-frames.

Pear, Hard (*Olínia cymósa* Thunb., var. *intermédia*: Order *Oliniéœ*). Cape Colony. *Zulu* "Umnonono." Height 14—16 or 30 ft.; diam. 1 ft. Yellowish, very heavy, hard, compact and tough. Valuable for axles and carriage-poles and suitable for musical instruments.

Pear, Native in Australia (i) (*Xylomelum pyrifórmé* Knight: Order *Proteáceœ*). New South Wales and West Australia. Known also as "Wooden Pear." Height 20—40 ft.; diam. 6—8 in. W 46. Sapwood narrow, light; heart a rich dark reddish with a beautiful rich figuring on tangential sections, taking an excellent polish. Occasionally used for picture-frames, veneers and walking-sticks. (ii) (*Hákea aciculáris* R. Br. var. *lissospérma*: Order *Proteáceœ*). South-east Australasia. Height 20—30 ft.; diam. 8—10 in. Hard and used in turnery.

Pear, Red (*Scolópia Ecklónii* Benth. and Hook. fil.: Order *Flacourtiáceœ*). South Africa. Height 30—35 ft.; diam. 2—3 ft. Heavy, hard, close-grained. Used chiefly by wheel-wrights and in mill-work.

Pear, Thorn or **Wolf** (*Scolópia Zéyheri* Benth. and Hook. fil.). Cape Colony. *Zulu* "Igumza elinameva." Height 60—70 ft.; diam. 2—3 ft. Straight-growing, very hard and close-grained, very difficult to saw. Used for cogs.

Pear, Wooden. See **Pear, Native**.

Pear-wood, White (*Pterocelástrus rostrátus* Walp.: Order *Celastráceœ*). South Africa. *Zulu* "Umdogan." Height 20—25 ft.; diam. 1—2 ft. W 42·84. E 635 tons. *f* 6·95. *fc* 3·04. Moderately heavy, strong and durable. Much used for felloes and other waggon work.

Penagah. See **Poon**.

Pencil Cedar. See **Cedar, Pencil**.

Pencil-wood (*Pánax Múrrayi* F. v. M.: Order *Araliáceœ*).

North-east Australia. Height 50—60 ft. S.G. 348. Light-coloured, the lightest in weight of any wood in the district, soft. Pith large. The wood hardens externally in drying, so that the outside is often harder than the centre. Cuts well and is recommended for lining-boards.

Pepperidge. See **Gum, Black.**

Peppermint, a name applied in Australia to various species of *Eucalýptus,* including (i) *E. amygdalína,* which is known as "Brown, Dandenong, Narrow-leaved," or "White Peppermint" [See **Ash, Mountain**]; (ii) *E. capitelláta* [See **Stringybark, White**]; (iii) *E. microcórys* [See **Tallow-wood**]; (iv) *E. pauciflóra* [See **Gum, Mountain White**]; (v) *E. Stuartiána* [See **Gum, Apple-scented**]; (vi) *E. piperíta*; and (vii) *E. odoráta*

E. piperíta Sm. (Order *Myrtáceæ*). Eastern Australia. Known also as "Blackbutt, Redwood, Messmate, White" or "Almond-leaved Stringybark." Height 80—100 ft.; diam. 2—3 ft. S.G. 1109—922. W 69·22. Red, very heavy, but works with difficulty and is very subject to shakes, durable. Used for posts, shingles and rough house-building.

E. odoráta Behr. South-East Australia. Also known as "Red Gum, Box" and "White Box." Small. W 60—70. Yellowish-white or light-brown, heavy, very hard, tough, close and straight-grained, generally hollow. Used for fencing, wheels and fuel.

Peppermint, Bastard. See **Gum, Broad-leaved Water.**

Pernambuco-wood. See **Peach-wood.**

Peroba branca or **Peroba de campos** (*Sapóta gonocárpa* Mart.: Order *Sapotáceæ*). Brazil. A large tree, yielding straight timber, 60—70 ft. long, siding 24—40 in. S.G. 868—739. W 50. Yellow, moderately heavy, stronger than Teak, but not so heavy, close and fine-grained, easily worked, taking a high polish, very durable, even when in contact with iron. Used in building Brazilian ironclads, and is suitable for engineering or building work, or for furniture.

Peroba vermelha (*Aspidospérma* sp.: Order *Apocynáceæ*). Brazil. Red, moderately heavy, smooth, close and fine in grain, somewhat resembling Pencil Cedar. Used in shipbuilding.

Persimmon (*Diospýros virginiána* L.: Order *Ebenáceæ*). Eastern United States. Known also as "Date-plum." *French* "Plaqueminier de Virginie." *Germ.* "Virginische Dattelpflaume." *Span.* "Persimon." Height 80 ft.; diam. 2 ft. S.G. 790. W 49·28. R 879 kilos. Sapwood very broad, sometimes 60 rings, cream-colour; heart dark-brown or black, heavy, hard, close-grained, strong and tough, resembling Hickory, but finer in grain. Used for shuttles, shoe-lasts, plane-stocks, etc.

Pimento (*Piménta officinális* Lindl.: Order *Myrtáceæ*). Jamaica. Imported as walking-sticks.

Pin-bush (*Hákea leucóptera* R. Br.: Order *Proteáceæ*). Central and Eastern Australia. Known also as "Beef-wood," "Water-tree," and "Needle-bush." Height 15—25 ft.; diam. 4—6 in. S.G. 818. Heavy, coarse-grained, soft, taking a good polish. Used for tobacco-pipes, cigarette-holders and veneers.

Pine is the general name, originally applied in the Northern Hemisphere to the trees and wood of the coniferous genus *Pínus*, and subsequently extended—mainly in Australasia—to the allied genera *Ágathis*, *Frenéla*, *Araucária*, *Dacrýdium*, *Podocárpus* and *Pseudotsúga*. Curiously enough, however, the wood of the various local varieties of the Northern Pine (*Pínus sylvéstris*) imported from Baltic ports, especially Dantzic, Memel and Riga, is known in commerce as "Fir," or "Red" and "Yellow Deals," the name "Pine" being used for the timber of other species of the same genus imported from North America. The pines, often called firs, are known in French as "pin," in German as "Kiefer, Föhre," or "Pynbaum," in Italian and Spanish as "Pino," in Swedish as "Fura," in Danish as "Fyr," and in Russian as "Sosna."

The wood of Pines, as also of those other trees that are so-called and of all Conifers, is of the simple structure described in our first chapter, consisting only of tracheids, with the pith-rays, and, in most cases, resin-canals. That of the true Pines—the genus *Pínus*—has numerous resin-canals, uniformly scattered through the annual rings and has a distinct dark-coloured heart, though when the wood is freshly-cut this last is often not recognizable. The pith-rays are rarely more than one cell thick, and are, therefore, invisible to the naked eye, but vary in depth, having generally from three to eight radial rows of elements, of which the central rows are parenchyma or cellulose-walled cells, with simple pits on their radial walls, *i.e.* on those in contact with the tracheids of the xylem, whilst the upper and lower row or two consist of tracheids with bordered pits. The rings are rendered distinct by the darker and firmer zone of autumn wood in each, consisting of more compressed, thicker-walled elements. This simplicity of structure and resinous character renders the wood uniform and even in texture, easy to work and of considerable durability. It is also on the whole soft, light, elastic, stiff, and strong, characters which, coupled with its abundance in pure forests—forests, that is, mainly made up of a single species—combine to render the Pines the most generally useful and among the cheapest of woods.

The wood even of a single species of Pine varies very much, according to the conditions under which it is grown; but, though connected by intermediate cases, most of the species fall into two fairly well-marked groups, known in America as the "Hard" and "Soft" Pines. The Hard Pines are harder, heavier, and darker-coloured, ranging from yellow to deep orange or brown: their autumn wood forms a much broader proportion of the ring and is somewhat abruptly marked off from the spring wood; and the tracheids of their pith-rays have their walls very unevenly thickened with tooth-like projections. This group includes the Northern (*Pínus sylvéstris*), Austrian (*P. austríaca*), and Mountain (*P. montána*) of Europe, and the majority of the North American species. The Soft Pines are softer and lighter; range in colour from light red to white; have a narrow autumn zone gradually merging into the spring wood on its inner surface; and have smooth walls to the tracheids of their pith-rays, with no tooth-like projections. The group includes the Cembra Pine (*P. Cémbra*) of Europe, and the White (*P. Stróbus*), Sugar (*P. Lambertiána*) and a few other species in America.

Pine seasons rapidly and with but little shrinkage, this being, however, greater in the harder kinds: it is never too hard to nail;

and when once well seasoned, is not subject to the attacks of boring insects. It is not, therefore, to be wondered at that Pine has become by far the most extensively used of all woods. The straight-growing, tapering stem fits it for masts and spars: its strength and lightness recommend it for ships' timbers, planking, bridges, and carriage-building, its durability for sleepers, its resinous character for torches or fuel, the refuse yielding charcoal and lamp-black, and its cheapness for street-paving, general carpentry, common furniture and boxes and paper-pulp.

Pine, Adventure Bay. See **Pine, Celery-topped.**

Pine, Aleppo (*P. halepénsis* Mill.) Mediterranean region; introduced in Australia. Height 50—80 ft.; diam. 2—3 ft. Yellowish-white, fine-grained. Valued locally for telegraph poles, turnery, joinery, or fuel, and as a source of turpentine.

Pine, Austrian (*Pínus austríaca* Höss.). Lower Austria and the north of the Baltic peninsula. Known also as "Black Austrian Pine." *Germ.* "Schwarzkiefer." French "Pin noir d'Autriche." Height 80—120 ft., relatively slender. Wood very similar to that of the Northern Pine (*P. sylvéstris*); but, when grown in poor soil, apt to be knotty. Suitable for fencing or fuel.

Pine, Bastard. See **Pine, Cuban.**

Pine, Bhotan (*P. excélsa* Wall.). At altitudes of 6000 to 12000 ft., from Bhotan to the Kuram pass in Afghanistan. Also known as "Himalayan Pine." *French* "Pin pleureur." *Germ.* "Thränen Kiefer." *Chinese* "Tong-schi." Height 50—150 ft.; diam. 2—3 ft. Red, compact, close-grained, very resinous, durable. Used for torches; but the most valuable wood of its district for building or engineering work and second only in durability to the Deodar.

Pine, Big-cone (*P. Coúlteri* Don). Coast-range of California. Reported to be of small value as timber.

Pine, Bishops'. See **Pine, Obispo.**

Pine, Black, in North America (*P. Jéffreyi* Balf.). California and Oregon, above 6000 ft. Known also as "Bull Pine." Height 100 ft. or more, up to 300 ft.; diam. 4 ft. or more, up to 10 or

12 ft. Light, hard, strong, very resinous. One of the "Hard Pines," closely allied to the Bull Pine (*P. ponderósa*). Used locally, chiefly as coarse lumber. See also **Pine, Lodge-pole**.

Pine, Black, in New Zealand (i) (*Prumnópitys spicáta* Masters: Order *Taxíneæ*). *Maori* "Matai." Height 80 ft.; diam. 2—4 ft. Heavy, close, smooth, and even in grain, strong, easily worked, very durable. Used for piles, sleepers, house-building, mill-wrights' work, etc. (ii) (*Podocárpus ferrugínea* Don: Order *Taxíneæ*). *Maori* "Miro." *Germ.* "Mirobaum." Height 50—80 ft.; diam. 1—3 ft. S.G. 1214 when green, 752—660 when seasoned. Light to dark reddish-brown, sometimes nicely figured, moderately heavy and hard, close, straight, and even in grain, strong, elastic, planing up well and taking a good polish, durable in contact with sea-water, but not in contact with the soil. Used for piles and suited for house-building, cabinet-work or turnery.

Pine, Black, in Australia. See **Cypress-Pine.**

Pine, Bull (*P. ponderósa* Dougl.). Western North America. Known also as "Yellow" or "Heavy-wooded Pine." *Germ.* "Westliche Gelbkiefer." Height 100—150 ft. or up to 225 ft.; diam. 5—6 or 12 ft. Sapwood wide; heart very variable in weight, strength and durability, generally hard, brittle, strong, resinous, but not durable in contact with the soil. Furnishing most of the hard Pine of the West, being largely used for lumber, railway ties, mining-timber and fuel. See also **Pine, Black** and **Pine, Nut.**

Pine, Canadian Red (*P. resinósa* Sol.). Michigan and Minnesota to Newfoundland. Known in Canada as "Norway Pine" and in Nova Scotia as "Yellow Pine." *French* "Pin rouge d'Amerique." *Germ.* "Rothkiefer" or "Harzige Fichte." Height 60—100 ft. or more; diam. 2—2½ ft. S.G. 578—485. W 30—44. E 650—850 tons. *e'* 1·32. *p'* ·81. *f* 3·71. *ft* 5·1—6·3. *c* 2705. *c'* ·357. *fc* 2·4—2·76. *v'* ·62. *fs* ·22—·35. R 800 kilos. Sapwood yellowish-white; heart slightly reddish, light, harder than White Pine (*P. Stróbus*), tough, elastic, moderately strong, fine-grained, working up well, with a silky lustre, very resinous, durable, not shrinking or warping much in seasoning. Used for spars, shipbuilding, and piles; but chiefly for flooring, for which it is preferable to White Pine, with which it grows, it being in fact a hard Pine, resembling resinous examples of Scots Fir.

Pine, Canadian Yellow. See **Pine, White.**

Pine, Carolina. See **Pine, Short-leaf.**

Pine, Cedar. See **Pine, Lowland-Spruce.**

Pine, Celery-topped (*Phyllocládus rhomboidális* Rich.: Order *Taxíneæ*; = *Podocárpus asplenifólia* Labill.) Tasmania. Known also as "Adventure Bay Pine." Height up to 60 ft., usually too slender to be useful. Even-grained and easily worked. Occasionally used for spars.

Pine, Cembra (*P. Cémbra* L.). From Kamtschatka to the Urals, Carpathians and Alps. Known also as "Swiss Stone Pine." *French* "Cembrot, Tinier." *Germ.* "Zirbelkiefer, Zirbe, Arve." *Swiss* "Alvier, Arolla." Height 60—70 or 90 ft. Sapwood broad, yellowish-white; heart, when dry, white or yellowish-brown, light, soft, fine-grained, easily split, shrinking little, susceptible of a fine polish, fragrant and obnoxious to insects; annual rings regularly circular; narrow autumn wood scarcely distinguishable; resin-ducts numerous and very large; pith-rays with one row of smooth-walled tracheids above and below, with small bordered pits, and generally three rows of parenchyma in the middle, with large simple pits. A soft pine, in request for wainscotting, carved work, lining clothes' chests, turnery, etc.

Pine, Chile (*Araucária imbricáta* Pavon: Order *Araucaríneæ*) Southern Chile. Known also as "Monkey Puzzle." *Germ.* "Chilitanne, Schmucktanne." Height 70—100 ft.; diam. 5—7 ft. at base. Wood in English-grown specimens cross-grained and not seemingly of value; but in Chile yellowish, beautifully veined, and susceptible of a fine polish. Used in Chile for masts.

Pine, Cluster (*P. Pináster* Sol. = *P. marítima* Lam.) Mediterranean region; naturalized in South Africa, Northern India, Australia, etc. *French* "Pin de Bordeaux, Pin maritime, Pin des Landes." *Germ.* "Sternkiefer, Strandkiefer." Height 50—60 ft. Reddish, soft, coarse-grained, not durable. Used only for coarse carpentry, packing-cases and fuel; but of great value as a source of turpentine, charcoal and lamp-black being manufactured from the refuse.

Pine, Colonial. See **Pine, Moreton Bay.**

Pine, Common, of Australia. See **Cypress-Pine.**

Pine, Corsican (*P. Laricio* Poir.). Corsica and the Maritime Alps. Known also as "Larch Pine." *Germ.* "Schwarzkiefer." Height 80—100 ft. Creamy white when freshly cut, brownish-

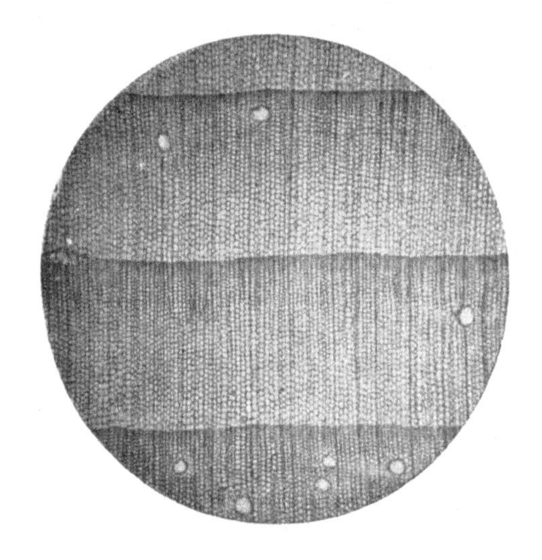

Fig. 56.—Transverse section of Corsican Pine (*Pinus Laricio*).

yellow when seasoned, tough, elastic, long but rather coarse in grain, very resinous, easily worked, susceptible of a fair polish, very durable, obnoxious to insects. Resembling Northern Pine (*P. sylvéstris*) of good quality in its structure and uses. (Fig 56.)

Pine, Cuban (*P. cubénsis* Griseb.). Southern United States. Known also as "Bastard, Meadow, Slash, or Swamp Pine," and in British Honduras as "Yellow" or "Pitch Pine." Height 75 ft. or more; diam. 2 ft. or more. Heavy, exceedingly hard, very strong, tough and durable, little inferior to Long-leaf Pine

(*P. palústris*), with which it is classed in Florida, but with wider sapwood and coarser grain. Used in carpentry.

Pine, Cypress. See **Cypress-Pine.**

Pine, Dantzic. See **Pine, Northern.**

Pine, Dark. See **Cypress-Pine.**

Pine, Digger. See **Pine, Nut.**

Pine, Dundatha (*Agathis robústa* Salisb.: Order *Coníferæ*). Queensland. Known also as "Queensland Kauri." Height 80—130 ft.; diam. 3—6 ft. Light yellow, soft, close-grained, easily worked. Largely used by joiners and cabinet-makers.

Pine, Flexible (*P. fléxilis* James). Rocky Mountains at altitudes of 4000—12,000 ft. Height 40—50 ft.; diam. 2—4 ft. Light clear yellow, turning red on exposure, light, close-grained, compact, very pliable, but very knotty and coarse-grained. Known and used locally as "White Pine."

Pine, Fox-tail (*P. Balfouriána* Murray). California at altitudes over 5000 ft. Known also as "Hickory" or "Awned Pine." Height 30—50 ft.; diam. 1—5 ft. Light, and apparently soft and not strong; but used in Nevada for mine-timbers.

Pine, Frankincense. See **Pine, Loblolly.**

Pine, Georgia. See **Pine, Long-leaf.**

Pine, Grey (*P. Banksiána* Lambert). Canada and Labrador from the Arctic Circle to Michigan and Maine. Known also as "Scrub, Jack, Yellow," or "Prince's Pine." *Germ.* "Strauchkiefer." Height 25—60 or 70 ft.; diam. 1—2 ft. Light, soft, not strong. Used chiefly for fuel and railway-ties.

Pine, Hickory (*P. púngens* Michx.). Alleghany Mountains. Known also as "Table Mountain Pine." *Germ.* "Stechende Kiefer." Height 25—40 ft. Light, soft, coarse-grained, not strong. Chiefly used for charcoal. [See also **Pine, Fox-tail.**]

Pine, Himalaya. See **Pine, Bhotan.**

Pine, Hoop, See **Pine, Moreton Bay.**

Pine, Huon. See **Huon-Pine.**

Pine, Jack. See **Pine, Grey.**

Pine, Japanese Black (*P. Thunbérgii* Parl.). Japan and

Korea. *Japan* "Omatsu, Kuro-matsu." Height, 80—90 or 120 ft.; diam. 2—4 ft. Used in house-building and for fuel.

Pine, Japanese Red (*P. densiflóra* S. and Z.). *Japan* "Me-matsu, Aka-matsu." Height 50—70 or 100 ft. Slender, coarse-grained, moderately strong, more ornamental than that of the preceding. Used for all kinds of carpentry and a favourite species in a dwarfed condition.

Pine, Jersey (*P. virginiána* Mill. = *P. ínops* Sol.). Eastern United States. Known also as "Scrub Pine." *French* "Pin chétif." Height up to 75—100 ft.; diam. up to 2—3 ft. East of the Alleghanies used only as fuel; to the west, where it reaches timber size, used in carpentry, especially in contact with water.

Pine, Kauri. See **Kauri.**

Pine, Lachlan. See **Cypress-Pine.**

Pine, Light. See **Cypress-Pine.**

Pine, Loblolly (*P. Tæda* L.). Southern United States. Known also as "Torch, Frankincense, Slash, Rosemary, Sap, Short-straw," or "Old Field Pine." *French* "Pin à l'encens," *Germ.* "Weihrauchkiefer." Height 80—100 or 175 ft.; diam. 2—5 ft. Sapwood wide; heart hard, though less so than in the Long-leaf Pine (*P. palústris*), with which it is generally confounded, lighter, coarser in grain, and with wider rings than that species, not strong or durable. Used for common lumber; but suited rather for fuel; rich in resin.

Pine, Lodge-pole (*P. Murrayána* Balf.). Mountains of Western North America. Known also as "Tamarack Pine, Black Pine," or, in the smaller form (*P. contórta* Dougl.) as "Oregon Scrub Pine." Height 70—80 or 150 ft.; diam. 4—6 ft. Light, hard, straight-grained, easily worked, but not strong or durable. Used locally for railway-ties and carpentry, and more generally for fuel.

Pine, Long-leaf, in America (*P. palústris* Mill. = *P. austrális* Michx.). Southern pine-barrens from North Carolina to Texas. Known also in the Northern States as "Southern, Georgia" or "Red Pine"; in the Southern States as "Turpentine-tree,"

"Yellow, Broom," or "Long-straw Pine"; and generally, especially in foreign trade, as "Pitch Pine," associated with the name of the port of origin, such as Darien, Pensacola, Savannah, etc. Height 50—100 ft.; diam. 1½—4 ft. S.G. 932—498. W 39—44. E 950 tons. *e'* 1·93—1·53. *p'* ·91—1·3. *f* 3·57. *ft* 4—5·09. *c* 4666. *c'* ·616. *fc* 3·99. *v'* ·847. Imported in logs and planks, 20—45 ft. long, squaring 11—18 in., or 3—5 in. thick and 10—15 in. wide. Reddish, resembling the Northern Pine (*P. sylvéstris*), but heavier and more resinous, owing to which latter character the broad zone of autumn wood appears greasy, tough, compact, clean, regular, straight, and sometimes fine and sometimes rather coarse in grain, susceptible of a high polish, rigid, rather difficult to work, but harder and stronger than other American Pine, liable to heart and cup-shake; but, I believe, very durable. "There are," says Mr. Stevenson, "numerous architects and civil engineers who rigidly adhere to the use of Memel and Dantzic Fir, and who will not allow the use of Pitch Pine, whilst there are others who rank it almost with the Oak, and state that in piling, and in jetties, exposed to the tides and weather, it will last double and treble the time allotted to Memel and Dantzic Fir." It is still the most abundant, and by far the most valuable Pine in the Atlantic States, occupying a belt from 80 to 125 miles wide, once covering 130,000 square miles. "Invaded from every direction by the axe, a prey to fires which weaken the mature trees and destroy the tender saplings, wasted by the pasturage of domestic animals, and destroyed for the doubtful profits of the turpentine industry, the forests . . . appear hopelessly doomed to lose their commercial importance at no distant day" (Sargent). Millions of feet of marketable timber are constantly destroyed by the carelessness of the turpentine-workers. As a source of turpentine it is the most important species in the world. Its timber is used in shipbuilding for spars, beams, and planking, the redder wood ("Red Pine" of the dock-yards in the Northern States) being specially valued for durability and a greater power of resisting the ship-worm than that possessed by Oak. In English shipbuilding it has almost extinguished the use of Baltic timber for spars, England importing, in all, over 500 million feet, or nearly 870,000 loads, more than a third of the entire export. In America it is largely used for fencing, railway-ties, mine-timbers, wood-paving, house-building, and fuel; whilst in this country it is largely used for wainscotting and church and school fittings, and to some extent in cabinet-making.

Pine, Long-leaf, of the Himalayas (*P. longifólia.* Roxb.). From Bhotan to Afghanistan at altitudes of 1500 to 7500 ft. *Hind.* "Chir." Height 100 ft. or more. Soft, not durable, easily worked. Used for tea-boxes, shingles and building; but chiefly valuable for its resin, one tree yielding 10—20 lbs. the first year. The wood is also used for torches and for charcoal, so that this is, on the whole, the most valuable Himalayan species.

Pine, Lowland Spruce (*P. glábra* Walt.). South-eastern United States. Known also as "Cedar" or "White Pine," and in Florida as "Old Field Pine." Height 80 ft. or more; diam. 3 ft. or more. Light, soft, brittle, easily worked, not strong or durable, resembling the Loblolly Pine (*P. Tǽda*), not resinous. Employed chiefly for inside work.

Pine, Maritime. See **Pine, Cluster.**

Pine, Meadow. See **Pine, Cuban.**

Pine, Monterey (*P. radiáta* Don = *P. insígnis* Loud.). South California. Height 80—100 ft.; diam. 2—5 ft. Light, soft, brittle, not strong. Used only for fuel.

Pine, Moreton-Bay (*Araucária Cunninghámi* Lamb.: Order *Araucarineæ*). North-east Australia and New Guinea. Known also as "Colonial" or "Hoop Pine." Height 150—200 ft.; diam. 3—5½ ft. S.G. 763—500. W 30—33·75. Yielding spars 80—100 ft. long, light-coloured, light, straight-grained, hard and strong, the sapwood liable to rot, but the heart durable if kept constantly dry or wet, working very easily, sometimes exhibiting a peculiar figure from groups of small knots. In request for flooring boards, carpentry, punt-bottoms, and to some extent for cabinet-work and spars, mountain-grown timber being preferred.

Pine, Mountain (*P. montána* Mill., including *P. Pumílio* Haenke, of Thuringia and the Carpathians, *P. Múghus* Wild. of the Tyrol, and *P. uncináta* Ram. of the Pyrenees). Central and Southern Europe, at altitudes of 500—8000 ft. *French* "Pin nain." *Germ.* "Bergkiefer, Krummholzkiefer, Zwergkiefer." Resembling the Northern Pine (*P. sylvéstris*), but small, often eccentric in growth, narrow-ringed, harder and heavier.

Pine, Murray and **Pine, Murrumbidgee.** See **Cypress Pine.**

Pine, New York. See **Pine, Short-leaf.**

Pine, Norfolk Island (*Araucária excélsa* R. Br.). *Germ.* "Norfolktanne." Height 150—200 ft.; diam. 5—7 ft., yielding excellent timber, but now scarce.

Pine, Northern (*P. sylvéstris* L.). Europe and Northern Asia, up to 700 ft. above sea-level in northern Norway and 6500 ft. on the Sierra Nevada of southern Spain. Known in Scotland as the "Scots," in England as the "Scotch Fir," and in the timber trade by various names according to its origin, such as White Sea, Baltic, St. Petersburg, Riga, Memel, Dantzic, Gefle, Soderhamn, Eliasberg, Saldowitz, Swedish or Norway Fir, Redwood, "Red" or "Yellow Deal." *French* "Pin sauvage." *Germ.* "Gemeine Kiefer, Föhre," or "Weissföhre." *Dutch* "Pynboom." *Danish* "Fyrre." *Russ.* "Sosna." Height 80—100 ft.; diam. 2—4 ft. S.G. 774—478. W 34—47. *e'* 1·3—1·69. *p'* ·74—1·087. *ft* 2·5—5·5. *c* 4051—3231. *c'* ·535—·427. *v'* ·93—·618. The characters and quality of the wood vary much according to climate and soil. Conversely to what is the case with Oaks, the more slowly-grown Pines of high latitudes or mountains, having narrower annual rings with a proportionally smaller amount of springwood, are heavier, denser and stronger than those of the south or of plains or from rich soils. English-grown Pine is thick-baited, carrying a great amount—often four inches—of sapwood, and is generally only used locally, not being nearly as durable as Larch. Scotch-grown wood is of better quality and is imported into the north of England, chiefly as mine timber. The pine from Prussia and central Russia is large, heavy, hard, resinous, and of excellent quality for sleepers, paving-blocks, masts, beams and planks; whilst from farther north, as from Gefle and Soderhamn in central Sweden, a wood of high quality is shipped, and from the White Sea a closer-grown, less resinous kind more suitable for joinery.

The sapwood is yellowish to reddish-white, the heart only becoming distinct as a brownish-red in drying, and the wood

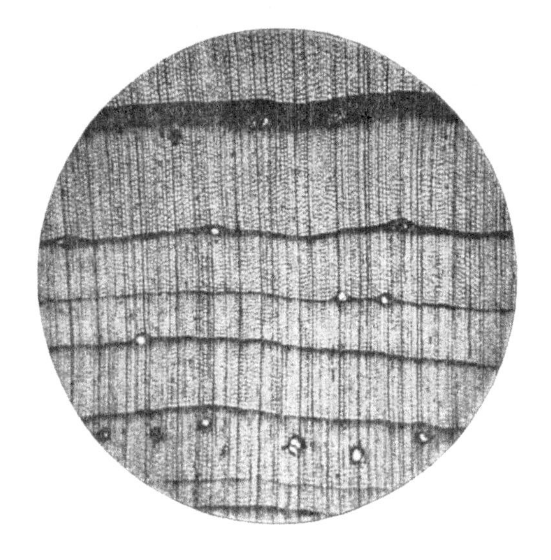

FIG. 57.—Transverse section of Northern Pine (*Pinus sylvestris*).

being on the whole whiter when grown on plains, redder when on hills. The annual rings are very distinct, owing to the broad sharply-defined zone of dark autumn wood that characterizes the "hard-pines," and they are slightly wavy. The resin-ducts are numerous, very large and distinct and mostly in peripheral zones near the outer margin of the rings (Fig. 57). The pith-rays consist of two or three rows of thin-walled parenchyma with large oval pits on their radial walls, each almost as wide as a tracheid, and two or three rows of tracheids above and below, with very unevenly thickened walls and small bordered pits. As the branches are in whorls, the knots serve to distinguish Pine from Larch, in which they are scattered (Fig. 58). In northern Europe the Pine is the chief timber used in house-building, both for framework of hewn logs, walls of logs in the round, and clap-boards: in Russia pine-logs are used for corduroy roads and the use of the wood for fuel is

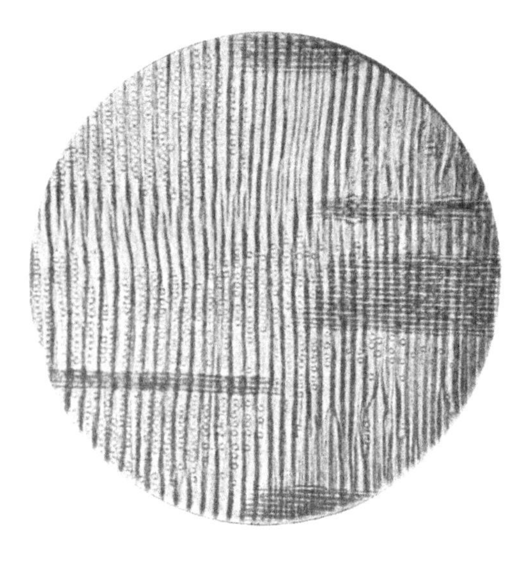

FIG. 58.—Radial section of Northern Pine (*Pinus sylvestris*).

universal. Baltic Pine was imported to and used in our east coast towns for flooring, wainscotting and joinery in the 15th, 16th and 17th centuries, when Oak was the chief building timber employed in England. Not till the beginning of the 18th was it recognized that the Scots Fir was the same species: the high price of foreign timber during the Napoleonic wars led to the clearing of the indigenous pine-forests of northern Scotland; and the excellence, easy working and great durability of the timber obtained from them broke down the prejudice in favour of Oak, and resulted in the great consumption of Baltic, White Sea and Canadian pine during the 19th century. Dantzic Fir, floated in rafts down the Vistula to that port, comes into the market in lengths from 18—50 ft., squaring from 11—20 in., in deals 2—5 in. thick, and as irregularly-grown logs for sleepers. The longest and straighest logs most free from knots are selected at Dantzic as "inch-masts," "hand-masts" and "spars" or "poles"; "inch-masts" being over 6 ft. in circumference and dressed octagonally or square; "hand-masts," from 2—6 ft. round, their length being measured in hands; and "spars" or "poles," less than 2 ft. round. This wood is valued for deck-planking, beams, joists, scaffolding, railway-work, etc. Its average specific gravity is given by Mr. Laslett as 582, and it is described by him as light, moderately hard, even and straight in grain, tough, elastie and easily worked. Riga Fir, with fewer knots but a slight tendency to heart-shake which makes it more wasteful in conversion into plank, averages about 541 in specific gravity, and so is on the whole inferior to Dantzic. Swedish Fir is yellowish-white, liable to heart, star and cup-shakes, and does not exceed 35 ft. in length, or 16 in. square. It furnishes cheap building material, deals for rough carpentry and much wood for matches and firewood. Norway Fir comes over in cheap prepared flooring and matchboarding, and as firewood.

Pine, Norway. See **Pine, Canadian Red**.

Pine, Nut (*P. Sabiniána* Dougl.). California, up to 4000 ft. Known also as "Digger" or "Bull Pine." *Germ.* "Nusskiefer, Weisskiefer." Height 40—50 or 80 ft.; diam. 1—4 ft. Light, very soft, resinous, cross-grained, not durable. Valuable only as fuel.

Pine, Obispo (*P. muricáta* Don.). Californian coast. Known also, by a corruption, as "Bishop's Pine" and as "Prickle-coned Pine." *Germ.* "Bischofs Kiefer." Height 25—50 ft.; diam. 1—2 ft. Light, resinous, coarse-grained. Used almost exclusively for fuel, except in the north, where it is employed in rough carpentry.

Pine, Old Field. See **Pine, Loblolly** and **Short-leaf**.

Pine, Oregon (*Pseudotsúga Douglásii* Carr.). Western North America from lat. 55° N. southward to lat. 23½° N., up to 11,000 ft. Known also from its discoverer as "Douglas Fir" or "Douglas Spruce," or as "Yellow" or "Red Fir." *French* "Sapin de Douglas." *Germ.* "Douglas-Tanne, Douglas-Fichte." Height 100—300 ft.; diam. 4—6 or 12 ft. S.G. 605. Though more nearly allied to the Firs and Spruces, resembling Larch or hard Pine in the general appearance, quality and character of its wood. Sapwood narrow, yellowish, heart variable, usually reddish-white, heavy, hard, strong, coarse-grained, with well-defined summer wood, but sometimes yellowish, lighter and finer in grain; close, straight and regular in grain, with few knots scattered as in Larch, firm, tough, elastic, not in the least liable to warp, more nearly resembling Canadian Red Pine (*P. resinósa*) than any other wood, rapid in growth, averaging 2 ft. in diameter at 100 years of age, very durable; annual rings well defined; resin-ducts small, often in groups of 8—30; pith-rays with tracheids with bordered pits as upper and under rows and parenchyma with simple pits in the centre; tracheids in the xylem with a spiral thickening, which distinguishes it from all allied wood. Coming to market in clean, straight spars 40—110 ft. in length and 9—32 in. in diameter, this timber is excellent for lower masts, yards, bowsprits, etc., though less adapted for top-masts, where there is much friction, than Riga, Dantzic, or Kauri timber. Used in its native country also for house-building, engineering work and fuel, its freedom from fungoid disease and durability even when grown rapidly in Scotland suggest that Douglas Fir may well supersede Larch for sleepers, etc. It is 25 per cent. cheaper than Archangel wood of the same quality.

Pine, Oyster-bay. See **Cypress-Pine.**

Pine, Pitch (*P. rígida* Miller). Ontario and New Brunswick to Florida. Height 40—80 ft.; diam. 2—3 ft. S.G. 515. W 32. R 739 kilos. Small, coarse-grained, resinous, knotty, light, soft, brittle. Used chiefly for fuel, for which it is unsurpassed in the North, and for charcoal; but formerly much used in New England for building The name in foreign trade belongs to the Long-leaf Pine (*P. palústris*).

Pine, Pond (*P. serotína* Michx.), little more than a southern variety of the last-named.

Pine, Port Macquarie (*Frenéla Macleayána* Parlat.: Order *Cupressíneæ*). North-east Australia. Height 20—30 ft.; diam. 6—12 in. Light, useful, and probably termite-proof, like other Cypress-Pines. Used in house-building.

Pine, Prince's. See **Pine, Grey.**

Pine, Pumpkin. See **Pine, White.**

Pine, Red, of America. See **Pine, Canadian Red.**

Pine, Red, of Australia. See **Cypress-Pine.**

Pine, Red, of New Zealand. See **Rimu.**

Pine, Rock. See **Cypress-Pine.**

Pine, Rosemary. See **Pine, Loblolly.**

Pine, Sand (*P. clausa* Vasey). South-east United States. Known also as "Upland Spruce" or "Scrub Pine." Height 15—20 or 80 ft.; diam. 1—2 ft. S.G. 557. W 34·75. Sapwood broad, nearly white; heart light orange, light, soft, brittle, not strong. Sometimes used for small masts.

Pine, Sandarac. See **Cypress-Pine.**

Pine, Sap. See **Pine, Loblolly.**

Pine, Scrub, of Australia. See **Cypress-Pine.**

" " of California or Oregon. See **Pine, Lodge-pole.**

" " of North-eastern America. See **Pine, Grey.** [See also **Pine, Jersey** and **Sand.**]

Pine, She. See **Cedar, Pencil.**

Pine, Short-leaf (*P. echináta* Mill = *P. mítis* Michx.). Eastern United States. Known also as "Carolina, Bull, Soft-leaved, Yellow, Slash, Old field," or "Spruce Pine." *French* "Pin-Sapin." *Germ.* "Glatte-Kiefer, Fichten-Kiefer." Height 80—120 ft.; diam. 3—4 ft. S.G. 610. W 34—38. *f* 3·68. *fc* 2·46. R 1038 kilos. Sapwood variable in amount, yellowish; heart light orange, rather heavy, hard, coarse-grained, compact, strong, not difficult to work, durable, much resembling Long-leaf Pine (*P. palústris*) and but little inferior to it as timber. Used in house-building, flooring, and interior finish.

Pine, Short-straw. See **Pine, Loblolly.**

Pine, Slash See **Pine, Loblolly, Cuban,** and **Short-leaf.**

Pine, Southern. See **Pine, Long-leaf.**

Pine, Spruce. See **Pine, Short-leaf.**

Pine, Stone (*P. Pínea* L.). Mediterranean region. *French* "Pin de parasol." *Germ.* "Italienische Steinkiefer." *Ital.* "Pino a pinocchi." *Span.* "Pino real, Pino de comer." Height 40—80 ft.; 1—2 ft. Whitish, very light, moderately resinous. Used in southern France and Italy for building and carpentry; but chiefly valued for its nuts. [See also **Pine, Cembra**].

Pine, Stringybark. See **Cypress-Pine, Mountain.**

Pine, Sugar (*P. Lambertiána* Dougl.). Oregon and California, at 2500—8000 ft. Known also as "White, Soft," or "Pumpkin Pine." *French* "Pin gigantesque." *Germ.* "Riesen-Kiefer, Zucker-Kiefer." Height 150—300 ft.; diam. 10—15 ft. The loftiest of all Pines. Wood very light, soft, coarse but straight grained, compact, very fragrant, easily worked, not cracking or warping. Much resembling White Pine (*P. Stróbus*) and used for indoor carpentry, cabinet-work, cooperage and wooden ware.

Pine, Swamp. See **Pine, Cuban.**

Pine, Swiss. See **Fir, Silver,** and **Pine, Cembra.**

Pine, Table Mountain. See **Pine, Hickory.**

Pine, Torch. See **Pine, Loblolly.**

Pine, Umbrella (*Sciadópitys verticilláta* S. and Z.: Order *Taxodíneæ*). Japan. *French* "Sapin à Parasol." *Germ.* "Japanische Schirmtanne." *Japan* "Kôya-maki." Height 100 ft.; diam. 2—3 ft. Wood nearly white, yellowish, or reddish-white, strong, straight-grained. Used at Osaka, to which port it is floated down the Kisiogaiva.

Pine, Upland Spruce. See **Pine, Sand.**

Pine, Weymouth. Pine, White, of America.

Pine, White, of Australia. See **Cypress-Pine.**

Pine, White, of America (*P. Stróbus* L.). Newfoundland and Quebec to Georgia. Known also as "Soft, Apple, Sapling, New England," or "Pumpkin Pine," and in England as "Weymouth Pine," having been largely planted by Lord Weymouth at Longleat, or in the timber-trade as "Yellow Pine." *Germ.*

"Weymouths-Kiefer, Strobe," *French* "Pin du Lord, Pin blanc." Height 140—180 ft., sometimes 100 ft. to first branch; diam. 3—4 or 8 ft. S.G. 385—600. W 24—25. E 600 tons. *e'* 1·46—6·94, averaging 3·48. *p'* ·6—·78. *f* 3. *ft* 1·5—5·1. *c* 2027. *c'* ·267. *fc* 2·24—2·5. R 626 kilos. Straight-growing; sapwood yellowish-white; heart pinkish-yellow to pinkish-brown, light, very soft, straight-grained, compact, not strong, free from resin, easily worked, susceptible of a fine polish, but not durable in contact with soil, subject to cup and heart-shake, and in old trees to a slight sponginess at the centre, very closely resembling the Cembra Pine (*P. Cémbra*), the narrow zone of autumn wood merging into the spring wood, the tracheids of the pith-rays having smooth walls, and the cells one or two large simple pits on their radial walls to each tracheid of the xylem. This is the most useful of American timbers, being very valuable for every description of joinery, doors, sashes, blinds, interior finish, laths, shingles, clap-boards, cabinet-work, and spars, and used also for fuel. Masts of this timber are much inferior to Baltic or Douglas Pine in strength, and cannot be relied upon for more than eight or ten years, especially if in the tropics. They should be very thoroughly seasoned before being painted and the paint then renewed almost annually. Trees of a size suitable for masts were protected in our American colonies at the beginning of the 18th century; but a century later seven-tenths of the houses in North America, except in the large towns, were built of wood, and of these about 75 per cent. were of this species. Reference has already been made to its reckless destruction by the axe and by fire.

Pine, White, of Western North America (*P. montícola* Don.). British Columbia to California, at altitudes of 2000—10,000 ft. Height 80—100 ft.; diam. 4—5 or 7 ft. Nearly white, very light, soft, close and straight-grained, but inferior to *P. Stróbus*, which it much resembles.

Pine, White, of New Zealand (*Podocárpus dacrydioídes* A. Rich.: Order *Taxíneæ*). *Maori* "Kahikatea." Height 80—150 or 180 ft.; diam. 4—5 ft., sometimes 60 ft. to the lowest branch. S.G. 490—428. Yielding timber 20—60 ft. long, squaring 1—2½ ft.; white, light, soft, straight and even in grain, tough, easily worked, not durable when exposed or in contact with soil, liable to become worm-eaten. Used in house-building and occasionally for canoes, but better adapted for indoor use or for paper-pulp.

Pine, Yellow. See **Pine, White, Bull, Grey,** and **Short-leaf.**

Piney Tree. See **Poon.**

Piney Varnish (*Vatéria índica* L.: Order *Dipterocarpáceæ*). Southern India and Ceylon. Known also as "Indian Copal" or "White Dammar." *Canarese* "Dupa maram." *Tamil* "Piney-maram." *Sinh.* "Hal." Height 30—60 ft.; diam. 2—5 ft. W 26. Sapwood reddish-white; heart grey, tough, moderately hard, porous, said to be termite-proof. Used on the West Coast of India for boat and house-building and masts, and in Ceylon for coffins, packing-cases, etc. It yields a fine copal or gum anime, used in Ceylon as incense, the finest specimens being sold as amber. Of allied species *V. acumináta* Hayne, of Ceylon is used for tea-chests.

Pink Ivory, *Zulu* "Mnini" (Order *Leguminósæ*). A beautiful, but as yet undetermined, wood, of an Acacia-like tree of moderate dimensions, growing in kloofs in south-western Natal, with yellowish broad sapwood and rose-pink heart, compact, fine-grained, moderately heavy and hard, and with indistinct rings.

Plane, Eastern, or **Oriental** (*Plátanus orientális* L.: Order *Platanáceæ*). Kashmir to Greece. *French* "Platane de l'Orient." *Germ.* "Morgenlandischer Platanus." *Arab.* "Doolb." *Pers.* "Chinar." Height 70 ft.; diam. 3—5 ft. Pale yellow or slightly reddish, resembling Beech, but softer, in very old trees becoming brown with black lines so as to resemble Walnut, fine, close, and smooth-grained, capable of a high polish, but very apt to warp and split, frequently worm-eaten and not durable, but improved by soaking for several years; annual rings finely but distinctly marked, bending outward at the pith-rays; pith-rays numerous and broad, occupying nearly half the surface and producing a pretty figure; vessels scarcely recognizable. Formerly used for "dug-out" boats at Mount Athos, and by the Turks for ship-building; in Persia and the Levant employed for cabinet-work, turnery, and carpentry, and in France as a substitute for Beech or Hornbeam.

FIG. 59.—Transverse section of Plane (*Plátanus occidentális*).

Plane, Western or **Occidental** (*Plátanus occidentális* L.) Eastern North America. Known also as "Sycamore, Buttonwood, Water-Beech, Button-ball Tree," or when cut radially as "Lace-wood" or "Honeysuckle Wood." *French* "Platane Americain, Platane de Virginie." Height 120 ft.; diam. 10—14 ft. or more. S.G. 568. W 51·5—35·4. R 635 kilos. Sapwood wider than in the Beech, yellowish; heart reddish-white, resembling Beech, except that the broad pith-rays are far more prominent; rather heavy, rather hard, compact, stiff, tough, not very strong, usually cross-grained, difficult to split, but, when dry, easy to cut in every direction, liable to warp, not durable if exposed; rings marked by a fine line bending slightly outward at the pith-rays; vessels evenly distributed (Figs. 59 and 60). Used considerably for cigar and tobacco boxes, wooden bowls, butchers' blocks, cooperage and blind-wood in cabinet work. The cabinet-makers of Philadelphia object to the wood when in plank from its tendency to warp; but when well seasoned it stands well

FIG. 60.—Tangential section of Plane (*Plátanus occidentális*).

and is imported into England for furniture. It is also cut radially as veneers, the "felt" or "silver grain" produced by the pith-rays being darker than the ground colour, which is just the converse of the arrangement of tint in Oak. Plane makes good fuel when dry, but the difficulty of splitting it hinders its use. The Californian species (*P. racemósa* Nutt.) has very similar wood.

By a tiresome confusion the name "Plane" is given in southern Scotland to the wood of the Sycamore or Great Maple (*Acer Pseúdo-plátanus* L.). See **Sycamore**.

Plum (*Prúnus doméstica* L.: Order *Rosáceæ*). Western Asia, cultivated elsewhere. *French* "Prunier." *Germ.* "Zwetschkenbaum." A small tree. Sapwood narrow, yellowish; heart deep brownish-red, resembling Mahogany, heavy, hard, not very durable; vessels much more numerous in the spring wood, so making a lighter-coloured zone; pith-rays numerous and very distinct. Used by cabinet-makers, turners, and instrument-makers.

Plum, in Australia. See **Acacia.**

Plum, Black (*Cargíllia austrális* R. Br.: Order *Ebenáceæ*). North-east Australia. Height 60—80 ft.; diam. 1½—2 ft. W 52. Close, very tough, firm, apt to split and discolour in seasoning and very liable to insect attacks. Used for whip-handles.

Plum, Burdekin (*Spóndias pleiógyna* F. v. M.: Order *Anacardiáceæ*). Queensland. Known also as "Sweet Plum." Dark brown with red markings, resembling Walnut, hard, close and straight in grain. Suitable for turnery or cabinet-work.

Plum, Grey (*Cápparis nóbilis* F. v. M.: Order *Capparidáceæ*). North-east Australia. Known also as "Caper Tree" and "Native Pomegranate." Height 20—25 ft.; diam. 6—14 in. Light coloured, hard, close-grained. Used for whip-handles and suitable for carving. [See also **Myrtle, Black**].

Plum, Hog (*Spóndias mangífera* Pers.: Order *Anacardiáceæ*). India and Burma. Known also as "Wild mango." *Sansk.* "Amrataca." *Hind.* "Amra." *Beng.* "Ambalam." *Telug.* "Ambara." Height 30 ft.; diam. 1 ft. Light grey, soft, valueless except as fuel. Yields a gum resembling Gum Arabic.

Plum, Kafir. See **Date, Kafir.**

Plum, Native or **Wild.** See **Apple, Black** or **Brush.**

Plum, Sebestan (*Córdia Mýxa* L. = *Sebestána officinális* Gaertn.: Order *Boragináceæ*). Egypt, Persia, Arabia, India, and the Malay Peninsula. Known also as "Nakkeru wood." *Sansk.* "Bukampadaruka." *Hind.* "Lesura." *Arab.* "Lebuk." *Tam.* "Vidi." *Telug.* "Nakkeru." Height 8—15 ft.; diam. 1—2 ft. W 28—42. Olive-coloured, greyish or light-brown, soft, coarse-grained, easy to work, strong, seasoning well, but liable to insect attack. One of the best woods for kindling fire by friction, used for boat-building, gun-stocks, etc., excellent for fuel and perhaps suitable for tea-chests, being said to have been used for Egyptian mummy-cases.

Plum, Sour (*Owénia venósa* F. v. M.: Order *Meliáceæ*). Queensland. Known also as "Tulip-wood." Height 30—40 ft.; diam. 1—3 ft. W 62. Highly coloured, with handsome figure and different shades from yellow to black, very heavy, very hard, very strong, easily worked, taking a good polish and durable. A valuable wood for cabinet-work. The allied *O. acídula* F. v. M., known by the same name and also as "Native Peach" and "Emu" or "Mooley Apple," which grows to about the same size and occurs also farther to the south and west, is reddish, but similar in texture and would be suitable for furniture.

Plum, Sweet. See **Plum, Burdekin**.

Plum, White. See **Ironwood** vi.

Pohutukawa. See **Ironwood** xxviii.

Pomegranate, Native. See **Orange, Native** and **Plum, Grey**.

Poon, an Indian commercial name, seemingly applied to the timber of several species used for masts and spars, especially species of *Calophýllum* (Order *Guttíferæ*). Of these the more important would seem to be (i) *C. inophýllum*, (ii) *C. tomentósum*, and (iii) *C. angustifólium*. *C. inophýllum* L. native to Madagascar, Mauritius, Ceylon, Southern India, Burma, Queensland and the Fiji islands. Known also as "Alexandrian Laurel," "Tatamaka," "Dilo." *Hind.* "Undi." *Telug.* "Punnaga." Apparently also the "Palo Maria," of the Philippines. Height 35—80 ft. or more; diam. 1½—5 ft. S.G. 579—647. W 63—35. E 755 tons. *c* 10,000—14,700. *c′* 1·3—1·9. Red-brown, with a pretty wavy figure, fairly hard, close but coarse-grained, very strong, durable. Said to be superior to Riga Fir for masts and spars: used in India for sleepers and suited for joinery and cabinet-work. *C. tomentósum* Wight, a native of Ceylon and of Queensland, is similar and is used in the former country for tea-chests. *C. angustifólium* Roxb., the "Piney tree" of Penang, which also attains large dimensions in the southern Ghats, and is apparently partly the source of

"Poons-pars." (iv) *Dillénia pentagýna* Roxb. (Order *Dilleniáceæ*), a native of India and Burma, in no way related to the species just mentioned, *Telugu* "Ravudana," seems also to be a source of these spars. It is a large tree sometimes 20 ft. to its lowest branch and 2 ft. in diam. W 69. Reddish-grey, heavy, very hard, strong, and durable in contact with the soil. Used for rice-mills, canoes, deck planks and house-building and yielding a good charcoal.

Poplar, a name applied, with few exceptions, to the woods of species of *Pópulus* (Order *Salicíneæ*), which are known in the United States, from their hairy seeds, as "Cottonwoods." *French* "Peuplier." *Germ.* "Pappel." *Span.* "Alamo." Like those of their allies the Willows, these woods are white or pale grey, yellowish, or brown, very soft, and light, with neither pith-rays nor vessels distinctly visible. They are used mainly for paper-pulp and cellulose; but to some extent for packing cases, blind-wood, sabots and other purposes, especially in France, at Ivry and elsewhere.

Poplar, Aspen. See **Aspen.**

Poplar, Balm of Gilead or **Balsam** (*Pópulus balsamífera* L.). North America. Known also as "Tacamahac." Height 70—80 ft.; diam. 5—7 ft. S.G. 363. W 22·6. R 550 kilos. Sapwood wide, nearly white; heart light reddish-brown, not strong or durable. Used only for paper-pulp, for which it is excellent; but as suitable for wooden-ware, etc., as other species.

Poplar, Black (*P. nígra* L.). Europe and Northern Asia. Height 50—60 ft.; diam. 1—2 ft. W 60·5 when green, 29 when dry. Sapwood wide, nearly white; heart light reddish-brown, shrinking more than one-sixth of its bulk in drying, not strong or durable. From its non-liability to splinter useful for the bottoms of waggons, sabots, clogs and turnery, and used also for carving and for charcoal.

Poplar, Black Italian (*P. monilífera* Ait.). Eastern United States, but now common in Italy, Switzerland and other parts of Europe. Known also as "Carolina" or "Necklace Poplar," "Big Cottonwood" or "Whitewood," or, in Europe, as "Swiss Poplar." *Germ.* "Wollpappel, Rosenkranz-Pappel." Height 150—200 ft.;

diam. 6—7 ft. S.G. 389. W 24·25. R 770 kilos. The quickest growing of Poplars. Sapwood very wide, nearly white, heart brownish, not durable if exposed to moisture, but of larger dimensions than, and equal in quality to any other Poplar. Used for flooring, clapboards, inferior fuel, and extensively for paper-pulp, for which purpose it is now largely and remuneratively planted in Britain. "Were every cottager to grow his own fuel . . . perhaps no tree would succeed so well" (*Loudon*). The polishing wheels used by glass-grinders are made of horizontal sections across the entire tree.

Poplar, Carolina. See **Poplar, Black Italian.**

Poplar, Grey (*P. canéscens* Sm.). Kashmir, Persia, Northern Africa and Europe. Height 60—100 ft.; diam. 2—4 ft. W 58 when green, 38·5 when dry. White, shrinking a quarter of its bulk in drying, and cracking; but not splitting when nailed. Said to be superior to White Poplar, and used on the continent for packing cases, rollers and boards for winding ribbon, silk, cloth, etc.

Poplar, Large-toothed (*P. grandidentáta* Michx.). Eastern Canada and the North-eastern United States. Known also as "Large Aspen" or "Whitewood." Height 60—75 ft.; diam. 2 ft. S.G. 463. W 29. R 721 kilos. Takes a smooth finish with a satiny lustre and shrinks but little. Used for clothes-pegs, turned ware, and formerly for ladies' high heels; but chiefly for paper-pulp.

Poplar, Lombardy (*P. dilatáta* Ait.). Kashmir, Persia and Mediterranean area. *French* "Peuplier pyramidal." *Germ.* "Pyramiden-Pappel." *Span.* "Alamo de Italia." Height 100—150 ft.; diam. 3—4 ft. Sapwood wide, nearly white; heart light reddish-brown, easily worked. Little used, chiefly for packing-cases; but, after some years' seasoning, also for churns and coach-panels.

Poplar, Necklace. See **Poplar, Black Italian.**

Poplar, Swiss. See " "

Poplar, White (*P. álba* L.). Central Europe, Northern Africa, Northern and Western Asia. Known also as "Abele." Height 60—100 ft.; diam. 2—4 ft. Sapwood white; heart at first yellowish, becoming browner and sometimes with reddish discolorations. Of little value.

Poplar, Yellow. See **Tulip-tree.**

Porcupine-wood (*Cócos nucífera* L.: Order *Palmáceæ*). Shores of India and throughout the tropics. Height 60—100 ft.; diam. 2 ft. W 70. The wood near the outside of this monocotyledonous stem, being crowded with dense dark-coloured fibro-vascular bundles resembling the quills of the porcupine, is very hard, strong and durable. It is used in India for rafters, beams, spear-handles and other purposes; but in England for walking-sticks, or as a veneer for work-boxes and other fancy articles.

Portia-tree. See **Umbrella-tree.**

Puriri (*Vítex littorális* A. Cunn.: Order *Verbenáceæ*). New Zealand. Height 60 ft.; diam. 3—5 ft. S.G. 1100 when green, 1000 when dry. Yielding timber 9—18 ft. long, squaring 10—18 in.; sapwood 2—3 in. wide, yellowish; heart dark-brown, very heavy, very hard, close-grained, very strong and durable. Much used for posts, piles, sleepers, etc.

Purple-heart of Guiana (*Copaífera pubiflóra* Benth., *C. bracteáta* Benth., and *Peltógyné venósa* Benth.: Order *Leguminósæ*). *French* "Amaranthe, Bois violet." *Dutch* "Purpuurhart." *Aborig.* "Kooroobovilli." *Germ.* "Amarantholz." Large trees yielding timber 20—50 ft. long, squaring 1½—2½ ft., brownish to blackish purple, especially when freshly cut, close-grained, very heavy, hard, strong, elastic, working easily, taking a fine polish, durable. S.G. 967-721. R 231 kilos. Used for furniture, gun-carriages and works of construction.

Purple-heart of Trinidad (*Peltógyné paniculáta* Benth.: Order *Leguminósæ*). Known also as "Zapateri." Yielding timber 20—25 ft. long and 12—15 in. wide, of a beautiful purple when freshly cut, but blackening with age, very durable. Sometimes used for furniture. The allied species *P. confertiflóra* Benth., the "Pao roxo" or "Guarabu" of Brazil, is similar.

Pyengadu and **Pynkado.** See **Acle** and **Mirabow.**

Quar (*Eúclea unduláta* Thunbg.: Order *Ebenáceæ*). Cape Colony. Height 20—30 ft.; diam. 12—15 in. Sapwood light brown; heart dark brown, heavy, very hard, close-grained, with beautiful transverse wavy figure. Suitable for furniture.

Queen-wood (*Daviésia arborea* W. Hill.: Order *Leguminósæ*). North-eastern Australia. Height 15—30 ft.; diam. 6—12 in. Streaked with pink, hard, close-grained, susceptible of a fine polish. An excellent cabinet-wood.

Raspberry Jam. See **Myall** iii.

Rassak or **Russock** (*Vática Rássak* Blume: Order *Dipterocarpáceæ*). Borneo. W 54. Light yellowish, becoming dark red on exposure, heavy, coarse-grained, durable. Used for piles, house-building, etc.

Rata (*Metrosidéros robústa*, A. Cunn: Order *Myrtáceæ*). New Zealand. Height 60—100 ft.; diam. 3—4 ft. S.G. 1228 when fresh, 786 when seasoned. Often 30—40 ft. to lowest branch, and yielding timber 20—50 ft. long, squaring 1—2½ ft. Red, very heavy, hard, close-grained, strong, easy to work, durable. Used in ship-building and for railway waggons.

Redwood, a name variously applied: (i) in the English timber trade to Dantzic Fir (*Pínus sylvéstris*) [See **Northern Pine**]; (ii) in Australia to *Eucalýptus piperíta* [See **Peppermint** vi.]; (iii) in Cape Colony to *Ochna arbórea* Burch. (Order *Ochnáceæ*). Known also as "Cape Plane." *Boer* "Roodhout." *Zulu* "Umtensema." Height 20—30 ft.; diam. 1½—2 ft. Red, hard, strong, durable. Used for waggon-building and furniture, and suitable for engraving.

Redwood, Andaman. See **Padouk.**

Redwood, Californian (*Sequóia sempervírens* Endl.: Order *Taxodíneæ*). Californian coast. *Germ.* "Immergrün Sequoie." "Eiben Cypresse." *Ital.* "Il Legno rosso di California." Height 180—250 or 300 ft.; often 75—100 ft. to lowest branch; diam. 12—20 ft. Sapwood light orange to dark amber, very soft and light, scentless; heart maroon to terra-cotta or deep brownish-red, darkening on exposure, light, soft, brittle, close-grained, not strong, without resin-ducts, very easily split, so that planks can be made from it without the use of the saw, susceptible of a high polish, in structure resembling Bald Cypress, very durable in contact with the soil; pith-rays very distinct (Fig. 61). The most valuable of Californian timber-trees, and the most used material for building and carpentry in the State, used also for sleepers, fencing, telegraph-poles, shingles and furniture, corresponding in quality and uses to White Cedar. Though suited for drawers or lining, it is somewhat too monotonous for ornamental furniture. In the London cabinet trade it is now known

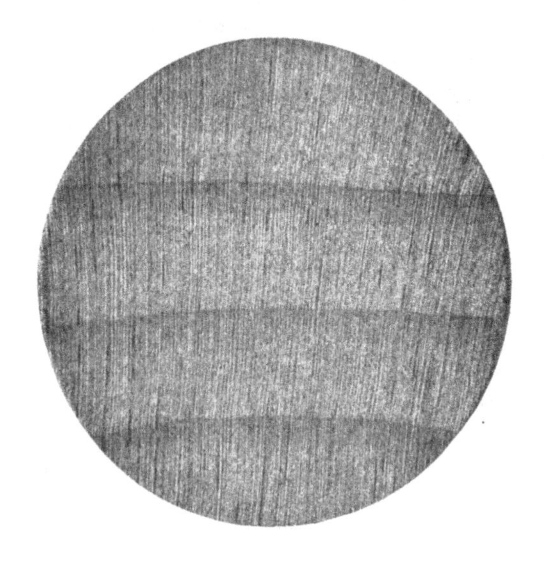

Fig. 61.—Transverse section of Californian Redwood (*Sequóia sempervírens*).

as "Sequoia." Though sending up vigorous coppice-shoots when felled "at the present rate of destruction not an unprotected Sequoia of timber-producing size will be left standing twenty years hence" (J. G. Lemmon in 1895).

Redwood, Coromandel or **Indian.** See **Mahogany, East Indian.**

Redwood, in Jamaica. See **Ironwood** xxiii.

Rewa-rewa (*Kníghtia excélsa* R. Br.: Order *Proteáceæ*). New Zealand. Known also as "Honeysuckle-wood." Height 100 ft.

On a radial section lustrous golden-yellow with pretty wavy warm red-brown silver-grain, perishable on exposure and becoming "foxy" unless thoroughly seasoned. Valued for inlaying and cabinet-work.

Rimu (*Dacrýdium cupressínum* Soland. : Order *Taxíneæ*). New Zealand. Known also as "Red Pine." Height 40—80 or 100 ft. ; diam. 2—5 ft. ; sometimes 40—50 ft. to the lowest branch. S.G. 678 when seasoned. Yielding timber 20—50 ft. long, squaring 10—30 in. Chestnut-brown near centre, lighter outwards, figured with light red or yellow streaks, moderately heavy and hard, very strong, working well and taking a good polish, but not durable in contact with soil. Extensively used in building for beams, girders, etc., for panelling, fencing, railway-ties, native canoes and furniture.

Roble, the Spanish for Oak, used in Trinidad for *Platymíscium platystáchyum* Benth. (Order *Leguminósæ*), a hard, tough wood with an ornamental silvery transparent grain, used locally in shipbuilding.

Rose-chestnut, Indian. See **Ironwood** xviii.

Rosewood. *French* "Bois du rose." *Germ.* "Rozenholz." *Ital.* "Legno rodie." *Span.* "Leno de rosa." *Port.* "Pao de rosada." The name of a number of different species in various parts of the world, mostly heavy dense dark-coloured woods, many of which belong to the Order *Leguminósæ*, such as the genera *Dalbérgia, Machǽrium* and *Pterocárpus*, and one or two of which contain a fragrant resin or oil, from which the name has originated. They have nothing more to do with the Rose.

Rosewood, African (*Pterocárpus erináceus* Poir. : Order *Leguminósæ*). Tropical West Africa. Known also as "African Teak." *French* "Santal rouge d'Afrique." Sapwood white; heart red-brown, hard, very elastic. Valuable as timber and as yielding an astringent resin or Kino.

Rosewood, Australian (i) *Acácia glaucéscens* [See **Myall** v]; (ii) *Dysóxylon Fraseriánum* [See **Cedar, Pencil**]; (iii) *Eremophíla Mitchélli* [See **Sandalwood, Bastard**]; and (iv) *Synóum glandulósum* A. Juss. (Order *Meliáceæ*). North-eastern Australia.

Known also as "Dogwood," "Bastard Rosewood" and "Brush Bloodwood." Height 40—60 ft.; diam. 1½—2 ft. W 41—45. Deep red and rose-scented when fresh, resembling Cedar but heavier and deader in colour, taking a fine polish, firm and easily worked. Used for shipbuilding, the inside of houses and cabinet-work, for which it has long been valued. An allied form *S. Lárdneri*, without scent and with more open grain, is known as "Pencil Cedar," or, from the smell of its bark, as "Turnipwood."

Rosewood, Brazilian, including that of Rio, the best, Bahia, the second best, and San Francisco, is probably *Dalbérgia nígra* Allem. (Order *Leguminósæ*), *Brazil.* "Jacaranda cabiuna," or in part also species of the allied genus *Machærium*, such as *M. scleróxylon* Tul., known as "Pao Ferro," *M. fírmum* Benth., "Jacaranda roxa," and *M. legálé* Benth., "Jacarando preto." S.G. 768—841. In half-round logs 10—20 ft. long, seldom over 14 in. in diam. Dark chestnut or ruddy brown, richly streaked and grained with black resinous layers, porous, open-grained, heavy, taking a fine polish, liable to heart-shake and frequently hollow and sold, therefore, by weight. Valuable, both solid and in veneers, for furniture and ornamental cabinet-work, especially pianoforte cases, and for turnery, realizing £10—£12 per ton for inferior, £20—£30 for good, and even up to £90 for the best qualities.

Rosewood, Bastard. See **Rosewood, Australian.**

Rosewood, Burmese. See **Padouk.**

Rosewood, Canary (*Convólvulus Scopárius* L., *C. virgátus* Webb, and *C. flóridus* L.: Order *Convolvuláceæ*). Canary Islands. "Lignum Rhodii." *French* "Bois des Rhodes des Parfumeurs." Derived from the rhizome and bases of the aerial stems, whence these species have been separated as a genus *Rhodorrhíza.* Rose-scented and distilled for the powerfully-scented oil "Oleum ligni Rhodii æthereum," used to adulterate attar of roses. Not otherwise used.

Rosewood, Dominica. See **Cypre, Bois de.**

Rosewood, Indian. See **Blackwood, Indian.**

Rosewood, Jamaica (*Linociéra ligustrína* Swartz.: Order *Oleáceæ*). See also **Granadillo**.

Rosewood, Moulmein, probably a species of *Millettia* (Order *Leguminósæ*), possibly *M. péndula* Benth., a dense, hard, dark-coloured wood.

Rosewood, Seychelles. See **Umbrella-tree**.

Rosewood, West Indian. See **Granadillo.**

Rowan (*Pýrus Aucupária* Gaertn.: Order *Rosáceæ*). Europe, Northern and Western Asia. Known also as "Mountain Ash." *Germ.* "Eberesche." Height 10—40 ft.; diam. 6—10 in. Sapwood reddish-white; heart reddish-brown, hard, tough, difficult to split; pith-flecks frequent; vessels and pith-rays indistinct; autumn wood slightly darker. Used to a small extent on the Continent in cabinet-work, carving and turnery.

Sabicu (*Lysilóma Sábicu* Benth.: Order *Leguminósæ*). West Indies, especially Cuba. Somewhat crooked in growth but yielding timber 20—35 ft. long, squaring 11—24 in. S.G. 899—957. *e'* 2·21. *p'* 1·6. *c* 5558. *c'* ·734. *v'* 1·161. R 435 lbs. Dark chestnut-brown, heavy, hard, strong, elastic, close-grained, free from shakes, though sometimes exhibiting on conversion a cross fracture of part of the inner wood, snapped, perhaps, by West Indian hurricanes, seasoning slowly, but shrinking but little and not splitting in the process, working up well, susceptible of a high polish, durable when exposed, and sometimes with such a curled figure as to be mistaken for Rosewood. Used in shipbuilding, especially for beams, keelsons, engine-bearers, etc., and for furniture. The staircases of the Exhibition of 1851 were of this wood and wore well. The allied species, *L. latisíliqua* Benth., native to the Bahamas and Florida, is similar.

Saffronwood (*Cassíné crócea* O.K., = *Elæodéndron cróceum* DC.: Order *Celastráceæ*). South Africa. Known also as "Safforan-wood." *Boer* "Saffranhout," *Zulu* "Umbomoana." *French* "Olivetier jaune, Bois d'or du Cap." Height 20—40 or 60 ft.; diam. 2—4 ft. W 47·5. E 510 tons. *f* 4·4. *fc* 3·18. Yellow, hard, close, fine-grained, tough, handsome. Used for beams, planks, waggon-building, furniture, etc.

Saj (*Terminália tomentósa* W. and A.: Order *Combretáceæ*). India and Burma. *Hind.* "Asan." *Tamil* "Maradu." *Mahrat.* "Eyn." A large tree yielding timber 18—28 ft. long and 1—2 ft. in diam. S.G. 892. R 462—602 lbs. Sapwood white, narrow; heart dark brown, finely variegated with darker streaks producing a wavy figure, heavy, hard, elastic, strong, difficult to work, but seasoning well and taking a high polish, liable to split on exposure and to dry-rot if not steeped. Its power of resisting termite attack is doubtful. Largely used for joists and rafters and in waggon and boat-building, and recommended for paving. Resembling the next.

Sal (*Shórea robústa* Gaertn.: Order *Dipterocarpáceæ*). Northern and Central India. Known also as "Saul," and formerly as "Morung Sal." *Sansk.* "Sala." *Philippine* "Guijo." A large tree yielding timber 20—60 ft. long and 1—2 ft. in diam. S.G. 458—842. W 28·6—52·6. R 1043 lbs. Sapwood whitish, narrow; heart light to deep brown, finely streaked with dark lines, heavy, hard, coarse and cross-grained, elastic, tough, comparing favourably as to strength with Teak, warping and splitting considerably in seasoning, but almost unrivalled for durability, its abundant whitish aromatic resin protecting it from termites. The most extensively used timber of Northern India for sleepers, piles, beams, bridges, planks, gun-carriages, wedges, tool-handles, blocks, cogs, etc., but too heavy to float and, therefore, expensive.

Saliewood (*Buddléia salvifólia* Lam.: Order *Loganiáceæ*). Cape Colony. *Zulu* "Unkaza." Height 15—20 ft.; diam. 10—15 in. Hard, tough, with a beautiful wavy grain. Used for cabinet-work, yielding veneers equal in appearance to Walnut, and for cogs; but recommended for wooden type and coarse engraving.

Sallow, in England, chiefly *Sálix Capréa* L. (Order *Salicíneæ*). Europe, Northern and Western Asia. Known also as "Goat Willow." *Germ.* "Sahlweide." A small tree. Sapwood reddish-white; heartwood a beautiful light red, light, very soft, easily split, lustrous, with wide annual rings, pith-rays indistinguishable, vessels minute and equally distributed, pith-flecks often present. Used chiefly for crate and hoop-making; but in France one of the most useful willows. [See **Willow.**]

Sallow, in Australia, or **Sally,** or **White Sally,** names applied to some species of *Acácia*, especially *A. longifólia* Willd., var. *floribúnda*, a brown black-streaked, light, tough, and hard wood, used for tool-handles; and to *Eucrýphia Moórei.* [See **Acacia.**]

Sandalwood, a name applied to the generally fragrant woods of *Sántalum álbum* and other species of the genus, to those of the other genera of the Order *Santaláceæ*, viz., *Fusánus, Exocárpus*, and *Osýris*, to some members of the Order *Myoporíneæ*, and a few other unrelated trees. True Sandalwood is *Sántalum álbum* L. (Order *Santaláceæ*). India, chiefly in the south, and perhaps also in the Malay Archipelago. Known also as "White" or "Yellow Sandalwood." *Sanskr.* "Chandana." *Hind.* "Chandana, Sandal." *Telugu* "Chandanam." *Burm.* "Sanda-ku." *Chinese* "Tan-mu." Height 30 ft., 8 ft. to lowest branch; diam. up to 2 ft. Sold in billets weighing 50—90 lbs. each. Yellowish-brown, very hard, close-grained and fragrant, the heartwood yielding on distillation about 2 drams of oil per lb., and increasing in fragrance with age, very liable to heart-shake. Used for carving, ornamental boxes, walking-sticks, fans, burnt as a perfume, ground into powder as a cosmetic, and distilled for its fragrant oil. Realizing £30 per ton in the Chinese market.

Sandalwood, Australian (i) *Fusánus spicátus* R. Br. = *F. cygnórum* Benth.: Order *Santaláceæ*). Southern and Western Australia. Known also as "Fragrant Sandalwood." Height 30 ft.; diam. up to 1 ft. Not very fragrant and scarce. Exported to Singapore and China at about £10 per ton. (ii) *Sántalum lanceolátum* R. Br. Height 15—25 ft.; diam 3—6 in. Yellowish, firm, close-grained, taking a good polish. (iii) *S. obtusifólium* R. Br. Eastern Australia.

Sandalwood, Bastard, of Australia. (i) (*Eremóphila Mitchélli* Benth.: Order *Myoporíneæ*). Eastern Australia. Known also as "Rosewood." Height 20—30 ft.; diam. 9—12 in. Brown, beautifully grained, very fragrant, very hard. Yielding handsome veneers. (ii) (*Myopórum platycárpum* R. Br.). [See **Dogwood** iv.].

Sandalwood, Bastard, of India (*Erythróxylon monógynum* Roxb.: Order *Erythroxyláceæ*). A small tree of Southern India and Ceylon. *Sansk.* "Devadara." *Arab.* "Shajr-ul-jin." *Hind.* "Deo dhari." *Tamil.* "Devatharam." *Telug.* "Devadari." Dark brown, very hard, taking a fine polish, very fragrant. Used as a substitute for Sandalwood.

Sandalwood, Bastard, of the Sandwich Islands (*Myopórum tenuifólium*).

Sandalwood, East African (*Osýris tenuifólia* Engl.: Order *Santaláceæ*). Portugese East Africa. "Mucumite." Length 6—8 ft.; diam. 8 in. Brown, with darker shades, heavy, crooked in growth.

Sandalwood, Fiji (*Sántalum Yási* Seem.).

Sandalwood, Fragrant. See **Sandalwood, Australian.**

Sandalwood, Indian. See **Mangosteen, False.**

Sandalwood, Native (*Fusánus persicárius* F. v. M.: Order *Santaláceæ*). Australia. Small and inferior.

Sandalwood, New Caledonia (*Sántalum austro-caledónicum* Vieill.).

Sandalwood, Red. See **Sanders-wood, Red.**

Sandalwood, Sandwich Island (*Sántalum freycinetiánum* Gaud. and *S. paniculátum* Hook. and Arn.).

Sandalwood, Scentless (*Eremóphila Stúrtii* R. Br.: Order *Myoporíneæ*). South-east Australia. Merely a shrub, with grey, nicely-marked, close-grained, hard wood.

Sandalwood, Scrub (*Exocárpus latifólia* R. Br.: Order *Santaláceæ*). North-east Australia. Known also as "Broad-leaved Cherry." Height 10—16 ft.; diam. 6—9 in. Dark-coloured, fragrant, very hard, coarse-grained, taking an excellent polish. Used in cabinet-work.

Sandalwood, White. See **Sandalwood.**

Sandalwood, Yellow. See **Sandalwood.**

Sandan (*Ougeínia dalbergioídes* Benth.: Order *Leguminósæ*). Northern India. Mottled brown and red, hard, tough, close-grained, taking a good polish, durable. Used for furniture, carriage poles, wheels, agricultural implements, etc.

Sanders-wood, Red (i) (*Pterocárpus santalínus* L. fil.: Order *Leguminósæ*). Southern and Further India, Ceylon, China, Java, etc. Known also as "Red Sandalwood." *French* "Santale rouge." *Germ.* "Ostindisches Santelholz, Caliaturholz." *Sansk.* "Rakta chandana." *Pers.* "Sandal surkh." *Hind.* "Chandana." *Tam.* "Chandanum." *Sinh.* "Rakt-chandan." Height 20-25 ft.; diam. 1 ft. S.G. 750. W 46·84. Deep red with lighter zones, heavy, very hard, fine-grained, taking a beautiful polish. Used for images, turnery, and occasionally building; but chiefly as a red dye, soluble in alcohol but not in water. (ii) *Adenánthera pavonína* L.: Order *Leguminósa*). India, Burma, Moluccas, North Queensland, and cultivated in Tropical Africa and America. Known also as "Red wood" or "Red Sandalwood." *Germ.* "Condoriholz." *Sansk.* "Cambhoji." *Hind.* "Ranjana, Ku-chandana." *Beng.* "Rakta-chandan." *Tam.* "Gandamani." A large tree. W 56. Yellowish-grey or light-brown, or in older trees a beautiful coral-red, sometimes with darker stripes when fresh cut, turning dark-brown, or purple like Rosewood, on exposure, rather heavy, hard, coarse but close-grained, durable, but apt to be worm-eaten. Used in house-building and for cabinet-work; or, ground into a paste by rubbing the wood against a stone with some water, as a dye.

Sanders, Yellow (*Ximénia americána* L.: Order *Olacíneæ*). Tropical America, Pacific, Australia, Asia, and Africa. *Brazil* "Ameixero," "Espinha de meicha." *Guiana* "Heymassoli." *San Domingo* "Croc." W 57·3. E 721 tons. *f* 4·5. *fc* 4. *fs* ·368. Yellow, fragrant, very hard, tough, close and even-grained. Employed in India as a substitute for Sandalwood, in the Fiji Islands for the peculiar pillows ("kali") used by the natives, and suggested for engraving.

Santa Maria (*Calophýllum Cálaba* Jacq.: Order *Guttíferæ*). Tropical America. Known also as "Galba," "Galaba," "Accite de Maria," and in Cuba as "Ocuje." Height 60—90 ft.; diam. 2—3 ft.; yielding logs 25—50 ft. long, squaring 12—22 in. S.G.

842. W 53. E 790 tons. *f* 5·14. *fc* 2·6. *fs* ·215. R 354 lbs. Pale red to orange-yellow, moderately heavy and hard, clean fine and straight in grain, flexible, with few knots, shrinking and splitting very little in seasoning, easily worked, durable. Has been used in our dockyards for beams and planks, and is equal to plain mahogany for interior finish.

Sappan-wood (*Cæsalpínia Sáppan* L.: Order *Leguminósæ*). India and South-east Asia. Known also as "Red wood, Brazil," or "Brasiletto wood." *Sansk.* "Patanga." *Hind.* "Bakam." *Malay* "Sapang." Height 30—36 ft.; diam. 8 in. Brownish-red. Used almost exclusively as a red dye for cotton goods, the roots, known as "Yellow-wood" or "Sappan Root," yielding an orange-yellow one.

Sapodilla (*Achras Sapóta* L.: Order *Sapotáceæ*). Tropical America. Known also as "Nispero," "Bully," or "Bullet-wood." Reddish-brown, very heavy, hard, and durable. Used for furniture, cabinet-work, and occasionally building.

Saquisaqui (*Bómbax mompoxénsé* H. B.: Order *Bombáceæ*). Venezuela. Known also as "Cedro dulce." S.G. 529. Rose-red, of better quality than other species of the genus, similar to the wood of the Cedar (*Cedréla odoráta*).

Sassafras in North America (*Sássafras officinálé* Nees: Order *Lauráceæ*). Known as "Sassafras" in Latin, Arabic, French, German, and Spanish, in German also as "Fenchelholz" and by the French in America as "Laurier des Iroquois." Canada to Florida and Texas. Height 50—90 ft.; diam. 3—7½ ft. S.G. 504. W 31·4. R 602 kilos. Sapwood yellow, narrow; heart orange-brown, with a slight characteristic aroma, light, soft, rather brittle, coarse-grained, very durable when exposed and partially insect-proof; with broad distinct annual rings, a marked pore-zone of springwood with 4—5 rows of vessels arranged radially in pairs, and very fine pith-rays, distinguished from the Red Mulberry (*Mórus rúbra*) by its lightness. Used for fencing, buckets, etc. The essential oil which brought the tree into notice in the 16th century is distilled from the bark of the roots. The name is applied in various parts of the world to other species of the Order

Lauráceæ and the closely-allied *Monimiáceæ* exhibiting the same characteristic smell.

Sassafras, Assam (*Cinnamómum glandulíferum* Meissn.). See **Camphor, Nepal.**

Sassafras, Australian (i) *Atherospérma moscháta* Labill.: Order *Monimiáceæ*). South-east Australasia. Height 100—150 ft. Dark-coloured, often well figured, close-grained, very tough, easily worked, taking a fine polish. Used for lasts, bench-screws, and cabinet-work, and suggested for sounding-boards and doors. (ii) (*Daphnándra micrántha* Benth.: Order *Monimiáceæ*). North-east Australia. Known also as "Satinwood" and "Light-yellow Wood." Height 50—80 ft.; diam. 1½—2 ft. W 43·5. Yellow when fresh, fragrant, soft, weak. Used for packing-cases and perhaps fit for cabinet-drawers. (iii) (*Dorýphora sássafras* Endl.: Order *Monimiáceæ*). North-east Australia. Height 50—100 ft.; diam. 2—3 ft. Light-coloured, sometimes neatly figured, light, fragrant, soft, weak, insect-proof, but probably not durable. Used like the last-mentioned. (iv) (*Nesodáphné obtusifólia* Benth.: Order *Lauráceæ*). North-east Australia. A large tree yielding light-coloured, close-grained wood, easy to work and suitable for joinery.

Sassafras, Black. See **Beech, She.**

Sassafras, Brazil (*Mespilodáphné Sássafras* C.: Order *Lauraceæ*).

Sassafras, Burmese. See **Camphor, Nepal.**

Sassafras, Cayenne (*Dicypéllium caryophyllátum* Nees = *Licaria guianénsis* Aubl.: Order *Lauráceæ*). Known also as "Cayenne Rosewood," "Licari," "Pepper-wood," "Bois canelle," "Rose male." Brazil and Guiana. An excellent wood. S.G. 1226—1108. R 360 kilos. Pale yellow, slightly fragrant, moderately hard. compact, and straight in grain, very easily worked, durable, The name "Sassafras" or "Rose femelle" is applied in French Guiana to *Acrodiclídium chrysophýllum* Meissn. (Order *Lauráceæ*). S.G. 806—688. R 184 kilos. Yellow, very durable, readily worked, yielding on distillation the essential oil known as "essence de roses." Excellent for shipbuilding or furniture.

Sassafras, Grey. See **Laurel.**

Sassafras, Indian. See **Camphor, Nepal.**

Sassafras, Nepal. See **Camphor, Nepal.**

Sassafras, Tasmanian. See **Sassafras, Australian** (i).

Satiné or **Bois de féroles,** of which there are two varieties, "Satiné rouge," a beautiful red-brown, and "Satiné rubanné," lighter-coloured, veined and lustrous, is apparently *Ferólia guianénsis* Aubl. and, perhaps, F. *variegáta* Lam., probably species of *Parinárium* (Order *Rosáceæ*). Guiana and Guadeloupe. Known also as "Bois marbré," and, in Demerara, as "Washiba." *German* "Feroliaholz." S.G. 877—825. Exported in logs 14—28 ft. long, squaring 13—15 in., hard, solid, and of good quality, working well, and susceptible of a beautiful polish. Used for furniture and cabinet-work.

Satin Walnut. See **Gum, Sweet.**

Satinwood (*Chloróxylon Swieténia* DC.: Order *Meliáceæ*). Central and Southern India and Ceylon. *Hind.* "Dhoura." *Tam.* "Mutirai, Porasham." *Sinh.* "Burutu." Height 30—60 ft.; diam. 12—15 in. W 64·3—55. E 699 tons. *f* 6·15. *fc* 3·37. *fs* ·85. R 329—510 lbs. Light orange, beautifully feathered, heavy, hard, close-grained, taking an excellent polish, durable, but liable to darken unless varnished, somewhat apt to split. Used in India for oil-mills, agricultural implements and furniture, the beautiful feathered variety being imported into England for the backs of hair-brushes, turnery, and cabinet-work.

Satinwood, in Australia (*Zanthóxylum brachyacánthum* F. v. M.: Order *Rutáceæ.*) North-east Australia. Known also as "Thorny Yellow-wood." Height 40—50 ft.; diam. 12—15 in. Bright yellow, silky, soft, close-grained, easily worked. Used in cabinet-work and said to be superior to some Satinwood in the English market. See also **Sassafras, Australian** (ii).

Satinwood, West Indian (*Fágara* (*Zanthóxylum*) *fláva* Krug. and Urb.: Order *Rutáceæ*). This appears to be the species imported in considerable quantities into England in logs 10 ft. long and 8 in. in width and thickness for ornamental purposes from the Bahamas and Porto-Rico, fetching from £3 to £10 per ton; but the name seems to be applied in Dominica to *Búcida capitáta*

Dow. (Order *Combretáceæ*), which is also known as "Yellow sanders," and in Brazil and the Guianas may be applied to other unascertained species.

Savicu. See **Sabicu.**

Schaapdrolletje (*Plectrónia ventósa* L.: Order *Rubiáceæ*) Cape Colony. Height 15—20 ft.; diam. 6—10 in. Heavy, hard, close-grained, tough, susceptible of a good polish and then handsome. Suitable for fancy-work.

Sequoia. See **Redwood, Californian.**

Securipa, an undetermined Brazilian wood, of considerable dimensions, straight growth, moderate weight and fair quality, brown in colour. Used for beams and planks in shipbuilding.

Serayah, probably a species of *Hópea* (Order *Dipterocarpáceæ*). Malay peninsula and Borneo. Known also as "White Cedar," "Borneo Cedar," "Majow," "Selangan." W 43. Reddish, resembling soft Mahogany, but strong, tough and easily worked. Used for house-building or furniture, being a substitute for Pine.

Service (*Pýrus torminális* Ehrh.: Order *Rosáceæ*). Europe, West Asia and North Africa. *Germ.* "Elsbeerbaum." A small tree, sometimes 30 ft. in height, with wood practically identical in character and uses with that of the Rowan.

Shad-bush (*Amelánchier canadénsis* Torr. & Gray). Eastern North America. Known also as "Shad-blow, Juneberry" and "Service-tree." *French* "Grand Amelanchier." *Germ.* "Traubenbirne." *Span.* "Nispero." Height 40 ft.; diam. 15 in. S.G. 784. W 48·85. R 1132 kilos. Sapwood thick, light brown, with red spots; heart reddish-brown, heavy, hard, strong, close-grained, taking a satiny finish. Little used, except for tool-handles and, under the name of "American Lancewood," for fishing-rods.

She-Oak. See **Oak, She.**

Shoondul (*Intsia bíjuga* O.K. = *Afzelia bíjuga* A. Gray = *Intsia amboinénsis* Thouars (?) = *Epérua decándra* Blanco: Order *Leguminósæ*). Seychelles, India, the Malay archipelago and the Pacific. Known also as "Pynkado." *Fiji* "Vesi." A moderate-sized tree. Reddish-brown, very hard, close-grained. Used in the

Pacific for war-clubs and in India for bridge and house-building, and exported as a furniture wood.

Silk-cotton tree (*Céiba pentándra* Gaertn. = *Eriodéndron anfractuósum* P. DC.: Order *Bombáceæ.*) The Tropics generally. Known also as "Cotton" or "White Cotton-tree." *Hind.* "Safed Simal." *Tam.* "Elava." *Sinh.* "Imbool." *Malay* "Paniala." *French* Arbre à coton." *Germ.* "Baumwollenbaum." *Cuba* "Ceiba." Height 150 ft. or more. S.G. 287. Straight growing, white, light, soft. Used for canoes, boats, rafts, floats, toys and packing-cases.

Silk-bark (*Celástrus acumindtus* L.: Order *Celastráceæ.*) Cape Colony. "Zybast." Height 20 ft.; diam. 1 ft. Prettily shaded, heavy, hard, even and close-grained, taking a good polish. Used in turnery and furniture and recommended for umbrella-handles.

Silverballi, Brown, Siruaballi or **Cirouaballi** (*Nectándra* sp.: Order *Lauráceæ*). British Guiana. Height 90 ft.; diam. 1½ ft. S.G. 830. Light brown, hard, easily worked, taking a fine polish, durable. Used for boat-planks and masts; but suited for furniture.

Silver-tree (*Tarriétia argyrodéndron* Benth.: Order *Sterculiáceæ.*) North-east Australia. Known also as "Ironwood, Stonewood, Black Stavewood." *Aborig.* "Boyung." Height 70—90 ft.; diam. 2—3 ft. White, hard, close-grained, tough, firm, a substitute for Beech. Extensively used for staves and suitable for piles.

Simarouba (*Simarúba amára* Aubl. = *S. officinális* DC.: Order *Simarubáceæ*). Northern Brazil, Guiana and the West Indies. "Maruba." "Acajou blanc" of Guadeloupe. A lofty tree yielding logs 13 or 14 ft. long, squaring 14 to 16 in. W 23. E 473 tons. *f* 3·36. *fc* 1·78. *fs* ·224. White, bitter, resembling Pine in quality, moderately hard, splitting in seasoning, easily worked, insect-proof.

Siris, Pink (*Albízzia Julibríssin* Durazz.: Order *Leguminósæ.*) Tropical and sub-tropical Africa and Asia from Afghanistan to China and Japan. Known also as "Sirsa" or "Sirissa."

"Cotton-varay" of Coromandel. Moderate sized tree. Dark brown to almost black, mottled, very heavy and hard, capable of a good polish. Valued for furniture and for house and boat-building.

Sissoo (*Dalbérgia Sissóo* Roxb.: Order *Leguminósæ*). Northern India. A large tree, 15 feet to its lowest branch, 1½ ft. in diam., yielding logs 10—15 ft. long. Greyish-brown, with dark longitudinal veins, very heavy, hard, close and even in grain, strong, elastic, seasoning well without warping or splitting, durable. One of the most valuable of Indian timbers, rapid in growth, sometimes almost as beautiful as Rosewood. Unrivalled for the naves and felloes of wheels, frames of carriages, boat-building, agricultural implements and furniture. Once extensively used for the wheels of gun-carriages; but not now plentiful.

Snakewood, in India, the wood of *Strýchnos colubrína* L. and *S. nux-vómica* L. (Order *Loganiáceæ*). The former is a climbing-plant 8—12 in. in diam.; light grey, hard and intensely bitter; the latter a tree reaching 15 ft. or 20 ft. to its lowest branch and 3 ft. in diam. S.G. 706. W 52. White, or ash-colour, hard, close-grained, strong, very bitter. Used for ploughshares and cart-wheels in Travancore. *Hind.* "Kuchila." *Telugu* "Naga-musada." *Portuguese* "Pao-de-cobra." *French* "Bois de couleuvre." The crooked and intensely bitter roots of *Ophiorrhíza Múngos* L. (Order *Cinchonáceæ*), a native of the Sunda Islands, and the twisted climbing stems of *Rauwólfia serpentína* Benth. = *Ophióxylon serpentínum* L.: Order *Apocynáceæ*), a native of the Malay archipelago, have also the same name. In the West Indies *Colubrína reclináta* Brongn. and *C. ferruginósa* Brongn. (Order *Rhamnáceæ*), from their twisted roots bear also the same English and French names, though known also as "West Indian Greenheart" or "Ironwood." Whilst all these woods get their names from their form or taste suggesting their use as remedies for snake-bite, the beautifully mottled Snakewood of British Guiana is *Brósimum Aublétii.* See **Leopard** or **Letter-wood.**

Sneeze-wood (*Pteróxylon útilé* Eck. and Z.: Order *Sapindáceæ*). South Africa. *Boer* "Neishout." *Zulu* "Umtati." Height

20—30 ft.; diam. 2—4 ft. W 61. E 782 tons. *f* 8·62. *fc* 5·96. Handsome, heavy, very hard, irregular in growth, difficult to convert, its dust producing violent sneezing, taking a fine polish, with a beautiful grain resembling Satinwood, and containing a gum-resin which renders it very inflammable and one of the most durable woods in the world, ranking with Jarrah and Greenheart, termite and teredo-proof, very slightly affected by water, and for bearings superior to brass, iron or Lignum-vitæ. Perhaps the most valuable of South African timbers. Used for engineering work, bridges, furniture, agricultural implements and carpentry.

Souari or **Schawari** (*Caryócar glábrum* Pers., *C. butyrósum* Willd., *C. tomentósum* Willd.: Order *Rhizoboláceæ*). Guiana. Known also as "Peki" or "Tatajuba." S.G. 932—820. R 211 kilos. Yellowish-red, moderately heavy and hard and very cross-grained. Excellent for carriage-building, but used chiefly in ship-building and cabinet-work.

Spearwood in Australia (i) *Acácia homalophýlla* [See **Myall** ii], (ii) *A. doratóxylon* and (iii) *Eucalýptus doratóxylon*. *Acácia doratóxylon* A. Cunn. (Order *Leguminósæ*) is known also as "Hickory, Brigalow," or "Caariwan." Height 20—35 ft.; diam. 6—12 in. S.G. 1215. Sapwood narrow, yellow; heart dark-brown, very heavy, hard, tough, close-grained, durable. Used for furniture, carriage-poles, gates, etc., and, by the natives, for spears and boomerangs. *Eucalýptus doratóxylon* F. v. M. (Order *Myrtáceæ*). South-west Australia. Height 60—80 ft.; diam. 2—3 ft. Straight-growing, hard and elastic, for which qualities its saplings are much valued by the aborigines for spears.

Spindle-tree (*Euónymus europǽus* L.: Order *Celástráceæ*). Europe, North Africa and Western Siberia. Height 5—20 ft.; diam. small. Clear yellowish-white with distinct annual rings but indistinguishable vessels or pith-rays, hard, tough, fine-grained, difficult to split, but easily cut. Used in turnery for spindles, shoe-pegs, etc., and yielding a fine crayon or gun-powder charcoal.

Spruce, a name applied originally to the Common or Norway Spruce (*Picéa excélsa* Link. = *Pínus Abies* L. = P. *Picéa* Duroi = *Abies excélsa* DC: Order *Coníferæ*) from Pruce or Prussia, whence it was obtained, and then extended to all the species of the genus *Picéa* and to a few other trees. Besides the fact of their cones falling off whole, and other botanical characters,

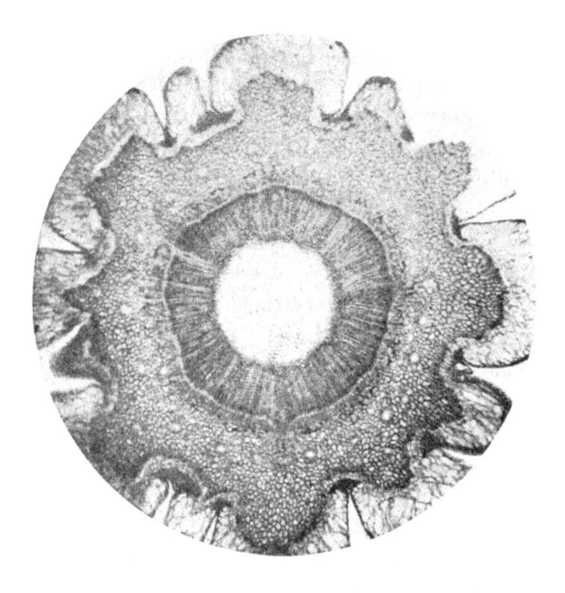

FIG. 62.—Transverse section of Spruce (*Picéa excélsa*), one year old.

by which the Spruces are distinguished as a genus from the Firs (*Abies*), their wood, though varying in durability according to the soil on which it is grown, has most of its characters common to all the species. There is no distinct heartwood, the whole being of a whitish colour: the resin-ducts are few and small; and the pith-rays have tracheids with bordered pits for their upper and lower rows of cells with four rows of parenchyma having simple pits in the middle (Figs. 62 and 63). The wood is less resinous than Pine, though equal to some Soft Pines and superior to Silver Fir as timber, superior to Pine for paper-pulp and much valued as a "resonance wood" for violins and sounding-boards. So similar are the Baltic and Canadian Spruce that in England each is used on that side of the country nearest to its origin and the price of one affects that of the other.

Spruce, American. See **Spruce, Black.**

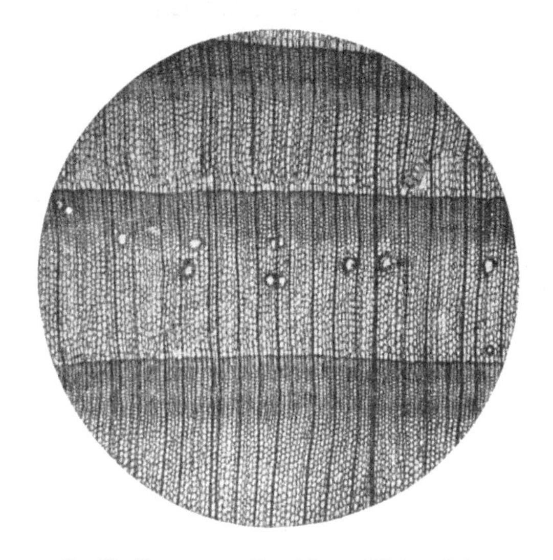

Fig. 63.—Transverse section of Spruce (*Picéa excélsa*).

Spruce, Bastard. See **Pine, Oregon.**

Spruce, Baltic. See **Spruce, Common.**

Spruce, Black (*Picéa nígra* Link). Alaska, Canada, Newfoundland and North-eastern United States. Known also as "American, Canadian, New Brunswick, St. John's, Double" or "Muskeag Spruce." *French* "Sapinette noir, Epinette noir, Epinette à la bière." *Germ.* "Schwarzfichte." Height 25—80 ft.; diam. 1—3 ft. S.G. 451—510. W 28·5. *e*′ 1·74. *p*′ ·83. R 747 kilos. Sapwood nearly white; heart slightly reddish, light, soft, elastic, strong, compact, with satiny lustre, being tougher, stronger, more elastic and more durable than White Pine, only slightly resinous and, therefore, not good as fuel. Trees with wide rings are known to lumbermen as "White Spruce." Used as "lumber," especially for flooring; for spars and other purposes in shipbuilding; for piles, paddles and oars; when "quarter-sawn," *i.e.* cut radially, in the manufacture of sounding-boards for pianos, violins, etc.; and very largely for paper-pulp. The "Canadian deals" largely imported to the West Indies and England are used, among other purposes, in Manchester and Birmingham for packing-cases. The tree yields a chewing-gum and its shoots are brewed into Spruce beer.

Spruce, Blue (*P. púngens* Engelm.). Rocky Mountains at altitudes of 6000—9000 ft. Known also as "Colorado" or "Rocky Mountain Spruce." Height 80—100 ft.; diam. 1—3 ft. Coarse, strong, useful lumber.

Spruce, Californian Coast. See **Spruce, Sitka.**

Spruce, Canadian. See **Spruce, Black.**

Spruce, Colorado. See **Spruce, Blue.**

Spruce, Common (*P. excélsa* Link). From the Urals and Lapland to the Pyrenees and Alps. Known also as "Spruce Fir, Norway Spruce," or "White Fir," and its wood as "White deal." *French* "Faux sapin, sapin-pesse, sapin gentil, serente, pinesse." *Germ.* "Fichte, Rothtanne, Pechtanne." Height 125—150 ft.; diam. 3—5 ft. W 64·7 when green, 28—32 when dry. E 715 tons. *f* 3·77. *ft* 5·5. *fc* 2·86. *fs* ·27. Stress requisite to indent it $\frac{1}{20}$ in. transversely to the fibres, 500 lbs. per sq. in. Straight-growing, white, reddish or yellowish, light, straight and even in grain, tough, elastic, easy to work except for the small hard knots, warping and shrinking slightly in seasoning, durable. Mostly imported from Norway with the bark on, in logs 30—60 ft. long, and 6—8 in. in diam., that from St. Petersburg being the best, that from the White Sea excellent and that from Riga, Memel and Dantzig large but coarser. These poles are used for scaffolding, telegraph-posts, ladders, roofs, fences, spars and oars. The largest wood is converted into deals and planking, chiefly for central and southern

Europe, for flooring, for toys, for which wide-ringed wood is preferred, for packing-cases, for sounding-boards, dressers and kitchen tables, on account of its whiteness, and to a very large extent for paper-pulp. Spruce is also largely used for charcoal and for fuel, while its resin is used in the preparation of Burgundy pitch.

Spruce, Double. See **Spruce, Black.**

Spruce, Douglas. See **Pine, Oregon.**

Spruce, Engelmann's. See **Spruce, White.**

Spruce, Hemlock. See **Hemlock Spruce.**

Spruce, Himalayan (*P. Moŕinda* Link = *Pinus Smithiána* Wall. = *Abies Smithiána* Loud. = *Picéa Smithiána* Boiss.). Bhotan to Afghanistan at 6000—11,000 ft. Known also as "Indian Spruce," "Morinda," or "Khutrow." Height 120—150 ft.; diam. 5—7 ft. White or nearly so, non-resinous, soft, straight-grained, easily worked, not durable, turning red and decaying rapidly on exposure. Used largely in Simla and its district of growth, for packing-cases, rough and indoor carpentry, planking and fuel.

Spruce, Hondo (*P. Hondoënsis* Mayr). Mountains of Central Japan. *Japanese* "Tohi." Perhaps identical with the Yesso Spruce. Very light and soft. Used in building and carpentry.

Spruce, Indian. See **Spruce, Himalayan.**

Spruce, Menzies'. See **Spruce, Sitka.**

Spruce, New Brunswick. See **Spruce, Black.**

Spruce, Norway. See **Spruce, Common.**

Spruce, Red (*P. rúbra* Link.). South-eastern Canada and Eastern United States. *French* "Sapinette rouge." *Germ.* "Rochfichte." Height 70—80 or 100 ft.; diam. 2—3 ft., being larger than the Black Spruce, with which it was confused. The most valuable timber of the district. Used for carpentry and paper-pulp.

Spruce, Rocky Mountain. See **Spruce, Blue** and **Spruce, White** (ii).

Spruce, St. John's. See **Spruce, Black.**

Spruce, Servian (*P. Omórica* Pancic). Mountains of Servia, Bosnia and Montenegro, at 2000—4000 ft. *Servian* "Omorica,

Morica." Said to have been largely exterminated for the sake of its timber.

Spruce, Single. See **Spruce, White.**

Spruce, Sitka (*P. sitchénsis* Carr.). Western North America from Alaska to California. Known also as "Menzies', Tideland" or "Californian Coast Spruce." Height 100—250 ft.; diam. 6—12 or 15 ft. Light brown, tinged with red, light, soft, straight-

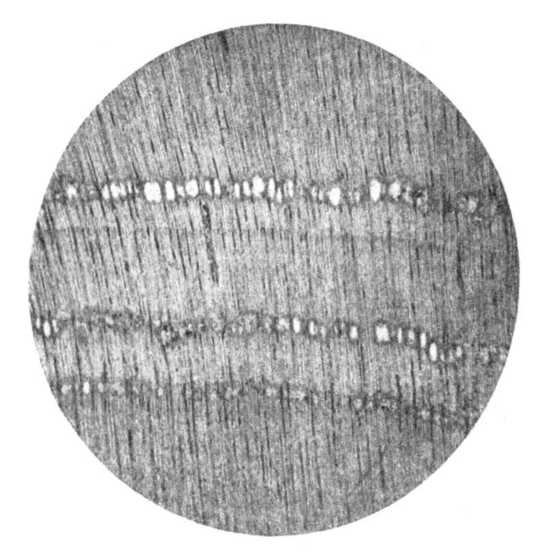

FIG. 64.—Transverse section of Sitka Spruce (*Picéa sitchénsis*).

grained, compact, not strong (Fig. 64). Said to be the best of American Spruces. Used for house and boat-building, carpentry, cooperage, woodenware, fencing, etc.

Spruce, Tideland. See **Spruce, Sitka.**

Spruce, White (i) (*P. álba* Link). Alaska to Newfoundland and the Northern United States. Known also as "Single Spruce." Light, soft, straight and even-grained, with a satiny lustre, compact but not strong. Confused with wide-ringed examples of the Black Spruce. (ii) (*P. Engelmánni* Engelm.). Rocky Mountains from Alberta to Arizona and New Mexico at

3000—11,500 ft. Known also as "Rocky Mountain Spruce." Height 100—150 ft.; diam. 3—5 ft. Very light, soft, straight and close-grained, not strong. Used locally for lumber.

Spruce, Yesso (*P. ajanénsis* Fisher). Northern Japan, Yeso, Saghalien, the Kurile Islands and Amurland. *Japanese* "Yezo-matsu, Kuro-matsu." Height 120—130 ft. Very light and soft. Much used in Yeso for carpentry.

Stavewood (*Stercúlia fœ́tida* L.: Order *Sterculiáceæ*). Deccan, Ceylon, Burma, East Tropical Africa, Moluccas, and, though doubtfully native, North Australia, and cultivated in Tropical America. Known also as "Horse Almond, Bastard Poon" or "Fetid Sterculia." *Beng.* "Jangli badam." *Tamil* "Pinnari." Exported from Cayenne as "Bois puant." Height to first branch 50 ft.; diam. 3—4 ft. W 26—33. Whitish, grey or reddish-brown, generally soft, open-grained and of little use; but said to be used for spars for small vessels. The name is also applied in Australia to other little-used woods.

Stinkwood (*Ocotéa bulláta* Benth.: Order *Lauráceæ*). South Africa. Known also as "Hard-black Stinkwood" or "Laurel-wood." Height 50—60 ft.; diam. 4—5 ft. Golden-brown, often mottled and resembling Walnut, sometimes iridescent, giving off a strong peculiar odour when worked, very tough and considered little and inferior to Teak in strength and durability. Used in house and waggon-building, for gun-stocks and furniture; but now very scarce.

Stinkwood, Camdeboo (*Céltis Kraussiána* Bernh.: Order *Ulmáceæ*). South Africa. Known also as "Soft Grey Stinkwood." *Zulu* "Umounari." Height 20 ft.; diam. 2 ft. Dark greenish, beautifully veined, very heavy, close-grained, liable to warp. Used in waggon-building and furniture.

Stinkwood, Red. See **Almond, Wild.**

Stringybark, a name, obviously descriptive, applied in various districts of Australia to a good many different species of *Eucalýptus* (Order *Myrtáceæ*), especially (i) *E. robústa* [See **Mahogany, Swamp**], (ii) *E. Sieberiána* [See **Gum, Cabbage**], (iii) *E. Stuartiána* [See **Gum, Apple-scented**], (iv) *E. macrorrhýncha*, (v) *E. oblíqua*, and (vi) *E. acmenioídes.*

E. macrorrhýncha F. v. M. South east Australia. Known also as Ironbark. *Aborig.* "Yangoora." Height 50—100 ft.; diam. 2—4½ ft. S.G. 1060—809. W 63·5. Tensile strength 11,700—23,400 lbs. per sq. in. Light brown, generally tinged with deeper red brown, sometimes figured with yellow and brown stripes, hard, strong, close-grained, tough, easily split, tearing under the plane, capable of a good polish, durable, furnishing a fair fuel. Used for fencing, flooring, wheelwright's work and house-carpentry, but suitable for furniture.

E. oblíqua L'Her. Tasmania and South-east Australia. Known also as "Black" or "Ironbark Box," and in Victoria, from its resemblance to, and association with, *E. macrorrhýncha*, as "Messmate." Height 100—150 or even 250—300 ft.; diam. 3—4 or even 15 ft. S.G. 1045—783. W 50—60·5. E 1202·tons. *f* 4·72. *fc* 2·9. *fs* ·476. Tensile strength 8200—8500 lbs. per inch. Straight-growing, light to dark brown, with a wavy figure near the base, heavy, hard, straight, close and even, but rather coarse in grain, strong, splitting very freely, somewhat liable to shakes and gum-veins, durable. Probably the most generally used of all Eucalypts, being employed for fencing, agricultural implements, joists, flooring, shingles, and for ships' beams and keels. It has been successfully introduced into India, especially in the Nilgiri hills.

E. acmenioídes Schauer. South-east Australia. Known also as "White Mahogany" and "Broad-leaved Box." Height 40—60 ft.; diam. 1½—2½ ft. S.G. 1066. W 67·25. Heavier and more durable than the preceding, sometimes prettily figured, strong, easily split, with a satiny lustre when planed. Used like the last-mentioned.

Stringybark, White (i) *Eucalýptus piperíta* [see **Peppermint**], (ii) *E. capitelláta* and (iii) *E, eugenioídes.*

E. capitelláta Sm. Eastern Australia. Known also as "Spotted Gum" and "Peppermint." *Aborig.* "Yangoora." Height 80—120 or 200 ft.; diam. 3—5 ft. S.G. 838. W 52·26. Moderately heavy, tough, strong, easily split, durable. Used for fencing, house-carpentry and fuel.

E. eugenioídes Sieb. South-east Australia. Known also as "Broad-leaved Stringybark." Height said to reach 200 ft. Light-coloured, said to be less easily split but more durable than the other Stringybarks, but inferior as fuel. Used for fencing, shingles and flooring.

Sugar-berry. See **Hackberry.**

Sugi (*Cryptoméria japónica* Don : Order *Taxodíneæ*). China and Japan; introduced into England in 1843. Growing at altitudes of 500—1200 or 3000 ft. Known also as "Japanese Cedar." Height 60—125 ft.; diam. 4—5 ft., tapering. Brownish-red, resembling *Sequoía* in texture. Used for common lacquer-ware. One of the most abundant and useful of Japanese forest-trees.

Sumach, Staghorn (*Rhús typhína* L. : Order *Terebinthácæ*). Canada and North-eastern United States. Known also as "Virginian Sumach." *Germ.* "Hirschkolben Sumach." Height seldom 20 ft.; diam. seldom 10 in. S.G. 436. W 27. Sap-wood very narrow, dingy yellowish-white; heart golden-orange to greenish, the autumn zones much darker, handsome, somewhat aromatic, light, soft, brittle, rather close-grained, difficult to split, lustrous, vessels larger and much more numerous in the spring wood, slightly dendritic, 1—7 together, pith-rays not visible. Used in dyeing, and occasionally in small pieces for inlaying in cabinet-work.

Sumach, Venetian (*Rhús Cótinus* L.). Southern Europe. Known also as "Wig-tree," "Wild Olive," "Young" or "Zante Fustic." A shrub, yielding crooked sticks 4—5 ft. long and 2—3 in. in diam. Sapwood narrow, white; heart golden-yellow or greenish, hard, easily split, lustrous; rings not distinct; vessels and pith-rays as in the preceding. Imported from Greece as a yellow dye for wool and leather.

Sundri (*Heritiéra fómes* Buch. : Order *Sterculiáceæ*). Sunderbunds of Bengal, Malay Peninsula, and Borneo. *Beng.* "Sundri." *Burm.* "Ka-na-zo." *Germ.* "Brettbaum." Not large, yielding timber 15 ft. long and 1 ft. in diam. S.G. 927—799. W 58—50. Brown, very hard, tougher than any other Indian wood, elastic, strong and durable. Used for boat and bridge-building, handles, naves and felloes of gun-carriages, and in Calcutta largely as fire-wood, and yielding gunpowder charcoal of the best quality. The close-allied *H. littorális* Dryand. is a native of East Africa, the Mauritius, Burma and Queensland, where it is known as "Red Mangrove." It is the "Looking-glass tree" of English gardeners. It yields timber 30 ft. long and 2 ft. in diam. W 102 when wet, 65 when dry. Dark-coloured, scented, firm, very tough, durable. Used in boat-building, for handles, gunstocks, planking and packing-cases.

Sycamore, a name that has been singularly and variously misapplied. Belonging originally to the Fig-mulberry of the Levant (*Ficus Sykomórus* L.: Order *Moráceæ*), a shade-tree yielding a very strong wood used for Egyptian mummy-cases, it is applied in England to the Great Maple (*Acer Pseúdo-plátanus* L.: Order *Acerineæ.*) Central Europe and Western Asia, almost naturalised in Britain. Known in the South of Scotland as "Plane." *French* "Grand Erable, Erable blanc de montagne, Fausse platane." *Germ.* "Bergahorn." Height 40—60 ft.; diam. 1—3 ft. W 64 when newly cut, 48—36 when dry. Without distinct heartwood, white, when young, becoming yellowish with age, or slightly brown in the centre, often beautifully figured, the fine but distinct pith-rays having a satiny lustre, which distinguishes it from Linden wood, compact, firm neither very heavy nor hard, fine-grained, tough, splitting evenly but with difficulty, easily worked, susceptible of a high polish, cracking and warping somewhat in seasoning and shrinking one-twelfth of its bulk, very durable if kept dry, and generally free from insect-attack; rings marked by fine evenly-circular lines; vessels indistinct, evenly distributed. Highly esteemed on the Continent by turners, cabinet-makers, carvers and toymakers, the figured wood being used for violins. Formerly much used for platters and spoons, it is still largely manufactured at Glasgow and elsewhere into bread-platters, butter-dishes and moulds and is also in demand for bobbins, reels, coach-panels, cutting-boards for shoemakers, shop-boards for butchers, and wooden type. Large wood is sought after for calico-printing rollers and when quartered for those of washing-machines. It is superior to Beech both as fuel and for charcoal; but is by no means plentiful or cheap.

In America the name is applied to *Plátanus occidentális* [See **Plane**]; in Australia to *Pánax elegans* [See **Laurel**] and to *Sterculia lúrida* F. v. M. (Order *Sterculiáceæ*), in the north-east, also known as "Hat Tree," a white wood occasionally used for shingles.

Sycamore, White (*Cryptocárya obováta* R. Br.: Order *Lauráceæ*). Queensland and New South Wales. Known also as "Bastard Sycamore, She-beech, Flindosa." *Aborig.* "Myndee." Height 70—100 ft.; diam. 2—5 ft. W 35. White, light, soft, fairly durable if kept protected, working well and darkening with age.

Tacamahac. See **Poplar, Balm of Gilead.**

Talura (*Shórea Talúra* Roxb.: Order *Dipterocarpáceæ*). Southern India. A large tree. Grey, very hard, smooth-grained. Used for housebuilding, especially in Madras.

Tallow-wood (*Eucalýptus microcórys* F. v. M.; Order *Myrtáceæ*). Eastern Australia. Known also as "Peppermint, Turpentine-tree" and "Forest Mahogany." Height 100—120 ft.; diam. 6—8 ft. S.G. 952. W 70·5—59·43. E 896 tons. *f* 5·48. *fc* 4. *fs* ·618. Light or dark yellow or yellowish-brown, straight or wavy in grain, strong, durable under or above ground, very greasy when freshly cut, liable to shakes and generally hollow when large. Used by wheelwrights and for ballroom floors. One of the best woods for paving.

Tamarack (*Lárix péndula* Salisb. = *L americána* Michx.: Order *Coniferæ*). Canada and North-east United States. Known also as "Hackmatack, American" or "Black Larch." *French Canadian* "Epinette rouge." Height 80, or rarely 100 ft.; diam. 2—3 ft. S.G. 263. W 38·86. R 901 kilos. Sapwood light; heart light brown or reddish-grey, moderately heavy, hard, rather coarse-grained, compact, very strong and durable in contact with soil, in microscopic structure resembling European Larch. One of the best American timbers for sleepers, valuable from its straight-growth for telegraph-poles and fence-posts, while naturally crooked pieces are used for knees in shipbuilding. Resembling Hard Pine in appearance, quality and uses, equal in durability to Oak and in strength to European Larch.

Tamarack, Western (*L. occidentális* Nutt). North-western United States. Known also as "Western Larch." Height 100—250 ft.; diam. 2—3 or 4—8 ft. Beautifully coloured, heavy, very hard, free from knots, strong and durable. The largest of Larches, harder and stronger than all other American conifers. Suitable for furniture or lumber; but used chiefly for sleepers, posts and fuel.

Tamarind (*Tamaríndus índica* L.: Order *Leguminósæ*). Throughout the Tropics. *Pers.* "Tamar-i-hindi." *Hind.* "Amli ka jhar." *Malay* "Asam, Kranji." *Sansk.* "Amlika." *Tam.* "Pulia." Slow-growing but large. Height 60—80 ft.; diam. 5—8 ft. S.G. 1323. Yellowish-white with an irregular heart of dark purplish-brown blotches in old trees, resembling Ebony or Lignum-vitæ, but apt to be hollow, very heavy and hard, difficult to work, durable and free from insect-attack. Used for blocks, mallets, rice-pounders, oil and sugar-mills, turnery, fuel and gunpowder charcoal.

Tamarind-plum. See **Kranji.**

Tampinnis (*Sloétia Sideróxylon* Teijsm. and Binn.: Order *Moráceæ*). Straits Settlements and Sunda Islands. Known also as "Ironwood." A large tree, yielding a dark-coloured, hard, valuable timber, used in making the large implements employed in stirring gambir.

Tanekaha (*Phyllocládus trichomanoídes* Don.: Order *Taxineæ*) New Zealand. Known also as "Celery-topped Pine." Height 50—80 ft.; diam. 2—3 ft., 30—40 ft. to lowest branch, yielding timber 18—45 ft. long, squaring 10—16 in. S.G. 1000—600. Yellowish-white, heavy, close and straight-grained, tough, very strong, working up well, very durable, especially in moist situations. Used for sleepers, piles, bridges, mine-props, masts, decks and building.

Taraire (*Beilschmiedia Taráiri* Benth. and Hook. fil.: Order

Laurineæ). New Zealand. Height 60—80 ft.; diam. 1—2 ft. Hard, compact, susceptible of a good polish, but not durable, if exposed. Suitable for cabinet-work.

Tatamaka. See **Poon.**

Tawa (*Beilschmiedia Táwa* Benth. and Hook. fil.). New Zealand. Height 60—70 ft.; diam. 1—2 ft. Compact, even-grained, but, like the allied Taraire, not durable if exposed. Suitable for furniture.

Tea, a name transferred in Australia, from the varieties of *Théa assámica* (Order *Camelliáceæ*) cultivated in China and now in India, to various species of *Melaléuca* and the allied genera *Kúnzea* and *Leptospérmum* (Order *Myrtáceæ*), the leaves having been used as a substitute for tea by Captain Cook's sailors. The name is applied unqualified (i) to *Leptospérmum lanígerum* Sm., seldom larger than a tall shrub, with light-coloured, heavy, hard, even-grained and durable wood, used by the aborigines for spears; (ii) to *L. flavéscens* Sm. similar, but reaching a height of 15—20 ft. and a diameter of 5—8 in.; and (iii) to *Melaleúca uncináta* R. Br., which reaches 70—90 ft. in height and 2—3 ft. in diameter.

Tea-tree, Black (*Melaleúca styphelioídes* Sm.). Eastern Australia. Known also as "Prickly-leaved Tea-tree." Height 20—30 or 80 ft.; diam. 1—3 ft. W 66·75—73·35 when seasoned. Heavy, hard, close-grained, difficult to work, splitting in seasoning, very durable even in damp situations. Used for posts.

Tea-tree, Broad-leaved (*Melaleúca leucadéndron* L.). Known also as "White" or "Swamp Tea-tree, Paper-bark Tree" or "Milkwood." Height 40—50 ft.; diam. 1—2 ft. Beautifully figured with ripple-like darker markings, heavy, hard, close-grained, very durable underground. Excellent for posts or boat-building.

Tea-tree, Mountain (*Kúnzea peduncuIáris* F. v. M.). South-east Australia. *Aborig.* "Burgan." Used for spears and boomerangs.

Tea-tree, Prickly-leaved (*Melaleúca armilláris* Sm.). Eastern Australia. Height 20—30 ft. Hard and durable under ground or water, but decaying on exposure. [See also **Tea-tree, Black.**]

Tea-tree, Soft-leaved (*M. linariifólia* Sm.). North-east Australia. Height 40—50 or 80 ft.; diam. 1—3 ft. Very heavy, hard, close-grained and imperishable under water, but splitting in seasoning. Used for piles, turnery and fuel.

Tea-tree, Swamp (i) (*M. ericifólia* Sm.). Eastern Australia and Tasmania. Small, very hard and durable. Used for hurdles or rafters, and, in Tasmania, for turnery. (ii) (*M. squarrósa* Sm.). South-eastern Australia and Tasmania. Height 6—10 ft. S.G. 713. Heavy, very hard, difficult to work, durable under water or when exposed. [See also **Tea-tree, Broad-leaved.**]

Tea-tree, White, in New Zealand, apparently (*Leptospérmum ericoídes* A. Rich.). Height 40—50 ft.; diam. 1—2 ft. Heavy and hard. Much valued for piles and using also in fencing and house-building. [See also **Tea-tree, Broad-leaved.**]

Teak (*Tectóna grándis* L.: Order *Verbenáceæ*). Central and and Southern India, Burma, the Shan States, Malay Peninsula, Sumatra, Java and Celebes, extensively planted by the Dutch in Ceylon. *Hind.* "Ságun." *Burm.* "Kuyon." *Malay* "Jati." *Tamil* "Teak." Height 80—100 ft. or more; diam. 2—4 ft., yielding logs 23—50 ft. long, squaring 10—30 in. S.G. over 1000 when green; but, being generally "girdled" three years before felling, 910—635 when seasoned. W 57 when green—37. E 1071—950 tons. *e'* 1·19. *p'* 1·08. *f* 6·92. *ft* 4—9, averaging 6·7. *c* 3301. *c'* ·436 or more. *fc* 4—5·4. *v'* ·832 or more. R 322—406 lbs. Straight-growing, light straw-colour to a brownish-red, when fresh, but darkening on exposure, some of the Teak of the Deccan is beautifully veined, streaked and mottled, whilst some old trees have burrs the wood of which resembles Amboyna-wood. It is very fragrant, so as to resemble Rosewood, owing to an oleo-resin which also renders the wood probably the most durable of known timbers, making it obnoxious to termites and keeping off rust from iron in contact with it. It is the general practice to "girdle" the trees, *i.e.* to cut a complete ring through both bark and sapwood, so killing the tree and rendering it light enough to float to the port of shipment; and, as usually a year elapses between the felling and its delivery in England, it arrives sufficiently seasoned, heavy, moderately hard, clean even and straight in grain, but little shrunken, split or warped in the process. The rapid drying, however, induced by girdling is said to render the wood inelastic, brittle and less durable, so that it splits too readily for use in gun-carriages. Teak varies very much according to locality and soil, that of Malabar being darker, heavier and rather stronger than, though not so large as, that of Burma. Though without shakes on its outer surface, Teak nearly always has a heart-shake, which, owing to a twist in the growth, may often at the top be at right angles to what it is at the butt, thus seriously interfering with conversion, though often little affecting the use of the timber in bulk. In the large Rangoon or Irrawaddy Teak there is also sometimes a close, fine star-shake. In these shakes an excretion of apatite or phosphate of lime consolidates in white masses, which will turn the edge of most tools. After girdling, the dead trees are often attacked by burrowing insects which may penetrate beyond the sapwood and so render the timber unfit for reduction to plank. Being a deciduous tree, Teak has distinct annual rings, with large and distinct vessels which are rather larger and more numerous in the spring-wood and are sometimes filled with the apatite. The pith-rays are distinct and light-coloured, as in Oak, but fine, the vessels in the spring-wood being 2—3 together between every two rays. Teak splits readily and is easily worked, somewhat like Oak, but it owes its superiority for shipbuilding over both Pine and Oak in part to its freedom from any change of form or warping, when once seasoned, even under the extreme climatic variations of the monsoons. In India, Teak is used for railway-sleepers, bridge-building and furniture. As the Indian Forest Department plant several thousand acres annually there is little fear of the exhaustion of the supply, whilst the timber from cultivated trees is said to be better than that grown in the natural forests. Teak is very largely exported, especially from Moulmein and Rangoon, that from the former port, drawn from the valleys of the Salwén and

Thungyen Rivers, being rather shorter but less shaky than that shipped at Rangoon from the Irrawaddy valley. Whilst it is the best timber known to us for shipbuilding, especially for the backing of armour-plates, Teak is also considerably used in England in the building of railway-waggons; but is comparatively little used in foreign dockyards. In the London market it is sorted into A, B and C classes, according to size, and has varied in price from £10 per load of 50 cubic feet in 1859 to £13 10s.—£15 10s. at the present time.

Teak, African. See **African.**

Teak, Bastard (*Pterocárpus Marsúpium* Roxb.: Order *Leguminósæ*). Central and Southern India. *Hind.* "Bibla." *Beng.* "Bija," "Bija Sâl." A large tree, yielding timber 18—30 ft. long and 1—2 ft. in diam. S.G. 820. R 518—378 lbs. Sapwood narrow, white soft, heart reddish-brown, handsomely streaked with a darker shade, very hard, requiring thorough seasoning, susceptible of a fine polish and very durable. Darker-coloured and harder than the allied Padouk, it is heavier than most Teak, equally strong, and less liable to split, but more expensive to work and not durable if exposed to wet. It is largely used for door and window-frames, posts, beams, agricultural implements, cart and boat-building, and furniture.

Teazle, the name in the walking-stick trade for *Vibúrnum Ópulus* L. (Order *Caprifoliáceæ*). A native of Europe, Northern and Western Asia, and North America, known also as "Guelder Rose" or "Balkan Rose," reaching a height of 6—12 ft., and imported as walking-sticks from the Balkans.

Tendu. See **Ebony, Bombay.**

T'eng-li-mu (*Pýrus betulæfólia* Bunge: Order *Rosáceæ*). China. The best wood in Wuchang for engraving purposes, being a tolerable substitute for Box, occurring in the market in planks 6 in. wide and 1½ in. thick, costing 150 cash, or 5½d.

Tewart or **Touart** (*Eucalýptus gomphocéphala* DC.: Order *Myrtáceæ*). Western Australia, where it is stated it covers 500 sq. miles. Known sometimes as "White Gum." Height

100—150 ft., yielding timber 20—45 ft. long, squaring 11—28 in. S.G. 1194—1000. W 73—66. *c* 10,284. *c′* 1·398. *v′* 1·229. R 257·25 lbs. Straight-growing, pale yellow or light brown, very heavy, hard, tough, strong and rigid, close, twisted or even curled in grain so as to be difficult to cleave or work and with no liability to split, with a slight heart and star-shake militating against its reduction into planks, shrinking very little in seasoning and apparently imperishable under any climatic changes. Used in shipbuilding for beams, keelsons, capstans and windlasses, strongly recommended for the woodwork in engine-rooms, where it is exposed to great heat, and for piles and dock-gates, and well suited also for naves and spokes of wheels, but, though one of the strongest known woods, too heavy for general use.

Thingan (*Hópea odoráta* Roxb.: Order *Dipterocarpáceæ*). Further India. *Burm.* "Thingan." *Anam* "Sao." Height up to 250 ft., 80 ft. to lowest branch; diam. 3—4 ft. S.G. 652—608. W 64—38. R 800 lbs. Yellowish-brown, heavy, hard, close and even-grained, not liable to insect-attack and very durable under water, but liable to split in the sun. Used for house-building, canoes, and cart-wheels, being one of the most valuable woods of its district.

Thitka (*Pentácé burmánica* Kurz.: Order *Tiliáceæ*). Burma, Pegu, Malacca and Java. Known also as "Kathitka." Very large. W 42. White or yellowish-red, light, soft, even-grained, taking a good polish; pith rays moderately broad, wavy, red, equidistant; rings visible. Used in Burma for boatbuilding and tea-chests, and exported in considerable quantities to Europe for furniture, resembling inferior Mahogany.

Thitkado. See **Cedar, Moulmein.**

Thitya (*Shórea obtúsa* Wall.: Order *Dipterocarpáceæ*). Further India. *Burm.* "Thitya," "Theya." A large tree, 50 ft. to its lowest branch; diam. 2 ft. W 75. Dark-coloured, handsome, very hard and durable wood. Used in house and canoe-building and for handles.

Thorn. See **Blackthorn** and **Hawthorn.**

Thorn, in Cape Colony (*Acácia hórrida* Willd.: Order *Leguminósæ*). Known also as "Mimosa." *Boer* "Doorn boom, Kamulboom." Height 20—25 ft.; diam. 1—1½ ft. Hard, tough. Used for building, agricultural implements, wheels, etc.

Tochi (*Ǽsculus turbináta* Bl.: Order *Sapindáceæ*). Japan. Height 20 ft.; diam. 2 ft. Used in house-building, box-making and lacquer-work.

Tonka-bean (*Coumaroúna odoráta* Aubl. = *Dípteryx odoráta* Willd.: Order *Leguminósæ*). Brazil and Guiana. Known sometimes as "Tonquin Bean," "Gaïac," "Cuamara." Height 60—70 or 90 ft.; diam. 1—2½ ft. S.G. 1213—1032. R 385 kilos. Dark yellow or reddish-brown, very heavy, hard, tough, cross-grained, difficult to work, taking a fine polish, very durable and said to bear a greater strain than any wood in the Colony. Used for cogs, shafts, mill-wheels, and to a small extent for turnery and furniture, and medicinally as a substitute for Guaiacum.

Toon. See **Cedar, Moulmein.**

Torreya, Japanese. See **Kaya.**

Totara (*Podocárpus Totára* A. Cunn.: Order *Taxíneæ*). New Zealand. Height 40—70 or 100 ft., 35—40 ft. to the lowest branch; diam. 2—6 or 10 ft., yielding timber 20—45 ft. long, squaring 10—22 in. S.G. 1230 when fresh cut, 600 when seasoned. Sapwood 2—3 inches wide, light-reddish; heart deep red, heavy, moderately hard, close, straight, fine and even in grain, strong, easily worked, not warping or twisting, very durable, teredo-proof. With the exception of Kauri, the most valuable timber in New Zealand, and far more abundant than Kauri. Used for piles in the sea, sleepers, telegraph-poles, fencing, shingles, bridges and general building purposes.

Trincomalee-wood (*Bérrya Ammonílla* Roxb.: Order *Tiliáceæ*). Ceylon, Philippines and Tropical Australia; introduced in India. *Sinh.* "Halmilla." *Tam.* "Katamanakku." *Germ.* "Halmalilleholz." Height 20—40 ft.; diam. 1—2½ ft. W 48. Straight-growing, dark red, light, very hard, straight and fine-grained, tough, strong, durable. Imported from Trincomalee to Madras.

Specially valued in Ceylon for staves for oil-casks, and in Madras for boat and carriage-building. Used for spokes and shafts, for handles, capstan-bars, etc.

Trumpet-tree (*Cecrópia peltáta* L. and *C. palmáta* Willd.: Order *Moráceæ*). The former in Jamaica, the latter in Brazil and Guiana. Height 50 ft.; diam. 1 ft. Very light and resonant. Used for floats for fishing-nets, razor-strops, for producing fire by friction, for trumpets and drums made from the hollow branches or stems.

Tsuga (*Tsúga Siebóldii* Carr.: Order *Coníferæ*). Japan. Known also as "Japanese Hemlock Fir." *Jap.* "Tsuga Araragi." Height 80—90 ft.; diam. 3—6 ft. Reddish-white, durable. Little used, owing to its inaccessibility.

Tulip-tree (*Liriodéndron tulipífera* L.: Order *Magnoliáceæ*). Eastern North America. Known also as "Saddle tree, Poplar, Yellow, White" or "Virginian Poplar, Whitewood, Canary Whitewood, Canary-wood," or "Canoe-wood." *French* "Tulipier." *Germ.* "Tulpenbaum." Height 100—150 ft.; diam. 3—10 ft. S.G. 423. W 26·36. R 657 kilos. Sapwood nearly white; heart light lemon-yellow or brownish, light, soft, close and straight in grain, tougher than many woods equally soft, compact, not very strong or durable, easily worked, shrinking and warping somewhat in seasoning, taking a satiny polish. Vessels minute, evenly distributed; pith-rays fine but distinct. Much heavier and more valuable as timber than the true Poplars. Excellent for shingles and clapboards, as it does not split under heat or frost; used for rafters and joists, and generally as a substitute for White Pine or Cedar in building, especially for doors, panels and wainscot, the seats of American Windsor chairs, box-making, turnery and boat-building; and formerly for Indian "dug-out" canoes. Imported from New York to Liverpool as "American" or "Canary Whitewood" in large planks and waney logs at a price equal to that of the best Quebec Yellow Pine, this wood is valued, as easily worked, firm when fully dried and taking polish, stain or paint very well, by carriage-builders, shop-fitters, cabinet-makers, etc.

Tulip-tree, in Australia (i) (*Lagunária Patersóni* Don: Order *Malváceæ*). Known also as "White Oak" or "Whitewood." Height 40—60 ft.; diam. 1½—2½ ft. White, close-grained, easily worked. Used for building. (ii) (*Stenocárpus sinuátus*). See **Fire-tree.**

Tulip-tree, in India. See **Umbrella-tree.**

Tulip-wood (*Physocalýmma scabérrimum* Pohl.: Order *Lythráceæ*). Brazil. *Portug.* "Pao de rosa." *French* "Bois de rose." *Germ.* "Rosenholz." A rose-coloured, beautifully striped wood. Considerably used for inlaying and small turned ware.

Tulip-wood, in Australia (i) (*Harpúllia péndula* Planch.: Order *Sapindáceæ*). North-east Australia. *Aborig.* "Mogum-mogum." Height 50—60 ft.; diam. 1—2 ft. The outer wood light-coloured, very tough, easily worked, the best wood in Australia for lithographers' scrapers and suggested for engraving. The inner wood beautifully marked with black and yellow, close-grained, strong, and much valued for cabinet-work; (ii) the similar wood of *Owénia venósa* [See **Plum, Sour**]; (iii) the very different light-coloured wood of *Aphanánthé phillippinénsis.* [See **Elm.**]

Tupelo. See **Gum, Black.**

Turnip-wood (*Synóum glandulósum* A. Juss.: Order *Meliáceæ*). New South Wales and Queensland. Known also as "Dogwood, Brush Bloodwood" and "Bastard Rose-wood." Height 40—60 ft.: diam. 1½—2 ft. W 41—45. Deep red when fresh with a scent like that of roses, afterwards resembling Cedar, firm, easily worked, but apt to tear under the plane, taking a good polish. Used for cabinet-work, interior finish, etc. The bark has a turnip-like smell. [See also **Cedar, Pencil.**]

Turpentine-tree, American. See **Pine, Long-leaf.**

Turpentine-tree, in Australia (i) *Eucalýptus microcórys* [See **Tallow-wood**]; (ii) *E. Stuartiána.* [See **Gum, Apple-scented**]; (iii) *Syncárpia laurifólia* Ten. (Order *Myrtáceæ*). North-east Australia. Height 100—150 ft.; diam. 4—5 ft. W 63. Sap-wood light-coloured; heart dark brown, hard, apt to shrink and warp unless well seasoned, very durable underground, in damp, or in sea-water, termite and teredo-proof owing to the resin it contains, taking a high polish, very uninflammable. Much used in Sydney for piles, excellent for sleepers, fence-posts and uprights in buildings. (iv) *S. Hillii* Bailey, of North Queensland, a dark pink, close-grained, tough wood, appears to be also useful.

Turpentine, Brush (i) (*Syncárpia leptopétala*) [See **Myrtle**]; (ii) (*Rhodámnia trinérvia* Blume: Order *Myrtáceæ*). New South Wales and Queensland. Sometimes known as "Black-eye." Height 70—80 ft.; diam. 2—3 ft. W 50—52·6. Reddish-brown, moderately hard, close-grained, firm, not easy to season and almost always hollow when large. Not much used.

Umbrella-tree (*Thespésia populnea* Corr.: Order *Malváceæ*). Western Tropical Africa, Seychelles, Ceylon, Southern India, Burma, Tropical Australia, Fiji and other Pacific islands, and Tropical America. Known also as "Tulip-tree, Portia-tree, Rosewood of Seychelles." *Sinh.* "Suriya." *Telugu* "Gangaravi." *Mahratti* "Bendi." *French* "Faux bois de rose, Bois de rose de l'Océanie." Height 40—50 ft. W 50. Sapwood light red; heart dark-red, a deep claret colour when hollow, resembling Mahogany, hard, very strong, durable, especially under water. Used in India for furniture, spokes and carriage-building, in Ceylon for gunstocks and also in boat-building.

Umzimbit (*Toddália lanceoláta* Lam.: Order *Xanthoxyláceæ*). South Africa. Known also as "White Ironwood." Height 20 ft.; diam. 2 ft. W 60·37. E 772 tons. *f* 8·56. *fc* 4·33. White, hard, tough, elastic, resembling Ash or Hickory. Used chiefly for ploughs, axles and waggon-building.

Uroobie (*Nephélium tomentósum* F. v. M.: Order *Sapindáceæ*). North-east Australia. Height 20—30 or 80 ft.; diam. reaching 3 ft. Yellow, close-grained, hard. Occasionally used for building.

Vaivai (*Seriánthes vitiénsis* A. Gray: Order *Leguminósæ*). Fiji. Light, tough. Valued for planks in boat-building.

Venatico. See **Mahogany, Madeira.**

Vau, in Fiji. See **Corkwood-tree.**

Varnish, Black or **Burmese** (*Melanorrhœ́a usitáta* Wall.: Order *Anacardiáceæ*). Further India. Known also as "Lignum-vitæ of Pegu." *Burm.* "Thit-si." *Manipuri* "Kheu." Height 30—40 ft. to lowest branch; diam. 2—4 ft. W 54. Dark red or brown, with yellowish streaks, heavy, very hard, fine and close-grained, tough, durable, the black gum which exudes from it repelling termites. Used for anchors and tool-handles and recommended for blocks, gun-stocks or sleepers. The varnish is largely used in lacquering and gilding.

Vesi. See **Shoondul.**

Vesi-vesi. See **Beech, Indian.**

Vine (*Vítis vinífera* L.: Order *Vitáceæ*). Originally a native of Armenia. Small. Sapwood narrow, reddish-white; heart reddish-brown; pith large; autumn wood narrow; spring wood full of large vessels. Light, pliable and elastic. Suitable for walking-sticks.

Violet-wood. See **Purple-heart.** In Australia the name is applied, on account of perfume, to *Acácia péndula.* See **Myall.**

Vlier (*Núxia floribúnda* Benth.: Order *Loganiáceæ*). Cape Colony. Height 20—25 ft.; diam. 1—2 ft. Light-coloured, hard. Used in waggon-building.

Wacapou (*Andira Aubletii* Benth.: Order *Leguminósæ*). Guiana. *French* "Épi de blé." *Dutch* "Bruinhart." In logs 14—21 ft. long, squaring 13—16 in. S.G. 1113—900. R 304 kilos. Dark-coloured, straight-grained, strong, easily worked, very durable and not liable to insect attack, one of the best of the hard-woods of the country; but subject to serious heart-shake. Excellent for ship building, railway-sleepers or cabinetwork.

Wallaba (*Epérua falcáta* Aubl.: Order *Leguminósæ*). Guiana. *French* "Wapa gras." Height 40—50 ft.; diam. 1—2 ft. S.G. 945—930. W 52. E 1171 tons. *f* 8·1. *fc* 4·28. *fs* ·424. R 224 kilos. Deep red with whitish streaks, heavy, hard, and, owing to a resinous oil, very durable, but rather coarse-grained. Used for vat-staves, shingles, palings and, to a small extent, for furniture. The **Ituri Wallaba** (*E. Jénmani* Oliver) is finer in grain.

Wallang-unda (*Monótoca ellíptica* R. Br.: Order *Epacridáceæ*). Eastern Australia. Known also as "Beech." Height 20—30 ft.; diam. small. W 37—44·5. Resembling Beech, sometimes with a beautiful silvery grain, working well across the grain and taking a polish. Excellent for planes, etc., but not answering for engraving.

Walnut (*Júglans régia* L.: Order *Juglandáceæ*). A native of Northern China and Persia, introduced into Greece and Italy in early times from Persia, and thence into the rest of Europe. *Pers.* "Jaoz, Charmagz, Akrot." *Greek* "Carua, Persicon, Basilikon." *Latin* "Juglans." *French* "Noyer." *Germ.* "Walnuss." Height 30—50 ft.; diam. 2—3 ft. W 58·5 when green, 47—46·5 when dry. Sapwood broad, greyish-white, very liable to become worm-eaten; heart brown to black-brown, often "watered," *i.e.*, showing dark wavy lines and zones, moderately heavy, hard, fine and close in grain, splitting very little in seasoning, but readily split artificially, taking a beautiful polish and durable, if kept dry, especially when dark in colour and figured: pith large, chambered; pith-rays fine, indistinct; rings distinct; vessels uniformly distributed, few, very large and open, single or in pairs. (Fig. 65.) Burrs, often 2—3 ft. across and a foot or more in thickness, weighing 5—6 cwt., occur, and being generally beautifully mottled are highly valued for veneers. The sapwood may be rendered more durable by smoking over a fire of Beech or by boiling in the juice of the green fruit; but the liability to worm-attack prevents Walnut being used in building. English-grown Walnut, coming mostly from rich plains, is pale, coarse, little figured, and perishable; French is better; that from the Black Sea, sometimes known as Austrian and Turkish, which is imported in waney logs 6—9 ft. long, squaring 10—18 in., is still more valuable; whilst that from Italy, which comes in planks 4—9 in. thick, 10—16 in. wide, and 5—12 ft. long, is the best. At the beginning of the 18th century Walnut became very fashionable as a furniture wood, marking the first departure from the previous universal use of Oak. The severe winter of 1709 killed most of the Walnut trees in Central Europe, the dead trees being bought up by the Dutch, who thus secured a "corner" in this wood. So scarce was it in France that its exportation was prohibited in 1720, and Mahogany, imported by the Dutch and Spaniards, largely replaced it for furniture. No wood, however, equals it for the manufacture of gunstocks, so that the wars of the 18th century created a great dearth of this timber, and we read of France consuming 12,000 trees a year in 1806, and of as

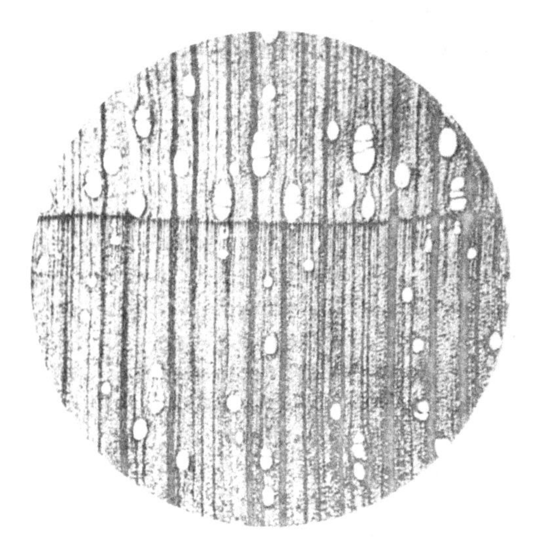

FIG. 65.—Transverse section of Walnut (*Júglans régia*).

much as £600 being paid for a single tree. European Walnut is still in use for the best gun-stocks. The burrs have realised £50 —£60 per ton, and veneers, some of which are of a beauty unsurpassed by any other wood, as much as two or three shillings per square foot. These are used in the pianoforte and furniture trades. Swiss carvings are mostly in Walnut, and the wood is also used in turnery, for screws for presses, musical instruments, sabots, etc.

Walnut, American or **Black** (*Júglans nígra* L.). Eastern North America. Height 60—150 ft.; diam. 3—8 ft. S.G. 611.

W 38·1. R 856 kilos. Sapwood narrow; heart violet-brown or chocolate-brown, blackening with age, heavy, hard, tough, strong, rather coarse in grain, easily worked, very durable in contact with the soil. More uniform in colour, darker, less liable to insect-attack, and thus more durable than European Walnut. Formerly used on the Wabash for "dug-outs," 40 ft. long and about 27 in. wide, and also largely for fence-posts, shingles, building, naves of wheels, etc., this wood has now become too valuable as a cabinet and veneer wood to be used for these purposes. Before the middle of the 19th century it was only used in England for carcase ends, frames for veneering, and other inferior purposes: it has now more than doubled in price and is more used than European wood, its uniform colour recommending it to shop-fitters and as a basis for painted or other ornamentation in the cabinet-trade. It is imported in logs 10—20 ft. long, squaring 15—30 in.; and, besides its use in cabinet-making, is employed for the stocks of our army rifles. It fetches from 1s. 8d. to 4s. per cubic foot in the London market.

Walnut, Satin. See **Gum, Sweet.**

Walnut, White. See **Butternut.**

Wandoo (*Eucalýptus redúnca* Schau.: Order *Myrtáceæ*). West Australia. Known also as "White Gum." Height up to 120 ft.; diam. up to 17 ft. W 70. Light-coloured, heavy, very hard, tough and durable. Valued for building, and especially for wheelwrights' work, being superior to Tewart for spokes and felloes and supplying the best naves and cogs; used also for furniture.

Water-tree. See **Pin-bush.**

Waterwood (*Chimárrhis cymósa* Jacq.: Order *Rubiáceæ*). West Indies. Known also as "Bois Rivière." Height 50—60 ft. Valued for furniture and joinery.

Wattle, the general name in Australia for species of *Acácia* (Order *Leguminósæ*), from their use by the early colonists in "wattling" their huts. *French* "Bois tressé."

Wattle, Black (i) (*A. binerváta* DC.), North-east Australia. Sometimes known as "Hickory." Height up to 30—40 ft.;

diam. 8—18 in. W 50·5—56·6. Dirty white to pinkish, close-grained, tough, light. Valued for axe-handles and bullock-yokes. (ii) (*A. decúrrens*) [See **Wattle, Feathery**]. (iii) (*A. mollíssima* Willd.). South eastern Australia and Tasmania. Known also as "Silver Wattle." *Aborig.* "Garrong" or "Currong." Height 20—30 ft.; diam. 6—9 in. S.G. 804—727. W 50·2. Light, tough, strong, liable to insect-attack. Formerly used for boomerangs, mulgas and spears, and now-a-days in Tasmania for cask-staves, treenails, etc.

Wattle, Feathery (*A. decúrrens* Willd.). New South Wales. Known also as "Black" or "Green Wattle." *Aborig.* "Wat-tah." Height 40—50 ft.; diam. 1—1½ ft. S.G. 773—727. W 62·8—45. Sapwood white, heart pinkish, light, tough, strong, easily worked, very liable to the attacks of boring beetles. Good for fuel or staves.

Wattle, Golden (i) (*A. pycnántha* Benth.). South-east Australia. Known also as "Green" or "Broad-leaved Wattle." S.G. 830. W 51·5. Tough, close-grained. (ii) (*A. longifólia*) [See **Sallow**].

Wattle, Green. See **Wattle, Feathery** and **Wattle, Golden.**

Wattle, Hickory (*A. aulacocárpa* A. Cunn.). Queensland. Dark-red, heavy, hard, tough. Useful for cabinet-work.

Wattle, Prickly (*A. juniperína* Willd.). South-east Australia and Tasmania. Height 8—12 ft.; diam. small, White, light, tough. Used for mallet-handles.

Wattle, Silver (i) (*A. dealbáta* Link.). Eastern Australia and Tasmania; established in India since 1840. Height 60—120 ft.; diam. 1—2 ft. Light-brown, moderately hard, warping considerably. Used for cask-staves, treenails, turnery and fuel. (ii) [See **Wattle, Black** (iii)]. (iii) [See **Blackwood**].

Wellingtonia. See **Big Tree.**

Whitethorn. See **Hawthorn.**

Whitewood, a name sometimes applied in the English timber-trade to the Norway Spruce. [See **Spruce.**] In the United States it refers mainly to *Liriodéndron* [See **Tulip-tree**]. In

Australia it is either (i) *Lagundria* [See **Tulip-tree** (i)], or *Pittospórum bícolor* [See **Cheesewood**].

Whitewood, American or **Canary** (*Liriodéndron*). See **Tulip-tree**.

Whitewood, Mowbulan. See **Laurel** (i).

Willow, a name restricted in Europe and North America to the numerous and variable species of the genus *Sálix;* but extended in Australia to several trees in no way related to the true Willows or to one another. These will be dealt with after the true Willows. Some species of *Sálix* are herbaceous and others mere prostrate undershrubs of no value. Several other species, and their numerous hybrids, are cultivated as coppice, in river eyots, or wet ground, under the name of "Osiers," for the manufacture of wicker-work. Of these the chief are *S. viminális* L., the Common Osier (*Germ.* "Korbweide"), with silky hairs on its young branches; *S. purpúrea* L., the Purple Osier, with red or purple bark; and *S. vitellína* L., the Golden Osier, with yellow bark. Other species are treated as pollards, the top being valuable for hurdles, clothes-props, hoops, handles for hay-rakes, etc. Those most important as timber-trees are the White Willow (*S. álba*) and the Redwood Willow (*S. frágilis*). *S. Capréa* has been already described. [See **Sallow.**]

Willow, Black (*S. nígra* Marshall). North America. *French* "Saule noir." *Germ.* "Schwarze Weide." Height up to 50 ft.; diam. up to 1½ ft., but generally a shrub. S.G. 446. W 27·77. R 424 kilos. Branches yellow, brittle at base; sapwood nearly white; heart reddish-brown, very light, soft, close-grained, easily worked. Used mainly for fuel and charcoal.

Willow, Crack (*S. frágilis* L.). Europe, Northern and Western Asia; introduced in America. Known also as "Withy," "Bedford" or "Redwood Willow," or "Stag's-head Osier." *Germ.* "Bruchweide." Height sometimes 50—90 ft.; diam. 4—7 ft. Branches green, yellow-brown, orange or crimson, smooth, polished, brittle at the base; wood, when dry, salmon-coloured, light, pliable, tough and elastic. Said to be used in Scotland for boat-building: used also in cabinet-work and for toys. Said to be superior to other Willow.

Willow, White (*S. álba* L.). Europe, North Africa, North and West Asia to the north-west of India. Height up to 80 ft.; diam. 7 ft. S.G. 785 when fresh, 461 when dry. W 35—24. Branches olive, silky, not easily detached; sapwood white; heart brownish, light, soft, smooth in grain, not splintering, shrinking more than $\frac{1}{6}$ of its bulk in drying, very durable in water; vessels uniformly distributed, indistinguishable; pith-rays indistinct. Used in Pliny's time, on account of its lightness, for shields, and formerly for flooring; now-a-days for break-blocks on railway-waggons, since, owing to the absence of oil or resin, it will not take fire on friction; for wheelbarrows, especially at iron-furnaces, as it will not split or warp when heated; for the paddles of steam-boats and strouds of water-wheels; for shoemakers' lasts and cutting-boards; for whetting fine cutlery, and for toys; but especially for cricket-bats, for which purpose large sound trees fetch exceptional prices. The smaller wood is used for clothes-props, the handles of hay-rakes, hurdles, fencing and hoops, for druggists' boxes, for paper-pulp and for fuel.

Willow, Yellow (*S. vitellína* L.). Europe; introduced into North America. Height up to 60 ft.; diam. to 3—4 ft. Sapwood wide, nearly white; heart irregular, reddish-brown, light, soft, not strong, easily worked and taking a beautiful polish. Its yellow twigs, known as "Golden Osiers," are used for basket-work, and larger wood for fencing, fuel and charcoal.

In Australia the name "Willow" is applied (i) to *Eucalýptus piluláris* [See **Blackbutt**]; (ii) to *Geíjera parviflóra*; and (iii) to *Pittospórum phillyrœoídes*, which is sometimes termed "Native Willow," a name also given to (iv) *Acácia salicína*. (ii) *Geíjera parviflóra* Lindl. (Order *Rutáceœ*). Known also as "Dogwood." *Aborig.* "Wilga." Height 20—30 ft.; diam. 6—12 in. Light-coloured, fragrant, hard, close-grained, apt to split in seasoning and liable to gum-veins. Used for naves of wheels. (iii) *Pittospórum phillyrœoídes* DC. (Order *Pittospóreœ*). Known also as "Butter-bush" and "Poison-berry." Height 20—25 ft.; diam.

4—6 in. S.G. 767. Light-coloured, very hard, close-grained. Useful for turnery. (iv) *Acácia salicína* Lindl. (Order *Leguminósæ*). Height 30—50 ft.; diam. 1—1½ ft. S.G. 763. W 47·5. Dark brown, prettily figured, heavy, close-grained, tough, taking a high polish. Used for boomerangs and furniture.

Wood-oil tree. See **Gurjun.**

Woolly-butt (i) (*Eucalýptus longifólia* Link.: Order *Myrtáceæ*). South-east Australia. Sometimes known as "Bastard Box." Height 100—150 ft.; diam. 3—6 ft. S.G. 1187. W 68·5. Red, heavy, hard, straight and close in grain, strong and tough, liable to gum-veins and shakes, durable, especially under ground. When sound, much prized for the felloes and spokes of wheels; but on account of its gum-veins, more used as fuel. [See also (ii) **Gum, Apple scented,** (iii) **Mahogany, Bastard** and (iv) **Gum, Manna.**]

Yacca (*Podocárpus coriácea* Rich. and *P. Purdieána* Hook: Order *Coníferæ*). Jamaica. Small, pale brown. W 38·6—46·9. E 456—596 tons. *f* 4·3—5·25. *fc* 2·49—2·55. *fs* ·346—·486.

Yarrah (*Eucalýptus rostráta*). See **Gum, Red.**

Yate-tree (*Eucalýptus cornúta* Labill.: Order *Myrtáceæ*). South-west Australia. Height 100 ft. S.G. 1235. The heaviest West Australian wood, hard, elastic, somewhat of the character of Ash. Valued for shafts, boat-ribs and agricultural implements.

Yellow-wood, a name applied in South Africa to various species of *Podocárpus* (Order *Coníferæ*), and in Australia to four or five woods in no way related to these or to one another. [See also **Fustic.**]

Yellow-wood, Bastard (*Podocárpus pruinósus* E.M., or perhaps also *P. elongáta* L'Hérit. and *P. Thunbérgii* Hook.). *P. pruinósus,* a native of Natal, is a tree of considerable size, yielding a pale-yellow, tough and durable wood, much used for building. [See **Yellow-wood, Natal,** and **Yellow wood, Real.**]

Yellow-wood, Dark or **Deep** (*Rhús rhodanthéma* F. v. M.: Order *Anacardiáceæ*). North-east Australia. Known also as "Yellow Cedar" or inappropriately as "Light Yellow-wood." *Aborig.* "Jango-jàngo." Height 60—80 ft.; diam. 1½—2 ft.

W 47. Rich brownish or yellowish-bronze colour, darkening with age, often beautifully marked, soft, fine and close in grain, taking a fine polish, with a silky lustre, durable. A handsome and valued cabinet wood.

Yellow-wood, Light (i) inappropriately, (*Rhús rhodanthéma*) [See **Yellow-wood, Dark**]; (ii) *Daphnándra micrántha*) [See **Sassafras, Australian** (ii)]; and (iii) and most appropriately (*Flindérsia Oxleyána*) [See **Jack, Long**].

Yellow-wood, Natal or **Outeniqua** (*Podocárpus elongáta* L'Hérit.). South and Tropical Africa. Known also as "White" or "Bastard Yellow-wood." *Boer* "Geel Hout." *Zulu* "Umkoba." Height 30—70 ft.; diam. 3—7 ft. Pale yellow, soft, light, close-grained, easily split and worked. Neither so common nor so hard as the Real Yellow-wood (*P. Thunbérgii*); but used indiscriminately with it for roofs, beams, planks, flooring and furniture, and, when creosoted, for sleepers.

Yellow-wood, Real or **Upright** (*Podocárpus Thunbérgii* Hook.). South Africa. *Boer* "Geel Hout." *Zulu* "Umceya." Height 75 ft.; diam. 4—8 ft. Light yellow, straight-growing, light, soft, even-grained, fairly elastic and strong, easily worked but somewhat liable to split or warp. Excellent for shingles and used also for furniture, and, like the last mentioned, in building.

Yellow-wood, Thorny (*Zanthóxylum brachyacánthum*). See **Satinwood** in Australia.

Yellow wood or **Yellow Cedar,** in Guiana (*Aníba guianénsis* Aubl.: Order *Lauríneæ*). *French* "Cèdre jaune." S.G. 606—489. R 145 kilos. Very strong, easily worked and durable. Used for planks in building.

Yen-dike (*Dalbérgia cultráta* Grah.: Order *Leguminósæ*). Burma. Apparently sometimes known as "Blackwood" and confused with some species of Ebony. Height 35 ft. to the lowest branch; diam. 1—3 ft. W 64. Black, sometimes with white and red streaks, straight-grown, very heavy and hard, tough, not brittle, elastic, but full of shakes, very durable, not cracking any more after conversion and resisting sun or rain.

Excellent for spokes, bows, handles of ploughs, tools, planes and spears, and largely used for carving.

Yen-ju (*Sóphora japónica* L.: Order *Leguminósæ*). China and Japan. Height 40 ft. or more; diam. 2—4 ft. Hard, fine-grained, ornamental. Used for turnery, furniture and interior finish; but valued as a shade tree and for its buds, the Chinese "Wai-hwa," which are used as a yellow or green dye.

Yew (*Táxus baccáta* L.: Order *Taxíneæ*). Europe, up to altitudes of 6000 ft. in Southern Spain; Northern and Western Asia, up to 11,000 ft. in the Himalayas; and Northern Africa. *French* "If." *Germ.* "Eibe," "Eibenbaum." *Welsh* "Yw." *Ancient Greek* "Taxos, Melos." *Modern Greek* "Maurelatos." *Latin* "Taxus." *Ital.* "Tasso." *Span.* "Texo, Tejo." Height 15—20 or even 50 ft., and in the Himalayas 100 ft.; diam. 1—5 ft. or more, up to 19 ft. Reddish brown, resembling Mahogany, irregular in its growth, heavy, very hard, close-grained, tough, very elastic and flexible, susceptible of a high polish, insect-proof and more durable than any other European wood, especially in contact with soil, it being an old saying that "a post of yew will outlast a post of iron." On old trees burrs occur, figured and mottled like Amboyna-wood. Sapwood very narrow, yellowish-white: annual rings very narrow, wavy, well marked by the broad dark zone of autumn-wood: pith-rays indistinguishable and without tracheids: wood without resin-ducts, entirely composed of spirally-thickened tracheids (Fig. 66). "The eugh obedient to the benders will," as Spenser calls it, seems to have been used, owing to its combined toughness and elasticity, for bows from very early times. In England, though home-grown wood was used, that imported, by Venetian traders, from Italy, Turkey and Spain, was of better quality. At the close of the 16th century the practice of "backing" bows with some other kind of wood was introduced, and at the present day they are largely made of Lancewood and Hickory. At the present day Yew is employed to some extent at High Wycombe and Worksop in chair-making, and on the Continent in turnery. When stained black it is one of the woods known as German Ebony. Small branches are valued for walking-sticks and whip-handles. In the latter part of the 18th century veneers of Yew burrs were largely used for tea-caddies and other small articles. There is in the library of the India Office a Persian illuminated manuscript on thin sheets of Yew.

Yew, Californian, Pacific or **Western** (*Táxus brevifólia* Nutt.). Pacific slope from British Columbia to South California,

Fig. 66.—Longitudinal section of Yew (*Táxus baccáta* L.).

up to altitudes of 8000 ft. Height 40—50 or 80 ft.; diam. 1—2 ft. Sapwood pale yellow; heart orange-red, heavy, hard, fine-grained, extremely stiff and strong, seasoning well and durable.

Yew, Japanese (*Táxus cuspidáta* S. and Z.). Yeso, and long cultivated throughout Japan. *Japan* "Ichii, Momi-noki, Araragi." Height 40—50 ft.; diam. 2 ft. Dark reddish, handsome, close-grained, tough. Used by the aboriginal Ainu for bows and by wealthy Japanese for cabinet-work and interior finish. Used also for pencils.

Yoke (*Piptadénia peregrína* Benth.: Order *Leguminósæ*). Tropical South and Central America. A large tree. W 70·48. E 286 tons. *f* 3. *fc* 2. *fs* ·317. Reddish-brown, heavy, hard, close-grained.

Zebra-wood, a beautifully marked furniture-wood, chiefly obtained from *Cónnarus guianénsis* Lam. (= *Omphalóbium Lambértii* DC.: Order *Connaráceæ*). British Guiana. "Hyawaballi." Height 90 ft., squaring 10—12 in. S.G. 1032. Reddish-brown, beautifully marked, working well and taking a good polish; but very rare. The name is also applied to the woods of *Eugénia frágrans* Willd., var. *cunedta* (Order *Myrtáceæ*), *Guettárda speciósa* L. (Order *Rubiáceæ*), the "Ron-ron" of Honduras, and *Centrolóbium robústum* Mart. (Order *Leguminósæ*), the "Arariba" or "Araroba" of Brazil, exported from Rio, in lengths of 30—40 ft.; and to *Diospýros Kúrzii* (Order *Ebenáceæ*) from the Andaman Islands.

Zelkova (*Zelkóva crenáta* Spach. = *Plánera Richárdi* Michx.: Order *Ulmáceæ*). Caucasus. Height 70—80 ft. Sapwood broad, light-coloured, very elastic, used for the same purposes as Ash or Elm; heart reddish, heavy, very hard, taking a good polish. Used for furniture.

APPENDIX I.

EXPLANATION OF SOME TERMS USED WITH REFERENCE TO CONVERTED TIMBER, ETC.

It may be useful to give here the following definitions of terms used in the English timber trades. (See also p. 89 *supra.*)

A **balk** is a log roughly squared.

A **plank** is 11 in. broad, from 2 to 6 in. thick, and generally from 8 to 21 ft. in length.

A **deal** is 9 in. broad and not more than 4 in. thick.

A **batten** is not more than 7 in. broad.

A **square** is 100 ft. superficial.

A **hundred of deals** is 120.

A **load** is 50 cubic feet of squared timber, or 40 cubic feet of unhewn, or 600 superficial feet of inch planking.

In France wood, especially firewood, is measured by the **stère** or cubic metre = 35·32 cubic feet, or 1·31 cubic yards.

The simplest formula for measuring timber is:

$$C=L\left(\frac{G+g+g'}{3}\right)^2,$$

where C = the cubic contents in cubic feet; L = the length of the log in feet; G = one-fourth of the girth of the tree midway in its length, in feet; g = one-fourth of the girth at one end, in feet; g' = one-fourth of the girth at the other end, in feet, a deduction for bark being made from each quarter-girth.

APPENDIX II.

THE MICROSCOPIC EXAMINATION OF WOODS.

It is generally recognised that much more trustworthy evidence as to the identity of out-of-the-way woods can be obtained by a microscopic examination than from reports of native opinion or rule-of-thumb tests applied by lumbermen or unscientific traders. Such microscopic examination, it is true, will not always enable us to refer an undetermined wood even to its Natural Order, and will in many cases fail to discriminate between the species of one genus; but, on the other hand, it will often afford, in addition to the evidence of identity, much valuable information as to strength, durability, or other reasons for suitability or unsuitability.

No one need be deterred from having recourse to this method of examination by any alarm as to its technical difficulties or expense. Though even an examination of a planed surface of wood with a pocket lens may give useful suggestions, a thin transverse section examined under the slightly higher powers of a compound microscope will afford vastly greater information. Any ordinary microscope costing from three to seven guineas will suffice; but it is desirable to have a wide simple stage and a triple or double nose-piece carrying, preferably, objectives of $1\frac{1}{2}$ in., $\frac{1}{2}$ in., and $\frac{1}{6}$ in. focus.

The preparation of the sections is not difficult. It is quite unnecessary for the ordinary purposes of study to make them of anything like the superficial dimensions of the beautiful preparations of Herr Burkart or Mr. Hough. All that is requisite is to get a transparent section, across the grain, embracing a few annual rings, though it is well to have, either in one or in more than one section, the pith or structural centre and some of the sapwood as well as the heartwood. The best results will be obtained with well-seasoned wood. The end of the specimen having been roughly smoothed with a plane or chisel, a shaving can be easily cut with a well-sharpened plane which, though not equally thin all over, will serve all the purposes of study. It is best to drop the shaving at once into some liquid dye, such as an alcoholic solution of methyl violet, as this renders the structure far more readily visible under the microscope. After dyeing, the thinner part of the shaving can be cut square with scissors and mounted, for which purpose we personally prefer an extra wide glass slip and a square cover-glass. If carefully dried under moderate pressure the sections can be preserved unmounted or mounted on paper, like the series prepared by Herr Nördlinger.

Another method we have found efficacious is to immerse a small squared specimen in a basin of water, such as a deep photographic basin, by means of metal weights, and then to slice off thin shavings with a broad chisel. The sections floating to the surface may then be dyed, or may be simply floated on to slips of glass, lightly covered with thin cover-slips and then dried slowly and cemented down. It is important to use thin cover-slips.

Though the transverse section is most important, longitudinal ones, either radial or tangential, are often also of use, the latter, for instance, affording the readiest means of distinguishing Alder from Birch.

APPENDIX III.

SELECT BIBLIOGRAPHY.

[The works of other authors to which I am most indebted are marked with an asterisk.]

ANDERSON, SIR JOHN.—The Strength of Materials and Structures. 11th edition. 1892.

BAUSCHINGER, JOH.—Mittheilungen aus dem Mech. Tech. Laboratorium in München, 1883 and 1887.

BOULGER, G. S.—Economic Forestry, *Trans. Scottish Arboricultural Soc.* xi. 1887.

BOULGER, G. S.—The Uses of Plants. 1889.

BOULGER, G. S.—The Scientific Study of Timber, *Trans. Surveyors' Instit.* xxiv. 1892.

CHARPENTIER, PAUL.—Timber. Translated by Joseph Kennell. 1902.

DARWIN, FRANCIS.—The Elements of Botany. 1895.

FERNOW, B. E.—Report of the Division of Forestry, U.S. Department of Agriculture, for 1886.

FLETCHER, BANISTER F. and H. P.—Carpentry and Joinery. 1898.

GAMBLE, J. S.—Manual of Indian Timbers. 1881.

*HARTIG, ROBERT.—Timbers and how to know them. Translated, from the 3rd German edition, by Dr. William Somerville. 1890.

*HOUGH, ROMEYN B.—American Woods. 1888.

HOLTZAPFFEL, CHARLES and JOHN JACOB.—Turnery and Mechanical Manipulation. 1843-1884.

KENT, ADOLPHUS H.—Veitch's Manual of the Coniferæ. 2nd edition. 1900.

*KEW, ROYAL GARDENS.—Official Guide to the Museums of Economic Botany. No. 3.—Timbers. 2nd edition. 1893.

KIRK, T.—The Forest Flora of New Zealand. 1889.

*LASLETT, THOMAS.—Timber and Timber Trees, Native and Foreign. 2nd edition, revised by Prof. H. Marshall Ward, D.Sc., F.R.S. 1894.

*MAIDEN, J. H.—Useful native plants of Australia. 1889.

MASTERS, DR. MAXWELL T.—List of Conifers ... in cultivation, *Journ. Royal Hort. Soc.* xiv.

NÖRDLINGER, H.—Les bois employés dans l'industrie. 1872.

NÖRDLINGER, H.—Holzquerschnitte. 1862-82.

ROGERS, JULIAN.—Analysis of Returns . . . relating to Colonial Timber, presented to Parliament, 1878.

*ROTH, FILIBERT.—Timber. U.S. Department of Agriculture, Division of Forestry. Bulletin No. 10. 1895.

SARGENT, PROF. C. S.—Sylva of North America. 1891-8.

SARGENT, PROF. C. S.—Report on the Forest Trees of North America. Ninth Census of the United States, vol. ix. 1880.

SHARPLES, STEPHEN P.—In the preceding.

STEVENSON, W.—The Trees of Commerce. 1888.

TIMBER TRADES JOURNAL. 1876-1901 (in progress).

TREDGOLD, THOMAS.—Elementary Principles of Carpentry. 7th edition, edited by E. W. Tarn. 1886.

*UNWIN, PROF. W. C.—The Testing of Materials of Construction. 2nd edition. 1899.

*WARD, PROF. H. MARSHALL.—Timber and some of its diseases. 1889.

*WARD, PROF. H. MARSHALL.—The Oak. 1892.

*WOOD, PROF. H. MARSHALL.—Disease in Plants. 1901.

WIESNER, JULIUS.—Die Rohstoffe des Pflanzenreiche. 1901.

WOOD, H. T.—Colonial and Indian Exhibition Reports. 1887.

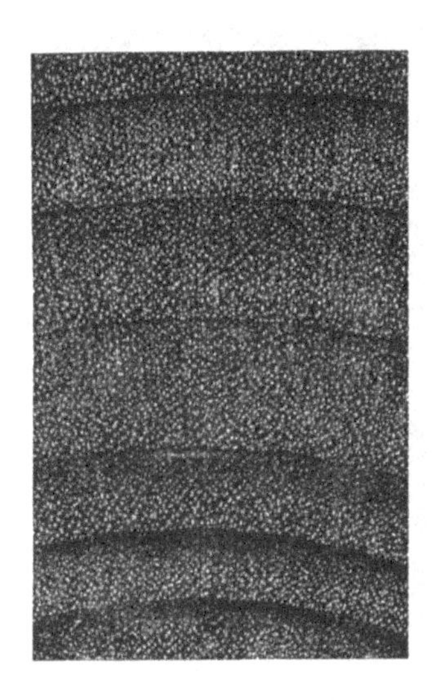

Acer platanoides, Norway Maple.

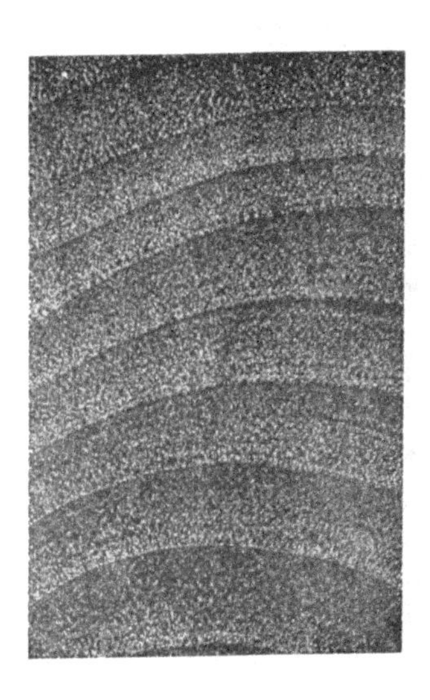

Æsculus Hippocastanum, Horse Chestnut.

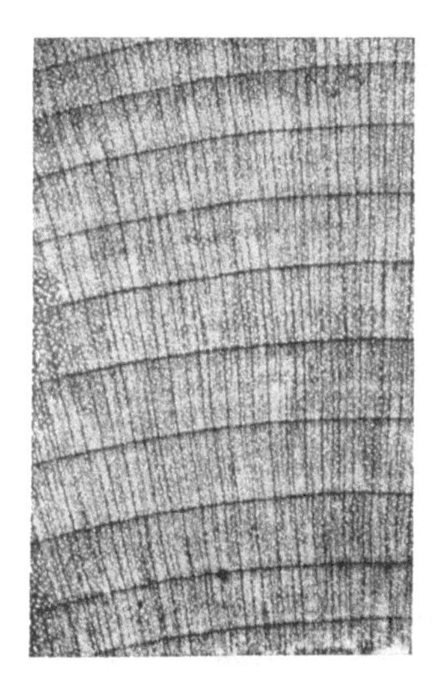

Acer Pseudoplatanus, Sycamore.

Sambucus nigra, Elder.

Prunus Avium, Cherry.

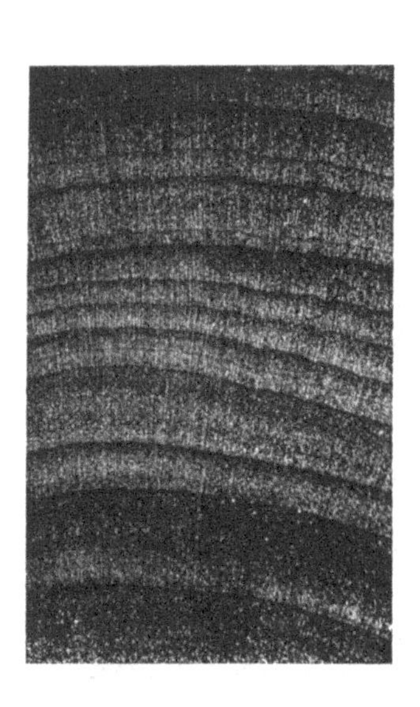

Pyrus Malus, Apple.

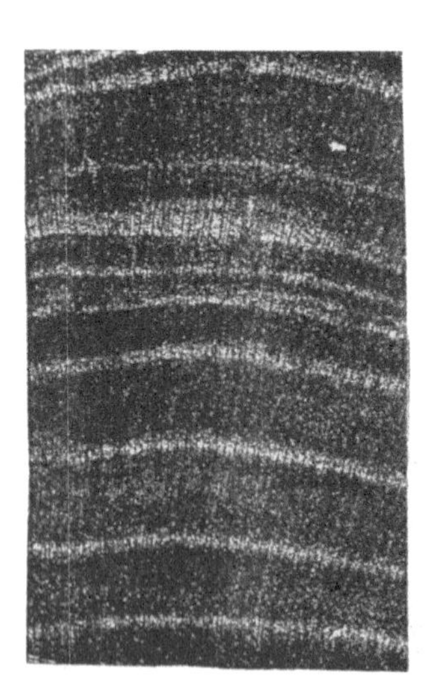

Prunus domestica, Plum.

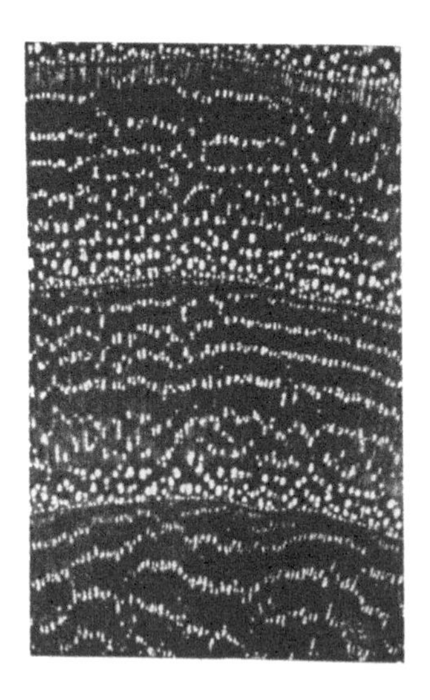

Robinia Pseudacacia, False Acacia.

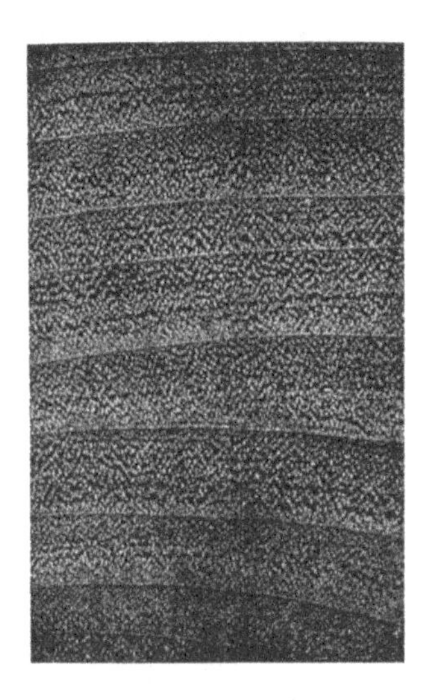

Populus tremula, Aspen.

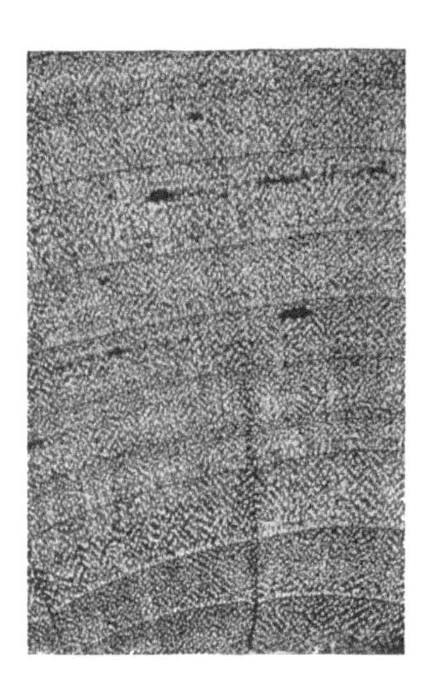

Betula alba, Birch.

Salix Caprea, Goat Willow.

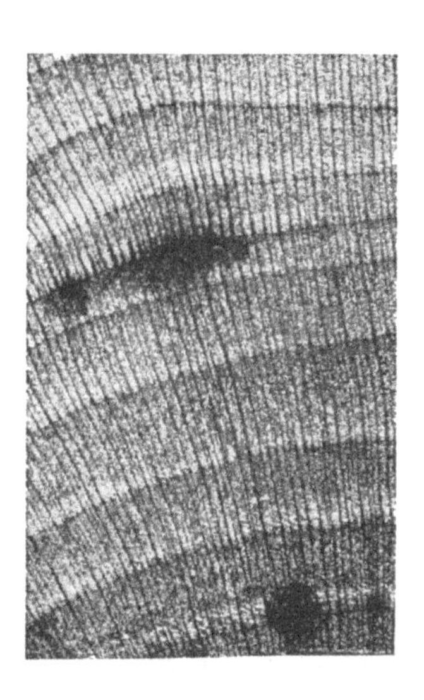

Platanus orientalis, Plane.

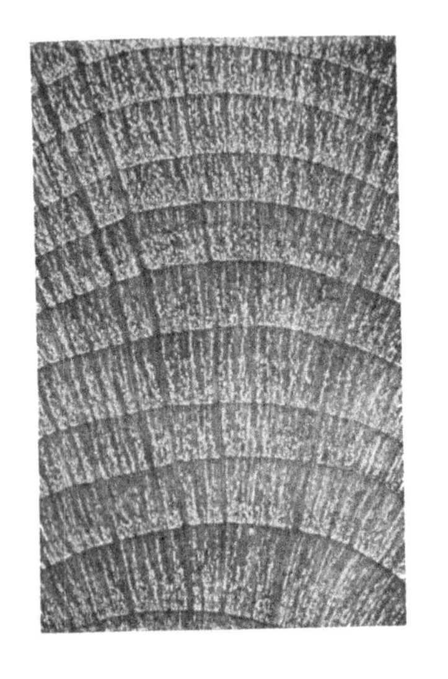

Corylus Avellana, Hazel.

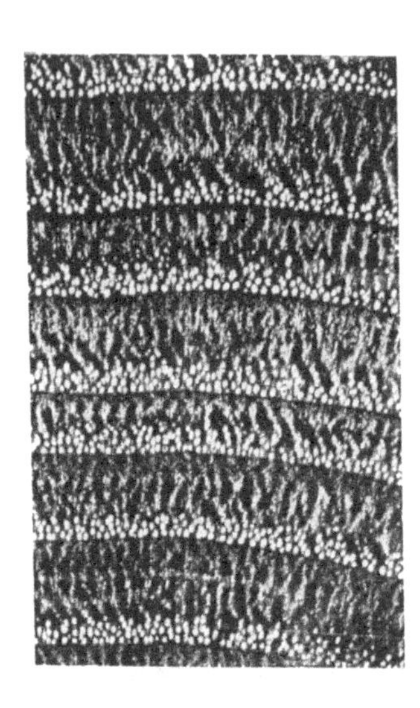

Castanea vesca, Spanish Chestnut.

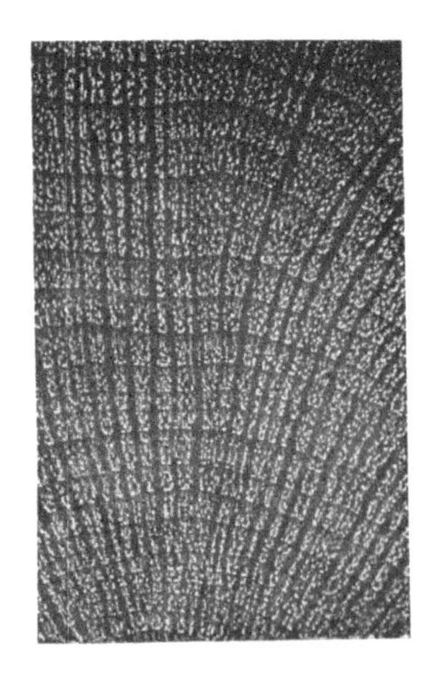

Carpinus Betulus, Hornbeam.

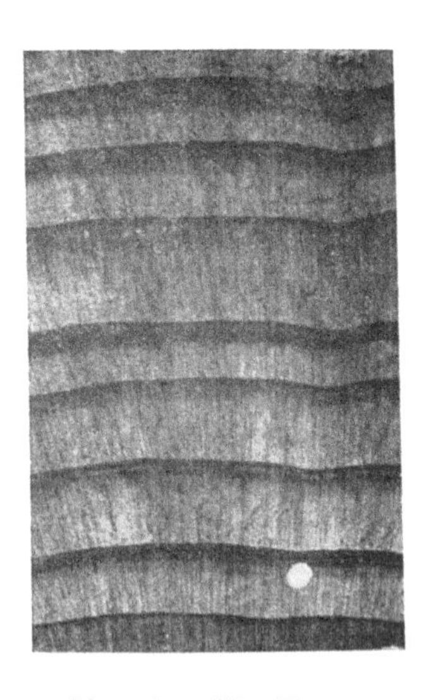

Abies pectinata, Silver Fir.

INDEX.

GLASGOW: PRINTED AT THE UNIVERSITY PRESS BY ROBERT MACLEHOSE AND CO.

Printed in the United States
By Bookmasters